Green Energy and Technology

More information about this series at http://www.springer.com/series/8059

Antonio Bonomi · Otávio Cavalett
Marcelo Pereira da Cunha
Marco A.P. Lima
Editors

Virtual Biorefinery

An Optimization Strategy for Renewable Carbon Valorization

Springer

Editors
Antonio Bonomi
Brazilian Bioethanol Science and Technology Laboratory (CTBE)
Brazilian Center of Research in Energy and Materials (CNPEM)
Campinas, São Paulo
Brazil

Otávio Cavalett
Brazilian Bioethanol Science and Technology Laboratory (CTBE)
Brazilian Center of Research in Energy and Materials (CNPEM)
Campinas, São Paulo
Brazil

Marcelo Pereira da Cunha
Institute of Economics
University of Campinas
Campinas, São Paulo
Brazil

Marco A.P. Lima
Institute of Physics Gleb Wataghin
University of Campinas
Campinas, São Paulo
Brazil

ISSN 1865-3529 ISSN 1865-3537 (electronic)
Green Energy and Technology
ISBN 978-3-319-79881-3 ISBN 978-3-319-26045-7 (eBook)
DOI 10.1007/978-3-319-26045-7

Springer Cham Heidelberg New York Dordrecht London

Softcover re-print of the Hardcover 1st edition 2016

Printed on acid-free paper

Springer International Publishing AG Switzerland is part of Springer Science+Business Media (www.springer.com)

Foreword

I am pleased and honored to be asked to write the Foreword for this book on the Virtual Sugarcane Biorefinery (VSB) project. I first learned of the VSB about six years ago and have been keeping track of the project ever since through meetings with my friends at the University of Campinas (UNICAMP) and the Brazilian Bioethanol Science and Technology Laboratory (CTBE). I was excited by the initial concept, and am delighted to see how much progress has been made by the CTBE and UNICAMP teams in realizing their objectives. They should be proud of what they have accomplished.

The VSB is unique; there is literally nothing like it in all the world. The VSB project embraces all phases of this system, from the agricultural sector through the final use of the biorefinery products. Nowhere else in the world are the tools of economic, environmental, and social modeling being combined to improve an existing industry (sugarcane refining) while also laying the foundation to understand and improve the sustainability of an emerging industry: second-generation ethanol (and other products) from sugarcane cellulosic biomass. The goal of the VSB is to focus research and development and increase our understanding so that a new industry can arise to sustainably use all of the sugarcane plant; sucrose, bagasse, and leaf matter, in order to provide food, liquid fuel, electricity, and other products. What an ambitious and important goal. The CTBE/UNICAMP team is leading the world in applying the latest in systems modeling to shape the emergence of a new, sustainable industry. I believe the world will increasingly appreciate and be grateful for what you are doing. Personally, I express my deepest thanks and admiration to my dedicated and capable Brazilian colleagues. I am grateful for your leadership in this effort.

Muito obrigado!

Bruce E. Dale, University Distinguished Professor
Michigan State University
Editor in Chief: Biofuels, Bioproducts and Biorefining

The first decade of this century was a vigorous moment for science, technology, and innovation policy in Brazil. In parallel with a fast economic growth characterized by a strong social inclusion policy focused on the expansion of the domestic consumer market, the federal government increased the amount of funds for research activities and, especially, for the development of some of the country's natural competitive advantages.

This book on the subject of virtual biorefinery brings illumining lights on the development of such an important subject in Brazil and enumerates the country's remarkable achievements on the development of this technology platform on its several dimensions and potential applications. At that time, benefited by a 30 years' experience since the inception of the 1970s National Ethanol Program, Brazil had a worldwide leadership on the production and use of bioethanol. The country had already been recognized by international studies as a Natural Knowledge Economy, a qualification resulting not only from its strong natural resources base, but also from a growing scientific performance in the period.

Within such an environment, a study organized by the Center for Strategic Studies and Management on Science, Technology and Innovation (CGEE) had identified the principal bottlenecks for the full exploitation of the Brazilian advantages on bioethanol from sugarcane as a substitute for gasoline. CGEE, a think-tank of the Ministry of Science, Technology and Innovation, had been created in 2001 with the objective to develop key studies that incorporate different views and actors (academy, industry, and government) of the innovation system to support policy decisions. In this study, a complete strategy to promote the increase of ethanol production with the use of sugarcane as the most important renewable organic matter was designed. Its main conclusion was that the expansion of the Brazilian capability of exploring the whole sugarcane potential for large-scale and sustainable ethanol production would depend on the cumulative scientific and technological advances along the entire chain of ethanol production, which ranges from the agricultural fields to the industrial stage. It was also suggested that without a strong basic research program, the challenges associated with the development of second-generation technology (i.e., one that allows the complete use of the full organic matter present on sugarcane) could not be surmounted.

The final decision of the federal government was for the creation of a national laboratory for ethanol technology development (Brazilian Bioethanol Science and Technology Laboratory – CTBE), using the well-established model of open laboratory already in operation in other areas in Brazil. The CTBE was to bring together, in a strong collaborative manner, the research efforts under way at the academic and industrial segments, as well as developing its own expertise. An interdisciplinary team from different areas of knowledge, including researchers in biotechnology, chemistry, agriculture science, bioscience and modeling engineering, among others, was recruited to start the Brazilian national effort toward the development of second-generation technology for ethanol production. Due to the complexity and challenges to be faced, such effort would require strong national and international collaboration. Each chapter in this book presents a special contribution relating the necessary efforts required for a well-balanced advancement of the knowledge needed to support the wise strategy for technology and innovation followed in pursuing a sustainable energy future for humanity.

Lucia C. P. de Melo
Pernambuco State Secretary of Science, Technology and Innovation
Former President of the Center for Strategic Studies and Management on Science, Technology and Innovation—CGEE

Energy policy is the continuation of war by other means and, only too frequently, the inverse is also true. The strategic, economic, financial, political, and social interests that determine and are determined by energy issues are at the core of the modern national industrial states. At the same time, and this is the reason we say energy policy is war pursued by other means, the consequences of the choices these nations make extrapolate their own borders. All sorts of international conflicts arise and become inevitable around energy issues. The fair number of international organizations, governmental or non-governmental, that exist to deal with these is a witness to their depth and breadth. In addition, over the past couple of decades, environmental issues, local and global, connected to the exploration of energy sources, transportation, transformation, distribution, and final uses of energy have become preeminent.

In 2015, we cannot say there is a global energy crisis, like the ones that occurred in the last decades of the twentieth century. The supply of fossil fuels is abundant and their prices are low. It is true that the world economy is not operating at full steam, and that we may be just experiencing one of those recurring moments in the economic cycle when supply overtakes demand. However, there is a latent energy crisis of another nature and almost unimaginable reach: anthropogenic global climate changes due to the emission of greenhouse gases into the atmosphere, mostly arising from the burning of fossil fuels. Global climate change, to the extent estimated by specialists and already partially indicated by real events, will have enormous negative economic and social consequences, especially from mid-twenty-first century onwards, that is, during the lifetime of the majority of human beings alive now. More and more, the questions appear to be not "if" there is trouble ahead, but "when, where, and how extensive, expensive and permanent" the damage will be. Dreams of a radiant future for mankind, of a poverty and disease free peaceful world, or more modestly, the digital utopia (the adjective delirious is superfluous) of bits and bytes prophets, are threatened by our undeniable success in securing, for the past two and a half centuries, abundant and cheap fossil energy supplies to power the modern industrial civilization.

The preceding paragraphs should underline the importance of the work reported in this book. The authors address in a systematic and quantified way the central question of the sustainability of biofuels. Biofuels are one of the several sources of renewable non-fossil energies that will have to be quickly deployed, if the worst consequences of global climate change are to be averted. The authors' focus is on the virtualization of bioenergy transformation centers—biorefineries—in particular for sugarcane feedstocks. As defined in Chap. 2, "Biorefinery is a facility that integrates biomass conversion processes and equipment to produce fuels, power, materials and chemicals from biomass." The novel simulation tools developed by the authors for the sugarcane biorefinery complement those available commercially for the chemical industry in general. More importantly, they cover the important upstream agricultural processes and the downstream transportation process, taking the Brazilian case as their paradigm. This is, as the authors themselves recognize, a work in progress. They point out the limitations and the required (ongoing) research to increase the reliability and accuracy of the simulations. However, the book comes at the right time to draw attention to the tools already developed to help researchers quantify the impacts of sugarcane biofuels (and co-products), to evaluate quantitatively the consequences of the gigantic production scale-up required if bioethanol is to have a real impact on the market and on the reduction of the risks of global climate changes, and to guide policy and investment decisions. There is one aspect of the life cycle analysis that the authors do not address directly, that is, the energy and environmental costs for building, operating, and decommissioning a biorefinery. An interesting question is, for instance, what is the energy payback time of a biorefinery? These issues will become relevant if the production scale of sugarcane bioethanol is to increase by tenfold or more. I am sure that the question is already on the authors' To Do list.

The team at the Brazilian Bioethanol Science and Technology Laboratory (CTBE) and UNICAMP is to be warmly congratulated for the original and painstaking work of designing and implementing a virtualization tool for sugarcane biorefineries. They have produced a reference work of great value that deserves to be widely known and used.

Cylon Gonçalves da Silva, Professor Emeritus
State University of Campinas—UNICAMP
Member of the Coordination for Special Programs, Scientific Directorate, FAPESP (São Paulo Research Foundation).

Preface

In the early 2000s, the Americans started a heavy investment for replacing part of the gasoline, used in light vehicles, by ethanol produced from corn. The global renewable energy sector incorporated a strong ally. The Brazilians, with long tradition on using ethanol from sugarcane (at that time, about half of the fuel used in light vehicles in Brazil was ethanol), saw, in this action, great opportunities to their country. If the idea would spread around, new ethanol markets would open for their economically competitive product. In 2005, a government-hired agency, Center for Strategic Studies and Management—CGEE, linked to the Minister of Science, Technology, and Innovation, financed a study evaluating the possibility of busting up the Brazilian production, to a point of being able to replace 5–10 % of the worldwide gasoline consumption in 2025, by Brazilian sugarcane ethanol. This study [a summary is in "Can Brazil replace 5 % of the 2025 gasoline world demand with ethanol?" (Leite et al. 2009)] pointed out the bottlenecks in sugarcane agriculture, in ethanol industry, in the sustainability strategy, and on the needed basic science, composing a necessary agenda to be developed, in order to accomplish this goal. As a result, a National Laboratory on Ethanol was created, initially with four scientific and technological programs designed to attack the four bottlenecks areas of this ambitious plan. I was the first director and responsible for implanting it. As scientists, we know how to measure the success of a research program: follow the publications, the leadership, the publication citations, the upcoming scientific research lines, and so on. For measuring success in technological innovation, follow the money (patent profits). Unfortunately, for a National Laboratory that needs to accomplish goals on all the above-mentioned programs and foment lines of research, it is not possible to follow the basic science reasoning for success or wait for the profit stage to grant success in technological developments. We needed a strategy for making decisions in a day-to-day basis. For this, we have hired a senior researcher, Dr. Antonio Bonomi, with great experience in computer simulations for comparing different technological routes for ethanol production. His responsibility in this new National Laboratory was to put a team together (the co-editors Dr. Cavalett and Dr. Cunha participate actively in this group) and run a Technological Assessment Program: the Virtual Sugarcane Biorefinery, a tool to link together all other programs and help the decision makers. This book is about this initiative—a strategy understood as an effective way to develop both science and technology looking for sustainable technological development.

Marco A.P. Lima

Reference

Leite RCC, Leal MRLV, Cortez LAB, Griffin WM, Scandiffio MIG (2009) Can Brazil replace 5 % of the 2025 gasoline world demand with ethanol? Energy 34:655–661

Acknowledgments

The authors would like to thank all those who have directly and indirectly contributed to the book and with the conception and development of the Virtual Sugarcane Biorefinery. Special mentions are to Rogério Cezar de Cerqueira Leite (University of Campinas and CNPEM), Luis Augusto Barbosa Cortez (University of Campinas), R. Maciel Filho (University of Campinas), Marcelo Zaiat (University of São Paulo), Carlos Alberto Labate (University of São Paulo), Carlos Eduardo Rossell (CTBE), Manoel Régis Lima Verde Leal (CTBE), Arnaldo Walter (University of Campinas), Marcos Buckeridge (University of São Paulo), Helena Chum (NREL), Guido Zacchi (Lund University), and Oscar Braunbeck (CTBE). Special recognition also for collaborators at CTBE and CNPEM, as well as the students and fellows who worked and have been working on Virtual Sugarcane Biorefinery. We thank all the present and past collaborators in the figures of Dr. Carlos Américo Pacheco (General Director of CNPEM), and Dr. Paulo Mazzafera (Director of CTBE). We acknowledge the Brazilian Center for Research in Energy and Materials (CNPEM); Ministry of Science, Technology and Innovation (MCTI); Coordination for the Improvement of Higher Education Personnel (CAPES); National Counsel of Technological and Scientific Development (CNPq), and São Paulo Research Foundation (FAPESP) for financial support for different parts of Virtual Sugarcane Biorefinery.

Contents

Editors and Contributors

About the Editors

Antonio Bonomi has Ph.D. in Chemical Engineering from the University of Minnesota (1977). He is the head of the Division of Integrated Assessment of Biorefineries at CTBE/CNPEM, since November 2008. His research areas include biotechnology processes, mathematical modeling and simulation, bioethanol technology, metrology in chemistry, among others.

Brazilian Bioethanol Science and Technology Laboratory (CTBE), Brazilian Center of Research in Energy and Materials (CNPEM), Campinas, São Paulo, Brazil.

Otávio Cavalett is a researcher in the Brazilian Bioethanol Science and Technology Laboratory (CTBE/CNPEM). His research areas and interests include sustainability assessment methods, Life Cycle Assessment, and technology assessment of biomass, bioenergy, and biorefineries.

Brazilian Bioethanol Science and Technology Laboratory (CTBE), Brazilian Center of Research in Energy and Materials (CNPEM), Campinas, São Paulo, Brazil.

Marcelo Pereira da Cunha has a M.Sc. in Applied Mathematics (2005) and a Ph.D. in Energy Systems Planning (2011) from the University of Campinas—Unicamp—Brazil. He is professor of the Institute of Economics at Unicamp since 2012. His research areas include evaluation of socioeconomic impacts in energy sectors, input–output analysis, general equilibrium modeling, and sustainability.

Institute of Economics, University of Campinas (UNICAMP), Campinas, São Paulo, Brazil.

Marco A.P. Lima is a full professor of Physics at UNICAMP. His undergraduation and master's degree are in Physics, both obtained at State University of São Paulo (USP). His Ph.D. is in Chemistry obtained at the California Institute of Technology (Caltech) in 1986. He was the first director of the Brazilian Bioethanol

Science and Technology Laboratory (CTBE), 2008–2012. In 2015, he was elected member of the State of São Paulo Academy of Science.

Institute of Physics Gleb Wataghin (IFGW), University of Campinas (UNICAMP), Campinas, São Paulo, Brazil.

Contributors

Oscar A. Braunbeck grew up at a family farm and graduated in engineering in Argentina. After doing graduate, MS, and Ph.D. programs in the USA, he kept working on teaching, research, and development of alternative processes for mechanized agriculture of sugarcane, leading engineering teams at CTC, Unicamp and CTBE/CNPEM. He produced research papers, books, patents, prototypes, and advised 25 graduate programs related to design of farm machinery.

Brazilian Bioethanol Science and Technology Laboratory (CTBE), Brazilian Center of Research in Energy and Materials (CNPEM), Campinas, São Paulo, Brazil.

João Luis N. Carvalho is a researcher in Brazilian Bioethanol Science and Technology Laboratory (CTBE/CNPEM). His research focuses on impacts of no-tillage, crop rotations, and straw management in the sugarcane sustainability.

Brazilian Bioethanol Science and Technology Laboratory (CTBE), Brazilian Center of Research in Energy and Materials (CNPEM), Campinas, São Paulo, Brazil.

Mateus F. Chagas is a Ph.D. student of Chemical Engineering in State University of Campinas (FEQ/Unicamp). He is a research assistant in the CTBE/CNPEM. His research focuses on Life Cycle Assessment of biofuels and techno-economic analysis of biomass production and processing chain.

Brazilian Bioethanol Science and Technology Laboratory (CTBE), Brazilian Center of Research in Energy and Materials (CNPEM), Campinas, São Paulo, Brazil.

Marina Oliveira de Souza Dias has a Ph.D. in Chemical Engineering from the University of Campinas (2011). Currently, he is a Professor at the Department of Science and Technology of the Federal University of São Paulo. Her research focuses on biorefinery process analysis and simulation, second-generation ethanol, and alcoholchemistry.

Institute of Science and Technology, Federal University of São Paulo (ICT/UNIFESP), São José dos Campos, São Paulo, Brazil.

Charles Dayan Farias de Jesus holds a Ph.D. in Chemical Engineering from the Federal University of São Carlos (2004). As a researcher of the Division of Integrated Assessment of Biorefineries, at Brazilian Bioethanol Science and Technology Laboratory, his research focuses on process modeling and simulation, development of sustainable bioprocesses, and application of methodologies and tools for estimating capital investment cost of industrial facilities.

Brazilian Bioethanol Science and Technology Laboratory (CTBE), Brazilian Center of Research in Energy and Materials (CNPEM), Campinas, São Paulo, Brazil.

Terezinha de Fátima Cardoso has a Ph.D. in Agricultural Engineering, University of Campinas (FEAGRI/Unicamp). She works in the Division of Integrated Assessment of Biorefineries at CTBE/CNPEM since august 2010. Her research focuses on technical and economic analysis of biomass production.

Brazilian Bioethanol Science and Technology Laboratory (CTBE), Brazilian Center of Research in Energy and Materials (CNPEM), Campinas, São Paulo, Brazil.

Bruna de S. Moraes has Ph.D. in Engineering Sciences (Hydraulics and Sanitation) from the University of São Paulo (2012). She is a postdoctoral researcher of the Division of Integrated Assessment of Biorefineries at CTBE/CNPEM, since May 2012. Her research areas include biological treatment of wastewater, biogas production from biowaste/biomass, energy recovery from biogas, environmental sustainability, among others.

Brazilian Bioethanol Science and Technology Laboratory (CTBE). Brazilian Center of Research in Energy and Materials (CNPEM), Campinas, São Paulo, Brazil.

Henrique Coutinho Junqueira Franco is agronomist and researcher leader at Biomass Crop Management Group at CTBE. He holds a Ph.D. in Soil Science and Plant Nutrition from ESALQ/USP/Brazil (2008). He is a postdoctoral fellow at CENA/USP/Brazil (2010) and visiting academic at University of Queensland—Australia (2012). He has been working with sugarcane crop since 2004 and at CTBE since 2010. His main research focus on nutrients management in biomass production aiming increase of nutrient use efficiency by crop with minimal environmental impacts. Further, he has been giving support for Mechanical Low Impact project (under CTBE coordination) within sugarcane planting part: precision planting.

Brazilian Bioethanol Science and Technology Laboratory (CTBE), Brazilian Center of Research in Energy and Materials (CNPEM), Campinas, São Paulo, Brazil.

Edgardo Olivares Gómez has Ph.D. in Agricultural Engineering in 2002 from the University of Campinas (UNICAMP). He is Professor of Mechanical Engineering Course at Engineering Faculty (FAEN) of the Federal University of Grande Dourados (UFGD), since June 2015. His research areas include thermochemical processes, such as pyrolysis, gasification, and combustion of biomass, fluid dynamics involving particles during separation and/or classification processes, and also cogeneration and electricity production technologies, among others.

Federal University of Grande Dourados (UFGD),Dourados, Mato Grosso do Sul, Brazil.

Vera L.R. Gouveia holds a doctorate in Chemical Engineering from the University of Campinas—UNICAMP. She is a researcher of the Division of Integrated Assessment of Biorefineries at CTBE/CNPEM. Her research areas and interests include bioprocess simulation, first- and second-generation ethanol production and technology assessment of biomass, bioenergy, and biorefineries.

Brazilian Bioethanol Science and Technology Laboratory (CTBE), Brazilian Center of Research in Energy and Materials (CNPEM), Campinas, São Paulo, Brazil.

Tassia L. Junqueira holds a doctorate in Chemical Engineering from the University of Campinas—UNICAMP. She is a process specialist at Division of Integrated Assessment of Biorefineries at CTBE/CNPEM, where she has been working since 2010. Her activities include process simulation and technical analysis of process alternatives involved in the integrated first- and second-generation bioethanol production process as well as other renewable products.

Brazilian Bioethanol Science and Technology Laboratory (CTBE), Brazilian Center of Research in Energy and Materials (CNPEM), Campinas, São Paulo, Brazil.

Bruno Colling Klein holds a master's degree in Chemical Engineering (State University of Campinas—Unicamp) and is a doctoral degree candidate in the same discipline. Working in the Division of Integrated Assessment of Biorefineries at CTBE/CNPEM since 2013, his areas of interest cover microalgae technology, biomass conversion processes, and production of biofuels, among others.

Brazilian Bioethanol Science and Technology Laboratory (CTBE), Brazilian Center of Research in Energy and Materials (CNPEM), Campinas, São Paulo, Brazil.

R. Maciel Filho
is a full professor of Chemical Engineering at UNICAMP and Ph.D. in Chemical Engineering from the University of Leeds, UK (1989). He is coordinator of the Laboratory of Optimization, Design and Advanced Process Control (LOPCA) since 1989, head of the Laboratory of Innovation in Biofuels—UNICAMP (LIB), and from 2010, Coordinator of the Brazilian Institute of Biofabrication (BIOFABRIS). The main research areas cover Modeling of Chemical and Biochemical Process: computer-aided design, operation and control and offline/online optimization, with special focus in Green Process Development and Biorefinery, specifically with bioethanol and by-products from fermentation as feedstock. Consideration is also given for the use of CO_2 and bioethanol as raw material for chemicals. He served as Head of Chemical Process Development, Director for Undergraduate Studies, Dean of Chemical Engineering School, and Pro-Rector at the State University of Campinas (UNICAMP). He is also a member of the Higher Scientific Research Council of Brazil Research Council (CA of CNPq.).

School of Chemical Engineering, University of Campinas (UNICAMP), Campinas, São Paulo, Brazil.

Paulo S. Graziano Magalhães has Ph.D. in Agricultural Engineering from the Cranfield University, Silsoe College, UK (1986). He works at Unicamp since 1980, where he became a professor in 2011. He is working at the Biomass Production Division at CTBE/CNPEM, being responsible for the Precision Agriculture program since 2010. His research areas include precision agriculture, soil dynamics, and biomass production, among others.

Brazilian Bioethanol Science and Technology Laboratory (CTBE). Brazilian Center of Research in Energy and Materials (CNPEM), Campinas, São Paulo, Brazil. School of Agricultural Engineering (FEAGRI). University of Campinas, (UNICAMP), Campinas, SP, Brazil.

Paulo Eduardo Mantelatto has a Ph.D. in Chemical Engineering from the Federal University of São Carlos (2011). He is a process specialist of the Division of Integrated Assessment of Biorefineries at CTBE/CNPEM, since October 2010. His research areas include biotechnology processes, downstream process, modeling and simulation, sugar and bioethanol technology, development of process, among others.

Brazilian Bioethanol Science and Technology Laboratory (CTBE). Brazilian Center of Research in Energy and Materials (CNPEM), Campinas, São Paulo, Brazil.

Edvaldo R. Morais has doctorate degree (2007) and postdoctorate (2012) in Chemical Engineering from the State University of Campinas. He is a researcher of the Division of Integrated Assessment of Biorefineries at CTBE since April 2012. He has expertise in mathematical modeling and computer simulation, with emphasis in chemical and biochemical reactors. His research areas include the technical assessment of biotechnology processes, first- and second-generation ethanol production and other chemicals from renewable sources. He is member of the team that developed the Virtual Sugarcane Biorefinery (VSB).

Brazilian Bioethanol Science and Technology Laboratory (CTBE). Brazilian Center of Research in Energy and Materials (CNPEM), Campinas, São Paulo, Brazil.

Lucas Gonçalves Pereira has Ph.D. in Food Engineering from the University of Campinas (Unicamp) (2012). He is currently a postdoctoral fellow at the University of Toronto. Focus of his research has been on the evaluation of the financial viability and environmental performance of bioenergy systems.

Brazilian Bioethanol Science and Technology Laboratory (CTBE). Brazilian Center of Research in Energy and Materials (CNPEM), Campinas, São Paulo, Brazil. Department of Civil Engineering, University of Toronto, Toronto, Ontario, Canada.

Mylene Cristina Alves Ferreira Rezende has doctorate degree (2007) and postdoctorate (2011) in Chemical Engineering from the University of Campinas. She is currently a researcher at the CTBE/CNPEM, since May 2012. Her research areas include process simulation and optimization, with special focus on thermochemical and catalytic conversion process.

Brazilian Bioethanol Science and Technology Laboratory (CTBE). Brazilian Center of Research in Energy and Materials (CNPEM), Campinas, São Paulo, Brazil.

Elmer Alberto Ccopa Rivera received his Ph.D. in chemical engineering from the State University of Campinas (UNICAMP) in 2006. He was a postdoctoral fellow at French National Institute for Agricultural Research (INRA) and is presently a researcher of the Division of Integrated Assessment of Biorefineries at CTBE/CNPEM, since June 2012. He effectively collaborated with other engineer and science professionals to create innovative modeling, optimization, and experimental procedures.

Brazilian Bioethanol Science and Technology Laboratory (CTBE). Brazilian Center of Research in Energy and Materials (CNPEM), Campinas, São Paulo, Brazil.

Isabelle Sampaio is graduated in Chemical Engineering from the State University of Campinas—Unicamp (2011). She works as a Technological Development Analyst on the Division of Integrated Assessment of Biorefineries at CTBE/CNPEM, since June 2012.

Brazilian Bioethanol Science and Technology Laboratory (CTBE). Brazilian Center of Research in Energy and Materials (CNPEM), Campinas, São Paulo, Brazil.

Alexandre Souza has Ph.D. in Food Engineering from the University of Campinas (2011). He is a postdoctoral researcher at the Division of Integrated Biorefinery Assessment at CTBE/CNPEM, since October 2013. His research areas include Social Impact Assessment of projects, Life Cycle Assessment, sustainable development, among others.

Brazilian Bioethanol Science and Technology Laboratory (CTBE). Brazilian Center of Research in Energy and Materials (CNPEM), Campinas, São Paulo, Brazil.

Marcos D.B. Watanabe has a Ph.D. in Food Engineering from the University of Campinas. Currently, his research areas are mainly focused on the techno-economic analysis of investment projects and hybrid input–output Life Cycle Assessment of sugarcane biorefineries.

Brazilian Bioethanol Science and Technology Laboratory (CTBE). Brazilian Center of Research in Energy and Materials (CNPEM), Campinas, São Paulo, Brazil.

Marcelo Zaiat is Chemical Engineer, Master in Chemical Engineering (Federal University of São Carlos—UFSCar, 1990 and 1992), and doctorate in Hydraulics and Sanitary Engineering (University of São Paulo—USP, 1996), with Free-Docency in Biological Wastewater Treatment (USP, 2004). He is associate professor at the São Carlos School of Engineering—USP since 1998 in the area of Environmental Engineering, member of the Coordination Committee of Engineering at São Paulo Research Foundation (FAPESP), and associate researcher at CTBE. He is specialized in Anaerobic Treatment Technology and published nearly 150 papers in double-refereed journals with h index of 21 (Web of Science, ISI). His research interest comprises hydrogen and methane production from wastewater, bioreactor design, and wastewater treatment with anaerobic reactors as the core technology.

University of São Paulo, São Carlos, São Paulo, Brazil.

Abbreviations

.NET	Microsoft .NET Framework
1G	First-generation
1G2G	Integrated first- and second-generation
2G	Second-generation
ABE	Acetone, *n*-butanol, and ethanol
ABNT	Brazilian Technical Standards Association
Adv	Advanced technology
AFEX	Ammonia fiber expansion
ALC	Alcoholchemistry
ALO	Agricultural land occupation
Anfavea	Brazilian Automotive Industry Association
ANOVA	Analysis of variance
ANP	Brazilian National Agency of Petroleum, Natural Gas and Biofuels
Anx	Annexed plant
ASPEN	Advanced system for process engineering
Aut	Autonomous distillery
B30	Baling system for 30 % straw recovery
B50	Baling system for 50 % straw recovery
B70	Baling system for 70 % straw recovery
BDO	1,4-Butanediol
BG	Biogas
BIG/GT	Biomass integrated gasifier/gas turbines
BIG–GT/CC	Biomass-integrated gasification–gas turbine combined cycle
BM	Biomethane
Bio-PE	Biobased polyethylene
Brix (Bx)	Sugar content of an aqueous solution
BRL	Brazilian real
BS	Baling system
But	*n*-butanol

C_{12}	Twelve-carbon sugar (sucrose)
C_2H_4eq	Ethylene equivalent
C4	Four-carbon molecule
C5	Five-carbon sugars (pentoses)
C6	Six-carbon sugars (hexoses)
CAPE	Computer-aided process engineering
CAPEX	Capital expenditures
CC	Conventional sugarcane
CCD	Central composite design
CCGT	Combined cycle gas turbine
CDM	Clean development mechanism
CER	Certified emission reduction
CETESB	Environmental Company of the State of São Paulo
CFC-11	Trichlorofluoromethane equivalent
CFD	Computational fluid dynamics
CH_4	Methane
CHP	Combined heat and power
CIP	Cleaning in place
CNPEM	Brazilian Center for Research in Energy and Materials
CO	Carbon monoxide
CO_2	Carbon dioxide
CO_2-eq	Carbon dioxide equivalent
COD	Chemical oxygen demand
Cog	Cogeneration
COLaN	CAPE-OPEN Laboratories Network
COM	Component Object Model
CORBA	Common Object Request Broker Architecture
CTBE	Brazilian Bioethanol Science and Technology Laboratory
Cur	Current technology
DAE	Differential algebraic equation
DAP	Diammonium phosphate
db	Dry basis
DCOM	Distributed Component Object Model
DDGS	Dried distillers grains with solubles
DECHEMA	Society for Chemical Engineering and Biotechnology
DEPG	Dimethyl ethers of polyethylene glycol
DHFORM	Standard enthalpy of formation
DHSFRM	Solid standard enthalpy of formation
DME	Dimethyl ether
DNIT	Brazilian National Department for Transport Infrastructure
DOE	US Department of Energy
EC	Energy cane
EH	Enzymatic hydrolysis
EIA	US Energy Information Administration
EIO-LCA	Economic Input–Output Life Cycle Assessment

Elec	Electricity
EML	EMSO modeling library
EMSO	Environment for modeling, simulation, and optimization
EU	European Union
FDP	Fossil depletion
FT	Fischer–Tropsch
FT-SPK	Fischer–Tropsch synthetic paraffinic kerosene
FWEP	Freshwater eutrophication
GBL	*g*-Butyrolactone
GHG	Greenhouse gas
GLUCOLIG	Glucose oligomers
GMO	Genetically modified organisms
Gproms	General process modeling system
Grid	Electric grid
GWP	Climate change (global-warming potential)
HAP	Hydroxyapatite
HC	Henry component
HDPE	High-density polyethylene
HMF	Hydroxymethylfurfural
HTP	Human toxicity
I30	Integral harvesting for 30 % straw recovery
I50	Integral harvesting for 50 % straw recovery
I70	Integral harvesting for 70 % straw recovery
IC	Internal circulation
IEA	International energy agency
INPM	Mass ethanol percentage
IO	Input–output
IOA	Input–output analysis
IRR	Internal rate of return
IS	Integral harvesting system
ISO	The International Organization for Standardization
LCA	Life Cycle Assessment
LCF	Lignocellulosic feedstock
LCI	Life Cycle Inventory
LCIA	Life Cycle Impact Assessment
LCM	Lignocellulosic material
LDPE	Low-density polyethylene
LGNSOL	Soluble lignin
LHV	Lower heating value
LLDPE	Linear low-density polyethylene
LP	Low polarization
LPMO	Lytic polysaccharide monooxygenase
MAP	Monoammonium phosphate
MARR	Minimum acceptable rate of return or hurdle rate

MIT	Massachusetts Institute of Technology
MIXCINC	Conventional and nonconventional components
MMA	Brazilian Ministry of the Environment
MPI	Message passing interface
MPS	Brazilian Ministry of Social Security
MTBE	Methyl tert-butyl ether
MTC	Million tonnes of cane
MTE	Brazilian Ministry of Labor and Employment
n.a.	Not applicable
N_2O	Nitrous oxide
NAE	Nonlinear algebraic equation
NC	Nonconventional components
NH_3	Ammonia
NMHC	Non-methane hydrocarbons
NO_3^-	Nitrate
NO_X	Nitrogen oxides
NPV	Net present value
NREL	National Renewable Energy Laboratory (USA)
NRTL	Non-random two-liquid model
NRTL-HOC	Non-random two-liquid activity coefficient model using the Hayden-O'Connell model for the vapor phas
Op	Optimized configuration
OpenMP	Open Multiprocessing
OPEX	Operational expenditures
P	Productivity
PACER	Process Assembly Case Evaluator Routine
PBS	Polybutylene succinate
Pc	Critical pressure
PE	Polyethylene
PHA	Polyhydroxyalkanoate
PHB	Polyhydroxybutyrate
PLA	Polylactic acid
PM	Particulate matter
PM10-eq	Particulate matter equivalent (10 μm or less in diameter)
Pol	Apparent sucrose content
PR-BM	Peng–Robinson equation of state with Boston–Mathias alpha function
PREP-EXT	Preparation and extraction
PSD	Particle size distribution
PSE	Process Systems Enterprise
PVM	Parallel Virtual Machine
Q	Quadratic
QLC	Queensland low color
R&D	Research and development

R	Straw recovery fractions
RCHO	Aldehydes
RKS-BM	Redlich–Kwong–Soave cubic equation of state with Boston–Mathias alpha function
S	Scenario
SC	Sugarcane
SHF	Separate hydrolysis and fermentation
S-LCA	Social Life Cycle Assessment
SNA	System of national accounts
SO2-eq	Sulfur dioxide equivalent
SPK	Synthetic paraffinic kerosene
SSCF	Simultaneous saccharification and co-fermentation
SSF	Simultaneous saccharification and fermentation
SSP	Superphosphate
St	Standard configuration
STEAMNBS	Steam tables
TAP	Terrestrial acidification
Tb	Normal boiling point
Tc	Critical temperature
TC	Tonne of cane (sugarcane stalks)
The	Thermochemical 2G ethanol
THF	Tetrahydrofuran
Tkm	ton–kilometer
TRS	Total reducing sugars
TSP	Triple superphosphate
UASB	Upflow anaerobic sludge blanket
UK	United Kingdom
US$	United States dollar
US	United States
USA	United States of America
V1	Bleed steam from first effect of multiple effect evaporator
V2	Bleed steam from second effect of multiple effect evaporator
VA	Value added
VHP	Very high polarization
VSB	Virtual sugarcane biorefinery
VVHP	Very very high polarization
WGS	Water-gas shift reaction
XYLOLIG	Xylose oligomers

List of Figures

List of Tables

Chapter 1
Background

Marco A.P. Lima, Antonio Bonomi, Otávio Cavalett and Marcelo Pereira da Cunha

Seven billion people living in the planet, with a nearly standard quality of life, demand a large-scale planning for sustainable production of energy, food, fresh water, and raw materials—which must be transformed in a sustainable way into all sorts of products that humans like and need to use. Taking into consideration the twenty-first century and the eight objectives of Millennium Development Goals—related to the Millennium Declaration adopted by United Nations in 2000—, that planning includes to eradicate extreme poverty and hunger, as well as to ensure environmental sustainability in a world with around 9 billion people.

All chemical elements, in principle, are limited in quantity and variety in our planet and they must be used in a rational way. When a particular element (lithium for batteries, for example) is no longer available, we have few choices: recycling it, searching and acquiring it from deeper mines, or developing new technologies to replace its role by other elements. The transformation process from raw materials to other products demands energy and, in the long term, the energy sources must be renewable and sustainable. Burning fossil fuels in large scale to produce electricity, to power engines, and to obtain heat is not sustainable, considering environmental issues and taking into account that its formation requires million years. In the timescale of human society, the ultimate sustainable source of energy is the Sun, a very diluted (in time and space) source of energy. Several sources of energy such as

A. Bonomi (✉) · O. Cavalett
Brazilian Bioethanol Science and Technology Laboratory (CTBE),
Brazilian Center of Research in Energy and Materials (CNPEM), Campinas, SP, Brazil
e-mail: antonio.bonomi@bioetanol.org.br

O. Cavalett
e-mail: otavio.cavalett@bioetanol.org.br

M.P. da Cunha
Institute of Economics (IE), University of Campinas (UNICAMP), Campinas, SP, Brazil
e-mail: marcelocunha@eco.unicamp.br

M.A.P. Lima
Institute of Physics Gleb Wataghin (IFGW), University of Campinas (UNICAMP), Campinas, SP, Brazil
e-mail: maplima@ifi.unicamp.br

A. Bonomi et al. (eds.), *Virtual Biorefinery*, Green Energy and Technology,
DOI 10.1007/978-3-319-26045-7_1

wind, solar, hydroelectric plants, and biomass belong to the sun sustainable cycle. Besides being renewable, sustainability imposes other requirements for energy and material sources and for the outcoming products and use. All these efforts must happen in a clean way without putting in risk our environment. Overloading the atmosphere with carbon dioxide and other radiative forcing agents, known to produce greenhouse effects and possible uncontrollable changes in the climate is too risked and not a clean strategy.

Sustainable processes are, nowadays, as a general rule, more difficult and expensive than the dirty and non-sustainable ones. This is because they impose boundary conditions and narrow down the number of practical solutions for the desired process. Needless to say, but mostly important, the full chain of production must be economically sustainable. This is a non-ignorable boundary condition.

How can we evaluate the impacts of boundary conditions in the energy and material sources production chain? By impacts we mean the social, environmental, and economic aspects of its sustainability. How can we evaluate the maturity of a new technology and the impact of its insertion in the production chain? How can we detect and solve the bottlenecks of a nearly mature technology? How can we indicate the basic science and public policies required for overcoming the bottlenecks? How can we evaluate suggestions for optimization of the production chain? How can we evaluate the potential value of a laboratory scale experiment for industrial applications? The purpose of this book was to show how some of these questions can be addressed and evaluated using computer platforms for a particular production chain: the Brazilian sugarcane biorefinery for sugar, bioethanol, bioelectricity, and other green chemistry products.

In this book, sugarcane biorefinery was chosen because it has played a strategic role in the Brazilian energy matrix as well as in the economy. For economic reasons, in 1975, Brazil created an ambitious program (Proalcool) for replacing part of the imported oil with ethanol from sugarcane, taking the advantage of an already-established sugar production sector. At the time, oil import was quite expensive and ethanol produced with Brazilian currency would balance the trade economy of the country. The automobile industry developed cars with engines designed to run exclusively with ethanol. As a result, quite rapidly, in the mid-1980s, more than 90 % of new light vehicles produced in the country were powered by ethanol and, from 1985 to 1995, fuel ethanol consumption (in volume) was higher than gasoline in the country. Due to the oil counter shock in the 1980s, gasoline consumption increased again, but fuel ethanol kept its share at least in 41 % from 1996 to 2003. From 2002 to 2008, the third oil shock motivated many countries to establish biofuels policies, including the USA, which creates a corn ethanol program. The introduction of flex fuel cars in Brazilian market in 2003, as well as an international market potential for fuel ethanol, boosted again Brazilian sugarcane industry. Sugarcane industry saw this as an opportunity for producing extra energy by burning all the bagasse and selling the surplus electricity to the grid. Many of the about 400 industries became a sort of biorefinery, producing three large-scale products: sugar, ethanol, and bioelectricity. A natural question arises: What can be done to turn this industry into a real high-tech biorefinery and look at the sugarcane not only as a source of energy, but also as a source of carbon? The virtual biorefinery described in this book was designed to assess and exploit these possibilities of carbon valorization.

Brazil has reached a situation in the use and exploitation of renewable fuels that it requires a well-planned development, considering the different routes of production mainly when the green chemistry and biorefinery concepts are incorporated in the analysis. The construction of a computer platform, like the Virtual Sugarcane Biorefinery, described in this book, has proven so far to be a good instrument to define priorities in science, technology, and public policies to be implemented by the Brazilian Government in order, for example, to set an industrial strategy aiming at boosting the ethanol production.

As it will become clear in the book, the technological assessment can serve as a guide for stimulating oriented basic science, pointing the need of new technologies, improving existing technologies, and even establishing a fully integrated view of existing industries. The latter can be useful to properly point out and solve technological bottlenecks or to evaluate the inclusion of newly integrated processes or new technologies in these industries, considering the existing regulations for preserving and implementing sustainable (and profitable) strategies.

Several companies are considering to use the same assessment strategy, in order to perform preliminary screening to set their business plans and research programs in the area, as well as using it to evaluate the obtained results in their efforts to develop new technologies, from the laboratories throughout the scale-up to pilot and demonstration plants. Therefore, such a tool can also be used for the bioenergy chain optimization considering also the valorization of renewable carbon.

Chapter 2
The Virtual Sugarcane Biorefinery Concept

Antonio Bonomi, Otávio Cavalett, Marcelo Pereira da Cunha and Marco A.P. Lima

Biorefinery is a facility that integrates biomass conversion processes and equipment to produce fuels, power, materials, and chemicals from biomass. The biorefinery concept is analogous to today's petroleum refineries, which produce multiple fuels and products from fossil oil. Industrial biorefineries have been identified as the most promising routes for the creation of a new domestic biobased industry (Kamm et al. 2006). Figure 2.1 shows a general scheme of biorefinery concept.

The Brazilian Bioethanol Science and Technology Laboratory (CTBE), one of the four National Laboratories that integrate the National Center for Research in Energy and Materials, was inaugurated by the Ministry of Science, Technology and Innovation of the Brazilian Government in 2010. The aim of CTBE foundation is to contribute in deepening the Brazilian leadership in the sectors of renewable energy sources and chemical industry raw material production, mainly by improving the sugarcane bioethanol production chain, through research, development, and innovation, integrating the productive sector and the Brazilian scientific-technology community.

Since the beginning of CTBE's construction, it was conceived that the idea of developing a "tool" (i.e., an applied methodology) that would allow comparing different technological routes and would contribute for the decision-making process.

A. Bonomi (✉) · O. Cavalett
Brazilian Bioethanol Science and Technology Laboratory (CTBE),
Brazilian Center of Research in Energy and Materials (CNPEM),
Campinas, SP, Brazil
e-mail: antonio.bonomi@bioetanol.org.br

O. Cavalett
e-mail: otavio.cavalett@bioetanol.org.br

M.P. da Cunha
Institute of Economics (IE), University of Campinas (UNICAMP),
Campinas, SP, Brazil
e-mail: marcelocunha@eco.unicamp.br

M.A.P. Lima
Institute of Physics Gleb Wataghin (IFGW), University of Campinas (UNICAMP),
Campinas, SP, Brazil
e-mail: maplima@ifi.unicamp.br

A. Bonomi et al. (eds.), *Virtual Biorefinery*, Green Energy and Technology,
DOI 10.1007/978-3-319-26045-7_2

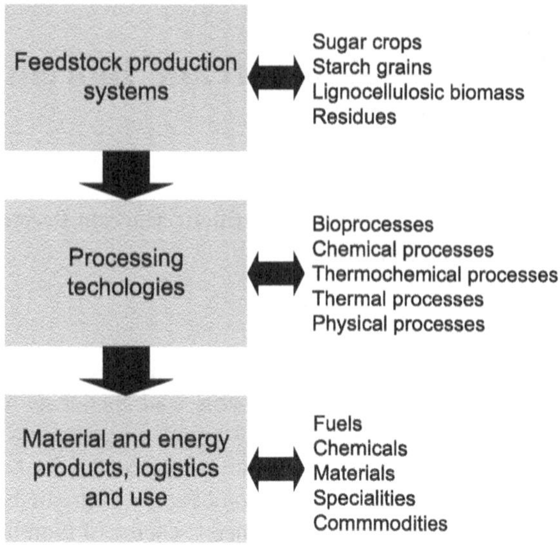

Fig. 2.1 Basic principles of a biorefinery (based on Kamm and Kamm 2004)

This technological assessment tool emerged from the need of setting a methodology to measure CTBE's success. With this purpose, the construction of a simulation tool was designed—the Virtual Sugarcane Biorefinery (VSB). A plan containing the idea, objectives, scope, and methodology for the VSB construction was submitted, evaluated, and approved by the bioethanol and sugarcane community (including academy and industry) in the workshop "VSB: Assessing success of new technologies," organized by CTBE in 2009.

The VSB is a simulation platform that allows the evaluation of the integration of new alternatives for the sugarcane sector—cellulosic ethanol and other products from the green chemistry in the biorefinery concept, new agricultural strategies for sugarcane production, as well as different strategies for ethanol and other products used as biofuels and biochemicals. In a first approach, technologies adopted today in the whole sugarcane production chain are used as reference for this evaluation. The results obtained with the VSB have been validated against existing plants or results from laboratory or pilot scales available experiments for the new technologies, in order to guarantee the accuracy of the sustainability impacts calculated with this simulation tool.

The VSB has also been used to assess the level of success reached by CTBE's Pilot Plant for Process Development in the development of new industrial technologies in the biorefinery concept, as well as the CTBE's Agriculture Program innovations.

This tool also considers the logistics and use of biorefinery products, therefore including their entire production and consumption chain. Methodologies used for the assessment are identified and developed together with the sustainability experts at CTBE.

Therefore, from the above-described features of the VSB, it is clear its importance as a tool for the continuous evaluation and improvement of CTBE's research activities, as well as to evaluate the potential of several possible alternatives and technologies covering all aspects of the program. Summarizing, the VSB strategy aims at developing an analysis tool for sugarcane biorefineries, including the agricultural, industrial, and product-use sectors, which will make it possible to:

- optimize the concepts and processes included in a biorefinery;
- assess different biorefinery alternatives referring to their sustainability impacts (economic, environmental, and social); and
- assess the stage of development of the new technologies included in the analysis.

The VSB is based on mathematical modeling and simulation of different processes and operations included in the sugarcane production chain (agricultural, industrial, and usage sectors), which allow the estimation and optimization of the economic, environmental, and social impacts obtained with the new technologies under development. These results allow assessing the stage of development of new technologies, as well as the interest in accelerating the implementation process, orienting the laboratories participating in the development, about possible optimum operating conditions, and looking for their experimental confirmation.

The development of models and the use of computational tools (including specific commercial software) make it possible to assess the impacts of new technologies on the Brazilian bioethanol production chain in the three pillars of the sustainability. Figure 2.2 illustrates the general concept of the VSB. Methods used for the sustainability assessment are summarized below and detailed in Chap. 6.

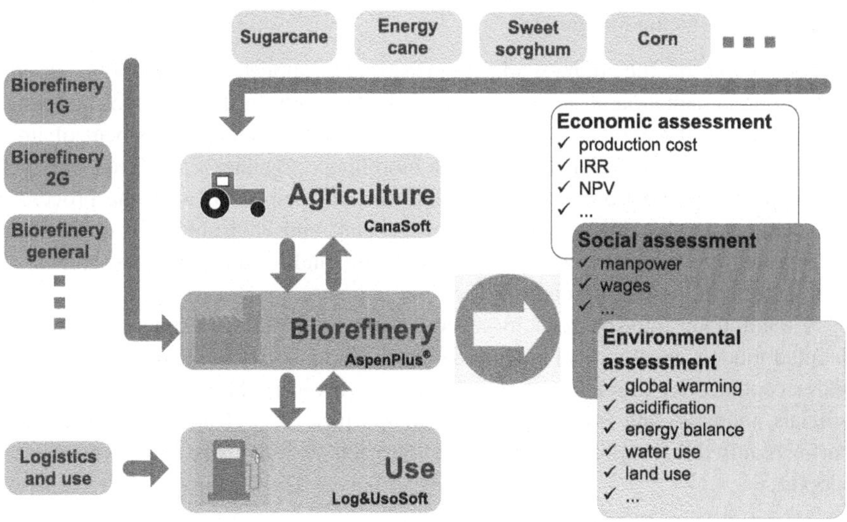

Fig. 2.2 General concept of the VSB

- **Economic**: Required investment, profitability (internal rate of return (IRR), net present value (NPV), and other parameters), revenues, production costs, and minimum selling prices of products and taxes among other parameters and their implications in the production chain are evaluated using economic engineering tools. At the same time, the sensitivity analysis of the most important technical parameters included in the technologies under development and test on the related costs and investments is performed, as well as a risk analysis related with the implementation of the new technologies.
- **Environmental**: Energy balance (relation between the renewable energy produced and the fossil energy consumed), greenhouse gas emissions balances, water consumption, and other environmental impacts included in the Life Cycle Assessment (LCA) methodology such as acidification, photo-oxidant formation, nitrification, eutrophication, and human toxicity are evaluated. New concepts and models such as land-use changes and improved methods for environmental impact assessment are being introduced in the environmental analysis of biofuels and biochemicals.
- **Social**: Some local impacts derived from the automation, plant scale, agricultural sector mechanization, among others, on the number and quality of created jobs (number of workers, occupational accidents, wages, and education profile), as well as land use, social relations with the community, and labor qualification are assessed. These impacts are established using a hybrid approach integrating Social Life Cycle Assessment (S-LCA) methodology and Input–Output Analysis (IOA). These economic models allow for the quantification of the changes in the activity level of each sector of the economy as a function of modifications on demand for products of one or more sector. The major advantage of the use of this hybrid methodology is the possibility of including the evaluation of indirect impacts that are normally unassessed in general Social Life Cycle methodologies.

The scope of the framework described and applied in this book is the construction/adaptation of a simulation platform aiming at helping the modeling, optimization and technical, socioeconomic and environmental assessment of integrated processes, major characteristic of a biorefinery, together with all the stages of the sugarcane production chain. This platform will help to identify the processes and parameters showing major economic, social, and environmental impacts, in order to help in the prioritization of the scientific and technological researches, as well as to help in the formulation of public policies.

A continuous improvement in several aspects of VSB fundaments, including detailed models development and validation of databases, is fundamental for its use and acceptance for the different stakeholders involved in the process (government officials, industry, and researchers). For this purpose, it is clear that financial support is required and should be provided by projects with industry and governmental support.

2.1 Agricultural Sector

VSB defines alternatives and includes the simulation of the agricultural operations required to produce and make the biorefinery feedstock—mainly the sugarcane, among alternative feedstock, available to the industry. These operations can be synthetically described as follows:

- preplanting operations;
- soil preparation;
- manual or mechanical planting;
- cultivation—sugarcane plant and ratoon;
- manual or mechanical harvesting; and
- sugarcane transport.

In order to evaluate the technical, socioeconomic, and environmental impacts of different technological scenarios, an agricultural spreadsheet-based model (including a detailed description of the above-mentioned operations), named "CanaSoft" was constructed, validated, and integrated to the simulation tools used to represent the other sectors of the sugarcane production chain (a detailed description of this platform is presented in Chap. 3).

2.2 Industrial Sector

In order to make the inclusion and the sustainability assessment of several biorefinery alternatives viable, it is necessary to define and technically evaluate different proposals and routes to transform biomass into products. The VSB focuses on sugarcane as the biomass to be used and the first- and second-generation bioethanol as the major product. It includes the analysis of other products such as sugar, electricity, other liquid fuels (obtained using the thermochemical route to convert the lignocellulosic material), materials (such as the polyhydroxyalkanoates obtained through sugars fermentation), and primers for the chemical industry (obtained from ethanol, sugar, or fractions of the lignocellulosic material), among others.

Therefore, some basic routes were designed and technically assessed, being a basis for the construction of the VSB, always using sugarcane as the major feedstock.

The simulation platform represents the different basic routes proposed in the construction of the VSB. Several commercial packages are available in the market, which are oriented to process simulation. ASPEN Plus (www.aspentech.com/products/aspen-plus.aspx), SuperPro Designer (www.intelligen.com/superpro_overview.html), and EMSO (www.enq.ufrgs.br/trac/alsoc/wiki/EMSO) are just some possibilities of commercial simulation packages. They were developed for a large spectrum of industries: oil, petrochemicals, pharmaceuticals, biotechnologies, fine chemistry, mineral processing, microelectronic, and effluent treatment, among others. For several reasons, which will be detailed in the topic related to the construction of the industrial sector of the VSB, Aspen Plus was selected as the platform for this simulation (a detailed description of the most important process simulations is presented in Chap. 4).

2.3 Logistics and Use Sectors

In order to complete the sugarcane production chain, the last sector to be simulated is the one that includes operations of commercialization and use of the different products produced in the biorefinery. If, for example, the ethanol is considered as the product to be assessed, the major operations to be considered for the simulation are as follows:

- transport of ethanol to/among the commercialization agents;
- mixture with gasoline (gasohol alternative); and
- final use (end of life) of ethanol in the vehicles.

A spreadsheet-based model so-called Log&UsoSoft was constructed detailing the operations involved with logistics, considering different transportation distances and modes; use, considering the product exercising a function; and end of life, considering the emissions and final disposal, of the different products of a biorefinery, allowing for the assessment of the sugarcane production chain completely (a detailed description is presented in Chap. 5).

Finally, it is clear that three development versions of the VSB are defined in order to characterize the quality and accuracy of the simulation performed during the use of the VSB for the assessment purposes. The descriptions of these versions are illustrated for the industrial sector, but they can be applied, with minor adjustments, to the other sectors of the production chain.

- **Preliminary Version**: All the simulations are performed based on preliminary flow diagrams and, in general, using data available in the literature.
- **Consolidated Version**: All the simulations are performed based on a conceptual design performed for the assessed technology or using operation description discussed in detail with specialists, when the other sectors of the chain are considered.
- **Validated Version**: The parameters used in the simulation are compared with the results obtained using data measured or obtained in commercial operations. Generally, three levels of validation, validated against one technology, validated against different technologies, and validated against different technologies and regional conditions, are considered.

The VSB has been successfully applied to compare different technological scenarios from government (public policies) and business aims, as well as to evaluate research bottlenecks to be prioritized. For instance, the VSB was used to perform a comprehensive assessment of evolution of 2G ethanol technologies in time to estimate the learning curve of 2G ethanol cost. These results were validated against dozens of stakeholders. The study counted with participation of the Brazilian Development Bank (BNDES), and it is a great example of the use of VSB for promoting public policies, considering and defining priorities and performing a periodic evaluation of the research progress (Milanez et al. 2015).

The VSB has also been used in several technology development projects in collaboration with companies, performing a periodic evaluation of the research achievements and directing efforts in the research and development.

References

Kamm B, Kamm M (2004) Principles of biorefineries. Appl Microbiol Biotechnol 64(2):137–145. doi:10.1007/s00253-003-1537-7

Kamm B, Gruber PR, Kamm M (eds) (2006) Biorefineries—industrial processes and products, vol 1. Wiley-VCH, Weinheim, p 441

Milanez AY, Nyko D, Valente MS, Sousa LC, Bonomi A, Jesus CDF, Watanabe MDB, Chagas MF, Rezende MCAF, Cavalett O, Junqueira TL, Gouvêia VLR (2015) De promessa a realidade: como o etanol celulósico pode revolucionar a indústria da cana-de-açúcar – uma avaliação do potencial competitivo e sugestões de política pública. BNDES Setorial 41:237–294

Chapter 3
The Agricultural Production Model

Otávio Cavalett, Mateus F. Chagas, Paulo S.G. Magalhães, João L.N. Carvalho, Terezinha F. Cardoso, Henrique C.J. Franco, Oscar A. Braunbeck and Antonio Bonomi

3.1 Biomass for Energy

Bioenergy has been proposed as a feasible strategy to mitigate greenhouse gas (GHG) emissions, because carbon emitted by combustion of biofuels can be partially offset by uptake through photosynthesis, promoting carbon recycling. Bioenergy is defined as all energy that is produced from biological mass that is available on a renewable basis and include liquid fuels (first- and second-generation biofuels for transportation), gaseous fuels (biogas), and solid fuels (for co-firing, heating, and electricity generation) (Don et al. 2012). Numerous crops have been used to produce bioenergy around the world, b'ut not all of them meet the requirements of a high yielding environmentally sustainable feedstock for bioenergy use. The ideal bioenergy crop would be the one that has fast growth, high biomass yield (Souza et al. 2013), with low energy input, low production costs, low levels of contaminants, and a low nutrient requirement (McKendry 2002).

Among the alternatives of biomass intended for renewable energy purposes, sugarcane is highlighted as a great alternative. It is a semi-perennial crop with potential for high accumulation of sugar and fibers. Therefore, this biomass is seen as a source of relative low-cost carbon for future biorefineries.

O. Cavalett · M.F. Chagas · P.S.G. Magalhães · J.L.N. Carvalho · T.F. Cardoso
H.C.J. Franco · O.A. Braunbeck · A. Bonomi (✉)
Brazilian Bioethanol Science and Technology Laboratory (CTBE), Brazilian Center of Research in Energy and Materials (CNPEM), Campinas, SP, Brazil
e-mail: antonio.bonomi@bioetanol.org.br

O. Cavalett
e-mail: otavio.cavalett@bioetanol.org.br

P.S.G. Magalhães
School of Agricultural Engineering (FEAGRI), University of Campinas (UNICAMP), Campinas, SP, Brazil

A. Bonomi et al. (eds.), *Virtual Biorefinery*, Green Energy and Technology,
DOI 10.1007/978-3-319-26045-7_3

Ethanol is one of the most widespread biofuel and the global production in 2014 was 93 billion liters, with the USA and Brazil producing 58 and 26 % of this amount, respectively (Renewable Fuel Association 2015). In Brazil, almost all ethanol is produced from sugarcane (Saccharum spp.), and recently, some initiatives have proposed to include other crops as raw materials in ethanol production in Brazil. Most of these crops should be produced in crop rotation with sugarcane, such as corn and sweet sorghum, while other are cultivated in dedicated lands, as energy cane.

In Brazil, the area cultivated with sugarcane is undergoing significant expansion due, mainly, to the growing demand for bioenergy production. Brazil is the largest producer of sugarcane, producing 634.8 million tonnes from an area of 9.0 million hectares in the 2014/2015 harvest. The south-central region accounted for 90 % of the national production of sugarcane, producing annually 37.9 million tonnes of sugar and 27.9 billion liters of ethanol (CONAB 2015).

Sugarcane is very productive C4 plant, which have a higher efficiency in converting solar energy in biomass and, therefore, it is expected that also produce a significant amount of primary energy per unit of cultivated area, when compared to other bioenergy crops. Historically, sugarcane has been cultivated for centuries to produce sugar, but in last decades escalating oil prices, energy security concerns, global warming problem, and agriculture products' low price increases interest in bioenergy and biofuel production (Leal et al. 2012). Sugarcane is a high-yield bioenergy crop and the primary energy contained in the whole plant is 7.4 GJ per Mg of cane stalks, based on the average Brazilian sugarcane quality. However, only 1/3 of this amount is used to produce sugar and ethanol (Leal et al. 2013) and the rest of the energy is contained in bagasse and in sugarcane straw, which potentially could be used to produce second-generation ethanol and electricity by cogeneration. However, sugarcane straw is currently used as soil amendment in sugarcane fields in order to improve soil quality indicators. Despite sugarcane already be a very productive crop in comparison with other bioenergy crops, studies have indicated a very high discrepancies between the real and potential yields in Brazilian conditions and most part of these differences are associated with the management practices used in the field (Goes et al. 2011).

Although sugarcane to be a highly efficient crop to accumulate biomass, to reach high levels of yield is necessary to modify several process of biomass production in the in agricultural phase, which represent around 50–60 % of total ethanol cost. (Cardoso et al. 2013; Braunbeck and Magalhães 2010). Thus, the knowledge of all steps of the agricultural phase of the sugarcane production is imperative in order to improve the yields and improve the sustainability of bioethanol production in Brazil.

Computer modeling and simulation of industrial chemical processes are recognized powerful tools to simulate, predict, and calculate mass and energy balances for industrial processes and can play an important role for a comprehensive evaluation framework and stimulate proper public policy for the bioenergy sector. For the evaluation of agricultural production systems, there is no software readily available for this purpose, due to its complexity, specificity, variability, interaction with environment and other inherent characteristics of agricultural systems. To overcome this lack, which in fact represents a challenge, this chapter presents an agricultural model, so-called CanaSoft, which is intended for simulation and assessment of the most important agricultural parameters of biomass production system alternatives and for integrated sustainability optimization of bioenergy chain. The complete description of the parameters considered in the model is provided as well as description of means to adapt it for other purposes. A case study of application of the model as well as its adaptation to other biomass is presented.

3.2 Description of Sugarcane Production System

Sugarcane is a semi-perennial crop sensitive to the climate, soil type, irrigation, fertilizers, insects, disease control, varieties, and the harvest period. The average yield of cane stalk is 60–100 Mg per hectare per year. However, it can vary depending on crop management practices and production region. After planting, sugarcane is harvested several times (usually 5 or 6 times in center-south of Brazil) before being replanted (Fig. 3.1).

The Brazilian sugarcane industry employs modern agricultural management practices including use of machinery, synthetic fertilizers, and pesticides. The main sugarcane operations are briefly described in the next sections of this chapter.

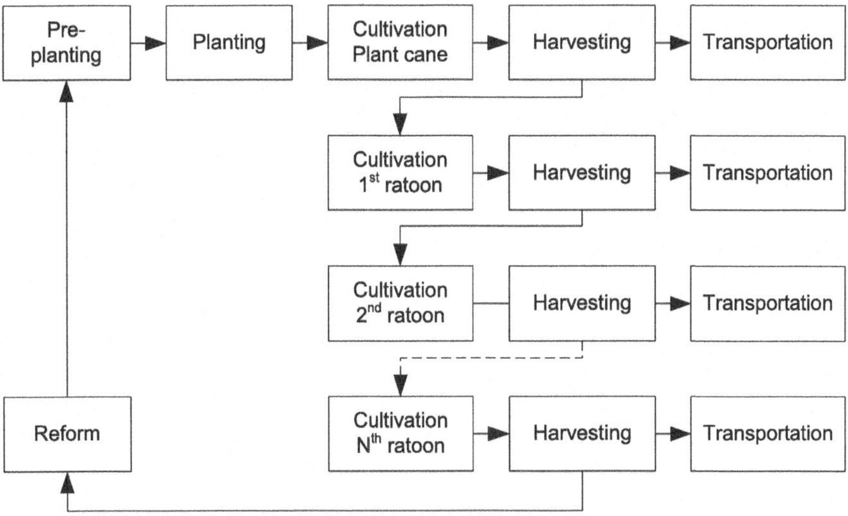

Fig. 3.1 Typical sugarcane production and transportation phases

3.3 The CanaSoft Model

Computer modeling and simulation are recognized powerful tools to simulate, predict, and calculate mass and energy balances for industrial processes. These tools play an important role for evaluation of technical aspects of industrial conversion alternatives for biomass.

However, agricultural production models are scarcer and there is no operational and readily available tool for modeling sugarcane production systems. The main reasons are the complexity, specificity, variability, interaction with environment, and other inherent characteristics of agricultural production systems.

To overcome this lack, which in fact represents a challenge, a computer model, so-called CanaSoft model, has been developed at CTBE. CanaSoft is intended for simulation and measurement of the most important agricultural parameters of biomass production system alternatives.

Although CanaSoft model contains "default" data for a typical sugarcane production system, its construction was planned to be customizable to other biomass production as well. Section 3.5 presents the main aspect of adapting CanaSoft for other biomass.

The CanaSoft model is developed as interconnected electronic spreadsheets and integrates several calculation modules. Starting from main agricultural production parameters that define a biomass production system (yields, type of planting, type of harvesting, use of fertilizers, use of agrichemicals, form of disposal of vinasse, transportation logistics of sugarcane stalks and straw, among others factors), the model provides economic, environmental, and social inventories.

Inventories calculation is linked to the calculation modules that contain the complete information about agricultural operations and inputs used in the biomass production system such as machinery performance parameters, fertilizer application areas, and emission factors. Calculation modules are connected to agricultural databases that contain basic information parameters such as machinery weight, initial investments, lifespan, and input prices. The calculation is therefore organized in interconnected modules representing different biomass production phases (e.g., preplanting, planting, cultivation, harvesting, and transportation).

The general framework used for development of the CanaSoft model is illustrated in Fig. 3.2.

The Canasoft model is intended to be useful to the sugarcane sector as supporting tool for strategic analysis, improvement programs, and optimal utilization of inputs and natural resources. It is expected that an improved decision-making process supported by assertive agricultural modeling will bring positive implications on productivity gains, profitability, and competitiveness for the sugarcane industry in the short and long term.

This quantitative assessment of sustainability impacts of alternative sugarcane production systems can also support new initiatives to add value and remuneration of this activity due to socioenvironmental benefits (positive externalities) that may be produced.

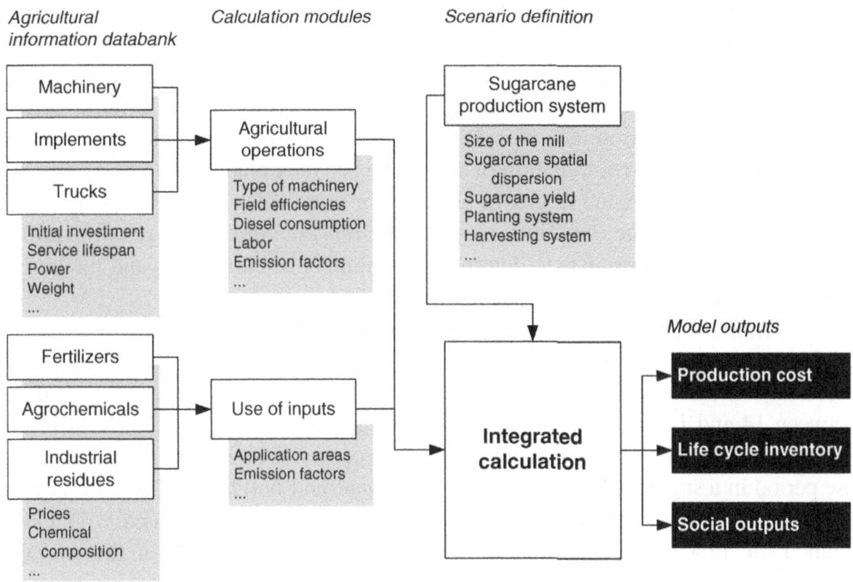

Fig. 3.2 Agricultural model scheme showing different calculation modules

3.3.1 Sugarcane Production Modeling

The biomass cultivation modeling comprises all cost and environmental impacts elements in all biomass cultivation steps, during the entire cultivation period. The CanaSoft model is designed to assess sugarcane production at mill scale, considering all the agricultural operations related to sugarcane production process, from preplanting to post-harvest, including transportation of raw materials to the field and of sugarcane from field to mill.

Each operation is modeled considering the consumption of infrastructure, raw materials, labor, and emissions to air, water, and soil, among other factors, related to one hectare of the operation. Since sugarcane is a semi-perennial crop, not all the operations are performed in the entire area of the mill every year. To solve this problem, the analytical unit for economic, environmental, and social assessments in CanaSoft model is one *representative hectare*, which includes all the agricultural practices needed for the crop system according to the proportion that each practice occurs in the total area of the mill. For example, if planting occurs once in a 5-year cycle, a representative hectare will include 20 % of one hectare of planting operation.

The parameters that define a sugarcane production system in CanaSoft model are presented in the subtopics of this section, as well as the mathematical relations of those parameters used to determine the representative hectare of analysis.

Modeling of hectare of agricultural operations, inputs use, and sugarcane transportation will be addressed in Sects. 3.3.2 to 3.3.4.

3.3.1.1 Sugarcane Production Cycle

Sugarcane production cycle is normally about 5 or 6 years long. It is renovated (replanted) when sugarcane yield is low enough according to the criteria of the producer. The production area of a steady operational mill is composed by plots of planting, plant cane (denomination of sugarcane between planting and first cut), and ratoon cane (the sugarcane re-grown after one or more cuts), in order to maintain a stable sugarcane production. Thus, to model the *representative hectare* of the mill it is of fundamental importance to determine the number of cuts (N_{cuts}) and the length of the average production cycle (L_{cycle}).

The length of the average production cycle depends on the number of cuts and on how much time after planting the first cut is performed. The first cut may be performed 12–14 months after planting, in the so-called year sugarcane system, or between 14 and 18 months, in the so-called year-and-half sugarcane system. In the last case, plant cane (first harvesting) yield is expected to be higher, just as the land use period in a sugarcane cycle. Either year or year-and-half first cut system may be adopted in planting areas, and fractions of systems used must sum 1 (Eq. 3.1). Once defined the first cut system and the number of cuts, the length of the average production cycle can be calculated using Eq. 3.2.

$$f_{year} + f_{year-and-half} = 1 \tag{3.1}$$

f_{year} fraction of the year first cut system in planting area, and
$f_{year\text{-}and\text{-}half}$ fraction of the year-and-half first cut system in planting area.

$$L_{cycle} = f_{year} N_{cuts} + f_{year-and-half}(N_{cuts} + 1) \tag{3.2}$$

L_{cycle} length of the average production cycle (years), and
N_{cuts} number of sugarcane cuts in a production cycle.

3.3.1.2 Crop Establishment

The establishment of sugarcane crop includes all the operations for land preparation and planting, which occur in the beginning of sugarcane cycle. The crop establishment includes the appreciation of the areas that will be reformed and prepared for a new crop cycle; incorporation of new areas that will be used for the production of sugarcane; preparation of production area maps as well as the planting rows usually using remote sensing tools. Crop establishment is often accompanied by construction or renovation of pathway, field roads, and water contention curves. Soil sampling, recommendation of fertilizers application, and sugarcane variety selection are also performed at this stage.

The area allocated for crop establishment depends on the expected sugarcane cycle length, Eq. 3.3.

$$F_{\text{planting}} = \frac{1}{L_{\text{cycle}}} \tag{3.3}$$

F_{planting} fraction of planting in a representative hectare, and
L_{cycle} length of the average production cycle (years).

A set of preplant operations are necessary in order to prepare the land for a new cycle of sugarcane cultivation and they depend on the previous land occupation. Sugarcane may be planted in new areas (not previously occupied with sugarcane) or in existent areas (reform of existing sugarcane fields); fractions of both types of planting areas must totalize 1 (Eq. 3.4).

$$f_{\text{reform}} + f_{\text{expansion}} = 1 \tag{3.4}$$

f_{reform} fraction of expansion area in planting area, and
$f_{\text{expansion}}$ fraction of reform area in planting area.

Fractions of each type of planting area in a representative hectare are given in Eqs. 3.5 and 3.6.

$$F_{\text{reform}} = F_{\text{planting}} f_{\text{reform}} \tag{3.5}$$

$$F_{\text{expansion}} = F_{\text{planting}} f_{\text{expansion}} \tag{3.6}$$

F_{reform} fraction of expansion area in a representative hectare, and
$F_{\text{expansion}}$ fraction of reform area in a representative hectare.

It is necessary to eliminate the previous sugarcane plants in the reforming areas. This operation may be performed with the use of herbicides (usually glyphosate) or mechanically (mainly for pest control) using a specific implement or harrowing.

Tillage operations are usually employed to prepare the soil for planting. The most common agricultural operations are subsoiling, harrowing, leveling harrowing, and application of limestone and gypsum to correct soil acidity. Adoption of some of these agricultural operations may be suspended whether reduced tillage or no-till practices may be adopted. Conventional, reduced, or no-tillage systems may be adopted for planting (Eq. 3.7), and their fraction in a representative hectare is expressed in Eqs. 3.8–3.10.

$$f_{\text{conv till}} + f_{\text{reduc till}} + f_{\text{no till}} = 1 \tag{3.7}$$

$f_{\text{conv till}}$ fraction of conventional tillage use in planting area,
$f_{\text{reduc till}}$ fraction of reduced tillage use in planting area, and
$f_{\text{no till}}$ fraction of no-tillage use in planting area.

$$F_{\text{conv till}} = F_{\text{planting}} f_{\text{conv till}} \tag{3.8}$$

$$F_{\text{reduc till}} = F_{\text{planting}} f_{\text{reduc till}} \tag{3.9}$$

$$F_{\text{no till}} = F_{\text{planting}} f_{\text{no till}} \tag{3.10}$$

$F_{\text{conv till}}$ fraction of conventional tillage in a representative hectare,
$F_{\text{reduc till}}$ fraction of reduced tillage in a representative hectare, and
$F_{\text{no till}}$ fraction of no tillage in a representative hectare.

The planting operation itself is mainly performed in two ways: semi-mechanized or mechanized. The semi-mechanized planting starts with furrow opening along with fertilizer application. The sugarcane seedlings are manually harvested in a nursery area and then transported to the planting area. The sugarcane distribution in the furrow and cutting of stalks are done manually, while the furrow opening and closing are mechanically performed. Closing operation is usually coupled with application of insecticide, nematicide, and micronutrients.

In the mechanized planting, collection of sugarcane seedlings is performed with an adapted harvester (rubberized coating of some internal parts). Sugarcane seedlings are then transported and discharged in the tractor-driven planter. These planters perform various operations including furrow opening, fertilization, distribution of seedlings, application of agrochemicals, and furrow closing.

The sum of the types of planting system fractions must be equal to 1 (Eq. 3.11). Fractions of each type of planting area in a representative hectare are given in Eqs. 3.12 and 3.13.

$$f_{\text{semi-mech-plant}} + f_{\text{mech-plant}} = 1 \tag{3.11}$$

$f_{\text{semi-mech-plant}}$ fraction of semi-mechanized planting in planting area, and
$f_{\text{mech-plant}}$ fraction of mechanized planting in a planting area.

$$F_{\text{semi-mech-plant}} = F_{\text{planting}} f_{\text{semi-mech-plant}} \tag{3.12}$$

$$F_{\text{mechanized planting}} = F_{\text{planting}} f_{\text{mech-plant}} \tag{3.13}$$

$F_{\text{semi-mech-plant}}$ fraction of semi-mechanized planting in a representative hectare, and
$F_{\text{mech-plant}}$ fraction of mechanized planting in a representative hectare.

3.3.1.3 Cultivation

Although there are different agricultural practices for cultivation of plant and ratoon cane, main operations are weeds and pests monitoring and control, preanalysis of sugarcane, fertilization, and subsoiling (to reduce soil compaction). In plant cane, leveling harrowing operation may be necessary to avoid losses of stalks during harvesting. This operation is carried out mechanically and fertilizer or herbicides application may be coupled to this operation.

Herbicides are usually applied after harvesting using a sprayed coupled tractor. The molecule(s) used depends on the period of the year (soil moisture condition), target plant, and amount of straw on the ground. If a pest infestation is detected, its control is done chemically (pesticides) or via biological control. The manual weed control is widely used in Brazilian sugarcane fields and has the advantage of low herbicide use.

Maturators may also be applied using airplanes in order to increase the sugar content in the stalks at the beginning of the crop season or its maintenance at the end of the harvesting season, depending on weather conditions. A wide range of molecules can be used depending on the country legislation but, in general, physiological maturators without herbicide action with low toxicity are preferred.

Fraction of plant cane in the representative hectare is equal to fraction of planting area (Eq. 3.14), and fraction of ratoon area can be defined as the relation between the number of ratoon cuts (number of cuts in a sugarcane cycle minus the cut of plant cane) and the length of the cycle (Eq. 3.15).

$$F_{\text{plant cane}} = F_{\text{planting}} = \frac{1}{L_{\text{cycle}}} \tag{3.14}$$

$$F_{\text{ratoon}} = \frac{(N_{\text{cuts}} - 1)}{L_{\text{cycle}}} \tag{3.15}$$

$F_{\text{plant cane}}$ fraction of plant cane in a representative hectare, and
F_{ratoon} fraction of ratoon in a representative hectare.

3.3.1.4 Harvesting and Transportation

Sugarcane harvesting and loading has been historically based on manual operations. It is a result from sugarcane field conditions such as amount of lodged stalks, length of stalks, slope conditions as well as socioeconomic factors. A first step toward mechanization was the introduction of mechanical loading of sugarcane stalks. The principle mechanical sugarcane harvest currently used in Brazil was invented in Australia in the 1950s. It combines the cutting and loading operations with transport. Mechanized harvesting of billet cane consolidated its position as the most versatile technology in terms of efficiency in a wide range of yield and plant land conditions. The harvesters available today process one or two rows of cane at a time.

Fraction of harvest area in the representative hectare is defined in Eq. 3.16. Harvest of sugarcane can be either manual or mechanized (Eq. 3.17), with or without preharvest burning (Eq. 3.18).

$$F_{\text{harvest}} = \frac{N_{\text{cuts}}}{L_{\text{cycle}}} \tag{3.16}$$

$$f_{\text{manual}} + f_{\text{mech}} = 1 \tag{3.17}$$

$$f_{\text{burned}} + f_{\text{green}} = 1 \tag{3.18}$$

F_{harvest} fraction of harvest area in the representative hectare,
N_{cuts} number of cuts in a sugarcane cycle,
L_{cycle} length of the average production cycle (years),
f_{manual} fraction of manual harvest in harvesting area,
f_{mech} fraction of mechanical harvest in harvesting area,
f_{burned} fraction of burned sugarcane in harvesting area, and
f_{green} fraction of unburned sugarcane in harvesting area.

Manual harvest is normally preceded by the practice of burning sugarcane fields to increase the efficiency of manual cutting. Manual harvest of green cane is less common; it is usually performed in the cutting of sugarcane seedlings. Mechanical harvest does not require preharvest burning. In spite of this, mechanical harvest of burned sugarcane is still practiced in some areas. Fractions of each type of harvesting in a representative hectare are given in Eqs. 3.19–3.22.

$$F_{\text{mech green}} = F_{\text{harvest}} \cdot f_{\text{mech}} \cdot f_{\text{unburned}} \tag{3.19}$$

$$F_{\text{mech burned}} = F_{\text{harvest}} \cdot f_{\text{mech}} \cdot f_{\text{burned}} \tag{3.20}$$

$$F_{\text{manual green}} = F_{\text{harvest}} \cdot f_{\text{manual}} \cdot f_{\text{unburned}} \tag{3.21}$$

$$F_{\text{manual burned}} = F_{\text{harvest}} \cdot f_{\text{manual}} \cdot f_{\text{burned}} \tag{3.22}$$

$F_{\text{mech green}}$ fraction of mechanical harvest of green cane in a representative hectare,
$F_{\text{mech burned}}$ fraction of mechanical harvest of burned cane in a representative hectare,
$F_{\text{manual green}}$ fraction of manual harvest of green cane in a representative hectare, and
$F_{\text{manual burned}}$ fraction of manual harvest of burned cane in a representative hectare.

Three types of trucks are available for sugarcane transport assessment in CanaSoft model: *Romeu e Julieta, Treminhão,* and *Rodotrem* (Eq. 3.23). Fractions of each type of sugarcane transport in a representative hectare are given in Eqs. 3.24–3.26.

$$f_{\mathrm{RJ}} + f_{\mathrm{treminhão}} + f_{\mathrm{rodotrem}} = 1 \tag{3.23}$$

f_{RJ} fraction of *Romeu e Julieta* truck use in sugarcane harvest area,
$f_{\mathrm{treminhão}}$ fraction of *treminhão* truck use in sugarcane harvest area, and
f_{rodotrem} fraction of *rodotrem* truck use in sugarcane harvest area.

$$F_{\mathrm{RJ}} = F_{\mathrm{harvest}} \cdot f_{\mathrm{RJ}} \tag{3.24}$$

$$F_{\mathrm{treminhão}} = F_{\mathrm{harvest}} \cdot f_{\mathrm{treminhão}} \tag{3.25}$$

$$F_{\mathrm{rodotrem}} = F_{\mathrm{harvest}} \cdot f_{\mathrm{rodotrem}} \tag{3.26}$$

F_{RJ} fraction of *Romeu e Julieta* truck use in a representative hectare,
$F_{\mathrm{treminhão}}$ fraction of *treminhão* truck use in a representative hectare, and
F_{rodotrem} fraction of *rodotrem* truck use in a representative hectare.

3.3.1.5 Straw Recovery

The large-scale use of sugarcane straw involves recovery, compaction, transportation, and preparation operations for combustion at boilers. Competitiveness restrictions of this process are frequently associated to high costs of these operations and biomass quality constraints (high mineral impurity levels). The most discussed systems for straw recovery that have been currently applied are integral and baling systems.

In the baling system, the straw is left in the field for about 15 days after sugarcane harvesting to decrease its water content. After that, straw is windrowed, collected, and compacted in bales, which are subsequently loaded and transported to the mill separately from the sugarcane stalks. Therefore, additional agricultural operations are required. The windrowing operation contributes with most of mineral impurities of straw. Cylindrical or prismatic bales have densities approximately two times higher than chopped straw. However, problems such as high content of mineral impurities and the complex sequence of operations remain. Also, processes with higher pressures and temperatures such as briquetting and pelletization achieve higher densities but significantly increase power consumption and reduce the efficiency of operation, which results in a significantly higher costs.

In the integral harvesting system, straw is processed by the sugarcane harvesters together with stalks. The integral harvesting system has an important advantage because sugarcane harvesters operate with the extractors shut down, which reduces stalks losses. The stalks and straw are jointly discharged into the infield wagon. This equipment transfers sugarcane stalks and straw to the trucks for transport to the industry. At the industry, sugarcane stalks–straw mixture is partially separated by a dry cleaning unit, involving pneumatic and mechanical separation. A second equally important advantage of the integral harvesting system is that straw is not left on the soil, which decreases its contamination with mineral impurities. The main limitations of this system are associated with the decreased loading density in transportation vehicles and the higher investment and low efficiency of the dry cleaning unit. The density becomes a major issue as the distances between the areas of production and industrial plant increase. In the case of bulk straw handling (integral system), density may be increased by reducing straw particles sizes by chopping processes. However, in the case of integral harvesting, the straw water content remains high, especially because sugarcane tops mixed with the load because of lodged sugarcane harvesting.

Both integral harvesting and baling systems are included in CanaSoft model. In the first one, straw is harvested and transported together with the sugarcane stalks, while in baling system, it is left in the field for about 15 days after sugarcane harvesting to decrease its water content.

Straw recovery can be performed only on green cane areas (Eq. 3.27), with one of the recovery systems (Eq. 3.28), in different recovery rates. Fractions of each type of straw recovery system in a representative hectare are given in Eqs. 3.29 and 3.30.

$$F_{\text{straw recovery}} = F_{\text{harvest}} \cdot f_{\text{unburned}} \cdot f_{\text{straw rec}} \tag{3.27}$$

$$f_{\text{integral}} + f_{\text{bale}} = 1 \tag{3.28}$$

$F_{\text{straw recovery}}$ fraction of total area with straw recovery,
$f_{\text{straw rec}}$ fraction of unburned sugarcane area with straw recovery,
f_{integral} fraction of straw recovery area that uses integral system, and
f_{bale} fraction of straw recovery area that uses baling system.

$$F_{\text{integral straw}} = F_{\text{straw recovery}} \cdot f_{\text{integral}} \tag{3.29}$$

$$F_{\text{baled straw}} = F_{\text{straw recovery}} \cdot f_{\text{bale}} \tag{3.30}$$

$F_{\text{integral straw}}$ fraction of total area with straw recovery using integral system, and
$F_{\text{baled straw}}$ fraction of total area with straw recovery using baling system.

Different fractions of straw recovery are possible to be obtained in integral harvesting adjusting the extractor of sugarcane harvester and, in recovery with the baling system, adjusting the straw windrow equipment. Straw recovery rates can be modified in CanaSoft model, with effects in the load density in the transport trucks when straw is recovery along with sugarcane stalks and in the efficiencies of agricultural operations for baling.

3.3.1.6 Industrial Residues Application

During the ethanol distillation process, large amount of a liquid waste, known as vinasse, is generated. For each liter of ethanol is produced from 8 to 15 L of vinasse. This material is rich in nutrients and it is usually used for fertirrigation. The nitrogen (N), and especially the potassium (K) in the vinasse is sufficient (depending on the application dose) to replace 20 and 100 % of N and K, respectively, of the mineral fertilizer of the ratoons.

Sugarcane mills adopt different vinasse distribution systems. The most common practices to vinasse distribution are based in channels and pipes and application with aspersion. Other types of vinasse distribution, such as, by trucks are less widespread. The type of distribution system and distance from the site of application strongly affects the cost of this operation, limiting the area of application of this important residue.

The use of filter cake in sugarcane is also an important source of nutrients to the crop. The filter cake mud is usually applied on the rows of sugarcane after harvesting. Depending on the available amount, filter cake mud can be a source of nutrients for crops, replacing up to 50 % of the sugarcane phosphorus (P) requirement. Usually this material is composted before being sent to the field. Its application in the field is similar to the fertilizer application operation. The industry also generates ash and soot, which generally are added to filter cake during the composting process.

The area covered with industrial residues depends on their industrial production and application rates. Industrial productions are quantified by process simulation (Chap. 4), usually in kilograms per Mg of sugarcane stalks. Application rates are informed as kilograms per hectare, and the fraction of total area with industrial residues use can be determined by using Eq. 3.31.

$$F_{\text{application}} = \frac{P_{\text{industrial}}}{A_{\text{rate}}} P_{\text{m,stalks}} \tag{3.31}$$

$F_{\text{application}}$	fraction of area with industrial residues (vinasse, filter cake mud, and ashes) application in a representative hectare,
$P_{\text{industrial}}$	Industrial production of industrial residues (kg Mg^{-1}),
A_{rate}	Application rate of industrial residues (kg ha^{-1}), and
$P_{\text{m,stalks}}$	Mean stalks productivity, considering the total area (Mg ha^{-1}).

3.3.1.7 Crop Rotation

Crop rotation in sugarcane field is performed only in the period between previous ratoon removal and sugarcane replanting, which is usually around six months long. It can be performed with crops of commercial interest, such as soybean (*Glycine max*), peanut (*Arachis hypogaea*), maize (*Zea mays*), and sweet sorghum (*Sorghum bicolor*), or with cover crops without commercial interest, such as sunn hemp (*Crotalaria juncea L.*) and *Mucuna* sp., for soil improvement, nitrogen supply, and pests and diseases control. The objective of cover crops is improve soil fertility, protect the soil surface from erosion, reduce pests infestation, break sugarcane diseases cycles, carbon sequestration, beyond other benefits.

In CanaSoft model, all the agricultural operations needed for the most common crops used in rotation with sugarcane are included. They can also be adapted to include other crops. Fraction of area available for crop rotation in a representative hectare is defined by Eq. 3.32.

$$F_{\text{rotation}} = F_{\text{planting}} \cdot f_{\text{year-and-half}} \tag{3.32}$$

3.3.1.8 Biomass Yield

In the case of sugarcane, yield is normally related to the gathered stalks from one harvested hectare of sugarcane. In the CanaSoft model, sugarcane yield is an exogenous variable. Typical yields for a specific region obtained from statistic databases or specific data for a biorefinery can be used to insert sugarcane yield in the CanaSoft model. However, CanaSoft model can be integrated with sugarcane growing models as DSSAT/CANEGRO (South African model) or APSIM-Sugarcane (Australian model) for modeling sugarcane yields as a function of region and climate variables, for example.

Real sugarcane production is used to calculate the gathered stalks, after accounting for harvesting losses (Eq. 3.33). Mean stalks yield in a representative hectare needs to be calculated considering the proportion of harvest area, according to Eq. 3.34.

$$P_{\text{stalks}} = P_{\text{crop}}(1 - h_{\text{loss}}) \tag{3.33}$$

P_{stalks} Mean stalks yield, considering the harvested area (Mg ha^{-1}),
P_{crop} Real stalks yield, considering the harvest area (Mg ha^{-1}), and
h_{loss} Harvesting losses.

$$P_{\text{m,stalks}} = P_{\text{stalks}} F_{\text{harvest}} \tag{3.34}$$

$P_{\text{m,stalks}}$ Mean stalks yield, considering the total area (Mg ha^{-1}).

The amount of straw produced is calculated as a fraction of the real yield of sugarcane stalks, usually around 140 kg of straw (dry mass) per Mg of stalks. Straw production in CanaSoft model is calculated according to Eq. 3.35. Straw available to be recovered from the field needs to take into account the fractions of manual and mechanized green cane harvest in a representative hectare (Eqs. 3.19 and 3.21).

$$P_{\text{straw}} = 0.140 \cdot P_{\text{crop}} \tag{3.35}$$

P_{straw} Straw yield, considering the harvest area ($t_{\text{dry mass}}$ ha^{-1}).

3.3.1.9 Mill Capacity, Production Area, and Transport Distance

Annual sugarcane milling capacity is an industrial parameter of fundamental importance to calculate the required sugarcane production area. The area required to produce the stalks for industrial processing is determined by Eq. 3.36.

$$A_{\text{milling}} = \frac{M_{\text{capacity}}}{P_{\text{stalks}}} \tag{3.36}$$

A_{milling} area required to produce the stalks for industrial processing (ha),
M_{capacity} Annual sugarcane milling capacity (Mg $year^{-1}$), and
P_{stalks} Mean stalks yield in harvest area (Mg ha^{-1}).

The area for harvesting includes the area to produce sugarcane for processing and to be used as seedlings in the crop establishment (Eq. 3.37). Fraction of area for seedling production in harvesting area depends on the quantity of seedlings used for planting, the productivity of stalks in harvested area and the number of cuts (Eq. 3.38). Harvesting area can be calculated using the milling area and the fraction of seedlings area, according to Eq. 3.39.

$$A_{\text{harvesting}} = A_{\text{milling}} + A_{\text{seedlings}} \tag{3.37}$$

$$f_{\text{seedlings}} = \frac{C_{\text{seedlings}}}{P_{\text{stalks}} \cdot N_{\text{cuts}}} \tag{3.38}$$

$$A_{\text{harvesting}} = \frac{A_{\text{milling}}}{1 - f_{\text{seedlings}}} \tag{3.39}$$

Additionally to the area of sugarcane harvesting, it is necessary to account for the land required for planting, if harvesting of plant cane is performed in a period longer than 12 months after planting (year-and-half sugarcane). Area for planting is calculated using Eq. 3.40, and total area for the mill can be finally calculated by using Eq. 3.41.

$$A_{\text{planting}} = \frac{A_{\text{harvesting}}}{N_{\text{cuts}}} \tag{3.40}$$

$$A_{\text{total}} = A_{\text{harvesting}} + A_{\text{planting}} \cdot f_{\text{year-and-half}} \tag{3.41}$$

Sugarcane transport
Specific trucks are used to transport sugarcane in Brazil. They can be divided into three types: *Romeu e Julieta, Treminhão,* and *Rodotrem* (Figs. 3.3, 3.4, and 3.5). According to the Brazilian National Department for Transport Infrastructure (DNIT 2009), these types are characterized by the number and design of trailers, as well as the length and the weight that they can support.

- Romeu e Julieta, with 19.80 m of length and up to 50 Mg.
- Treminhão with seven axes, up to 30 m of length and 63 Mg.
- Rodotrem with 9 axes, up to 30 m of length and 74 Mg.

The sugarcane transport loading capacity varies according to the harvest system. Manually harvested sugarcane has a small bulk density, causing smaller transport capacities compared to mechanically harvested sugarcane, as presented in Table 3.1.

Transport distance
The transport distance of sugarcane from farm to the mill is calculated based on the total area and on the spatial dispersion of sugarcane fields. The spatial dispersion factor (*s*) reflects the sugarcane concentration in the entire area confined into the circular perimeter around the mill, and its value was defined based on regression of several transport distances and milling capacities of Brazilian sugarcane mills.

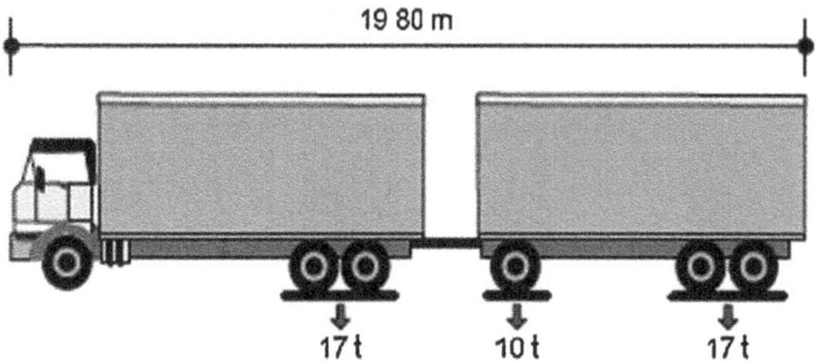

Fig. 3.3 Design of a Romeu e Julieta truck

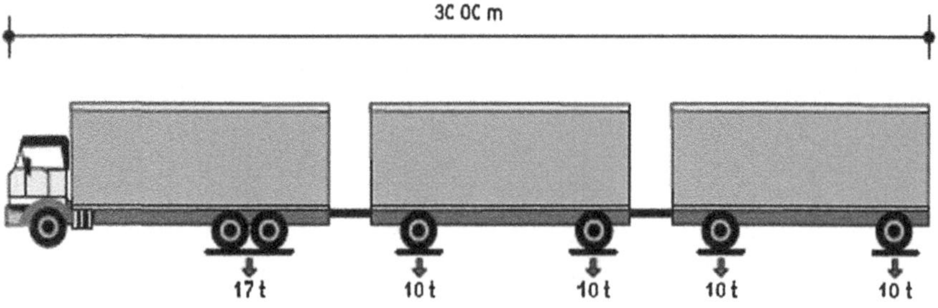

Fig. 3.4 Design of a Treminhão truck

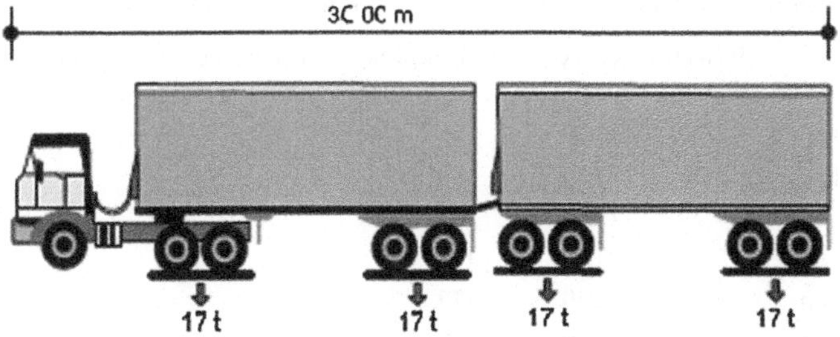

Fig. 3.5 Design of a Rodotrem truck

Table 3.1 Sugarcane transport capacity according to harvest system

Truck	Sugarcane transport capacity (Mg)	
	Mechanically harvested sugarcane	Manually harvested sugarcane
Romeu e Julieta	32	28
Treminhão	50	45
Rodotrem	60	50

Reference Figueiredo Filho (2011)

Using the CanaSoft model to compare different sugarcane production scenarios, spatial dispersion factor remains constant and transport distance varies according to milling capacity, sugarcane productivity, number of cuts, use of seedlings, among other factors that affect the total area of the mill. Mean transport distance is calculated with Eq. 3.42.

$$d_{\text{transport}} = \frac{2}{3}\frac{1}{10}\sqrt{\frac{A_{\text{total}}}{s \cdot \pi}} \tag{3.42}$$

$d_{\text{transport}}$ transport distance of sugarcane (km),
A_{total} total area of the mill (ha), and
s spatial dispersion factor.

3.3.2 *Agricultural Mechanized Operations*

3.3.2.1 Machinery Data Bank

The machinery data bank includes technical, economic, and performance information about the agricultural machinery used in the biomass cultivation system. Most of the information is obtained from machinery producer companies, consultants, and market data.

Table 3.2 Part of agriculture machinery data bank for tractors and harvesters

Machinery	Weight (kg)	Service lifespan (h)	Nameplate power (hp)	Op. width (m)	Op. speed (m s^{-1})	Investment (US$)
Tractor 75 hp	2880	13,500	75	–	–	40,217.39
Tractor 90 hp	3410	13,650	88	–	–	46,673.91
Tractor 125 hp	4780	11,750	125	–	–	58,843.48
Tractor 140 hp	5470	12,000	140	–	–	69,078.26
Tractor 175 hp	6412	13,250	175	–	–	74,913.04
Tractor 220 hp	9450	13,250	220	–	–	128,565.22
Tractor 280 hp	11,273	12,000	280	–	–	151,304.35
Harvester 1 row	19,000	12,000	358	1.50	1.50	413,043.48
Harvester 2 rows	21,000	12,000	358	3.00	0.93	456,521.74

The agriculture machinery data bank is the base for modeling the agricultural mechanized operations. Table 3.2 illustrates part of agriculture machinery data bank for tractors and harvesters and Table 3.3 for implements.

It is important to highlight that the parameters contain "default" data, but they can be customized for a certain scenario. New machinery is constantly added when necessary.

Weight and nameplate/suggested power are obtained from the machinery fabricant specification. Service lifespan is estimated based on technical information from sugarcane mills. Tractors operational widths and speeds depend on implement parameters (Table 3.3) that are also obtained from their respective fabricants and from technical information from sugarcane mills. The machine or implement initial investment (cost) is their market price for a specific year. Initial investments are constantly updated in CanaSoft model, either by applying specific price index or by price quoting on the market.

3.3.2.2 Machinery Field Capacities

Information from the machinery data bank, coupled with operational data for the machinery performing a specific operation, is used to calculate field efficiencies for the mechanized operations, i.e., the number of hectares they can work per hour.

Theoretical field capacity depends only on the operating width of the agricultural machine and on its average speed in the field (Eq. 3.43) and represents the maximum possible worked area per hour.

$$\mathrm{TFC} = \frac{3600}{10,000} l_{\mathrm{m}} \cdot v_{\mathrm{m}} \tag{3.43}$$

Table 3.3 Part of agriculture machinery data bank for tractor implements

Machinery	Weight (kg)	Service lifespam (h)	Suggested power (hp)	Op. width (m)	Op. speed (m s^{-1})	Investment (US$)
Leveler disk—28 disks	746	8000	70–75	2.53	2.50	4226.09
Leveler disk—40 disks	1175	8000	95–105	3.70	2.50	9652.17
Leveler disk—56 disks	1445	8000	120–130	5.26	2.50	10,873.91
Disk plow—14 disks	1197	8000	80–90	1.75	1.67	7069.57
Disk plow—24 disks	2134	8000	140–150	3.11	1.67	10,078.26
Disk plow—36 disks	2840	8000	230–250	4.72	1.67	13,691.30
Disk harrow—14 disks	2052	8000	120–130	2.21	1.39	13,230.43
Disk harrow—20 disks	3336	8000	220–240	3.23	1.39	16,260.87
Disk harrow—30 disks	4250	8000	280–300	4.93	1.39	21,308.70
Limestone spreader—5500 kg	1325	8000	80–100	9.00	2.22	10,086.96
Sugarcane billet planter—2 rows	7700	8000	160–200	3.00	2.00	106,443.48
Boom sprayer—18 m length	1730	10,000	70–90	18.00	2.92	25,478.26
Sugarcane farm trailer—30 m^3	7690	10,000	160–200	3.00	0.93	41,126.09

TFC Theoretical field capacity (ha h^{-1}),
l_m operating width of the machinery (m), and
v_m average operating speed of the machinery (m s^{-1}).

Actually, TFC is not achieved under real conditions, due to several factors that cause delays in scheduled operating time. These time delays include time spent on travels across sugarcane fields, waiting for management control and operator resting stops (management efficiency); time spent on checking and adjusting machine to fix damages and improve its performance (maintenance efficiency); and time spent in turning maneuvers and waiting for filling machines tanks (operational efficiency). Effective field capacity can be calculated from Eq. 3.44.

$$\mathrm{EFC} = \mathrm{TFC} \cdot \eta_{\mathrm{mg}} \cdot \eta_{\mathrm{mt}} \cdot \eta_{\mathrm{op}} \tag{3.44}$$

EFC effective field capacity (ha h^{-1}),
η_{mg} management efficiency,
η_{mt} maintenance efficiency, and
η_{op} operational efficiency.

Table 3.4 illustrates some default values of field capacities and efficiencies used in CanaSoft model for some agricultural operations. They can be customized for specific assessments of specific mills or regions and are used to calculate diesel, labor, and machinery use, among other technical, environmental, social, and economic outputs.

Table 3.4 Field capacities and efficiencies for selected agricultural operations in CanaSoft model

Mechanical operation	Machinery hp	Implement	η_{mt}	η_{mg}	η_{op}	TFC (ha h^{-1})	EFC (ha h^{-1})
Limestone application	Tractor 88	Limestone spreader—5500 kg	0.80	0.47	0.80	7.20	2.16
Subsoiling	Tractor 220	Subsoiler—7 shanks	0.80	0.78	0.80	1.68	0.84
Harrowing	Tractor 220	Disk harrow—18 disks	0.80	0.78	0.80	1.45	0.72
Leveling harrowing	Tractor 125	Leveler disk—40 disks	0.80	0.78	0.80	3.33	1.66
Mechanical planting	Tractor 180	Sugarcane billet planter—2 rows	0.80	0.60	0.80	2.16	0.83
Agrochemical application	Tractor 75	Boom sprayer—18 m length	0.80	0.78	0.80	18.90	9.45
Broadcast incorporated fertilizer application	Tractor 125	Cultivator with land leveler	0.80	0.63	0.80	1.20	0.48
Manually harvested sugarcane loading	Sugarcane loader 180	–	0.95	0.80	0.90	0.38	0.26
Mechanical harvesting	Harvester 1 row 360	–	0.80	0.70	0.79	0.81	0.36
Straw windrowing	Tractor 75	Straw windrower	0.80	0.94	0.80	7.50	4.50
Straw baling	Tractor 175	Big square baler	0.80	0.60	0.80	2.66	1.02

3.3.2.3 Diesel Use in Mechanical Operations

Diesel use in the agricultural operations is calculated using the agricultural machine power, power usage factor, field efficiencies, field capacities, and specific diesel consumption.

The power usage factor represents the average percentage of the nameplate power used by the machine, since the machine does not operate always at its highest power. This factor is obtained from literature review, consultancy to experts, and technical information from the mills.

The specific diesel consumption is the rate of fuel consumption divided by the power produced. It is a measure of the efficiency of fuel burning to produce power. In CanaSoft model, the default values considered for specific diesel consumption are 0.26 L hp^{-1} h^{-1} for the engines up to 120 hp and 0.20 L hp^{-1} h^{-1} for engines with high power, since tractors with higher engines power are generally newer and designed to be more efficient on fuel consumption. These values were obtained from the literature review and can be customized for specific assessments.

Equation 3.45 is used for calculating the diesel consumption per hour. Fuel consumption per hectare of a certain operation is calculated based on the time spent doing one hectare of work (effective engine working time) and the consumption per hour, according to Eq. 3.46.

$$FC_h = MP_n \cdot \eta_{power} \cdot sc_{diesel} \tag{3.45}$$

FC_h diesel consumption per hour (L h^{-1}),
MP_n nameplate machinery power (hp),
η_{power} power usage factor, and
sc_{diesel} specific diesel consumption (L hp^{-1} h).

$$FC_{ha} = \frac{FC_h}{TFC \cdot \eta_{op}} \tag{3.46}$$

FC_{ha} diesel consumption per hectare (L ha^{-1}),
TFC theoretical field capacity (ha h^{-1}), and
η_{op} operational efficiency.

Table 3.5 illustrates the parameters used for diesel consumption per hectare for selected agricultural operations in CanaSoft model.

3.3.2.4 Labor Use in Agricultural Operations

Labor use is calculated in CanaSoft model considering every mechanized or manual activity involving human work, always considering one hectare. Labor use in mechanical operations is calculated by summing up all the machine operation time.

Table 3.5 Diesel consumption parameters for selected agricultural operations in CanaSoft model

Mechanical operation	MP_n (hp)	η_{power}	FC_h (L h^{-1})	η_{op}	TFC (ha h^{-1})	sc_{diesel} (L hp^{-1} h^{-1})
Limestone application	88	0.35	7.90	0.80	7.20	1.37
Subsoiling	220	0.6	26.04	0.80	1.68	19.37
Harrowing	220	0.6	26.04	0.80	1.45	22.45
Leveling harrowing	125	0.6	14.79	0.80	3.33	5.55
Mechanical planting	180	0.55	19.53	0.80	2.16	11.30
Agrochemical application	75	0.3	5.77	0.80	18.90	0.38
Broadcast incorporated fertilizer application	125	0.5	12.33	0.80	1.20	12.84
Manually harvested sugarcane loading	180	0.3	6.92	0.90	0.38	20.35
Mechanical harvesting	360	0.7	49.43	0.79	0.81	77.51
Straw windrowing	75	0.6	11.54	0.80	7.50	1.92
Straw baling	175	0.6	20.71	0.80	2.66	9.72

This value is derived from the machinery use tables, considering the occurrence areas in a representative hectare of the mill.

Labor use in manual operations is obtained knowing the amount of time used per hectare in each operation. The most important manual labor in sugarcane production cycle is the manual planting and harvesting. To exemplify, typical values for labor use for seedlings distribution and cutting in semi-mechanized planting areas are 94.4 h ha^{-1} (considering seedlings use of 12 Mg ha^{-1}), and for manual harvesting is 74.8 h ha^{-1} in areas with manual harvesting of burned sugarcane (considering the yield of 80 Mg of collected stalks per hectare). Other manual operations include pest-monitoring, collection of soil and biomass samples for analysis, planting, and harvesting monitoring (even in mechanized operations). The total manual labor used is calculated summing up all the labor time used for manual operations, considering the occurrence areas of manual operations in a representative hectare of the mill.

3.3.2.5 Machinery Use in Mechanical Operations

Agricultural machinery usage is calculated based on their total weight and expected lifetime (Tables 3.2 and 3.3), and annual using time, based on agricultural efficiencies for each operation (Table 3.4). Those factors were adjusted based on literature review and consultancy to experts. The most common mechanized operations for the sugarcane production system are already inserted with "typical data" in the agricultural operations calculation module. A "default" agricultural machinery (and implement, if any) is already defined for all mechanized agricultural operation using the options available in the agriculture machinery data bank. Again all the parameters can be customized according to the system under evaluation.

Machinery use is calculated as its weight divided by its lifespan. Tires consumption is calculated the tires weight in the total machinery weight (2.75 %) and the lifespan of tires (2500 h) (ecoinvent center 2007).

3.3.3 *Use of Inputs*

3.3.3.1 Seedlings

It is considered that seedlings requirement varies according to planting system. Normally, semi-mechanized planting requires less seedlings per hectare and seedlings are manually harvested and, while mechanized planting generally uses a large amount of mechanically harvested seedlings, to try to avoid the problem of uneven distribution in the furrow.

3.3.3.2 Fertilizers and Soil Amendments

Due to soil characteristics in most of the sugarcane production areas in Brazil (such as low base saturation and acidity), limestone and gypsum are applied to correct soil acidity, increase base saturation, and eliminate aluminum toxicity. The limestone and gypsum application is normally done before sugarcane planting, and doses will vary depending on soil chemical properties.

Chemical fertilizers are supplied to the sugarcane crop in the form of basic compounds of several commercial formulations. Ammonia, urea, ammonium nitrate, monoammonium phosphate, single superphosphate, potassium chloride, and several combinations of these are the most commonly used fertilizers in the sugarcane production systems. The typical commercial formulas and fertilization rates recommended for sugarcane are comprised in the CanaSoft model. Default formulations and application rates can be customized depending on the system under study.

The fertilizer data bank of CanaSoft model is comprised of detailed information about chemical composition and market prices of the basic compounds used as fertilizers, whether isolated or in a mixture. These basic compounds can provide a single macronutrient (N, P, or K) or a binary of them (NP, NK, or PK). The nitrogen-based fertilizers in CanaSoft data bank are ammonia, urea, ammonium nitrate, and ammonium sulfate. The straight phosphate fertilizers are the single superphosphate (SSP) and triple superphosphate (TSP), while potassium fertilizers comprise potassium chloride and potassium sulfate. Binary fertilizers include monoammonium phosphate (MAP) diammonium phosphate (DAP), and potassium nitrate. NPK formulations can be easily added in data bank by combining basic compounds.

3.3.3.3 Industrial Residues

Industrial residues of sugar and ethanol production processes are normally recycled and used as nutrient source in the agricultural sugarcane production system. Filter cake mud and ashes from bagasse (and straw) burning in industrial boilers are normally used in planting area. Vinasse is spread in the field normally during ratoon cultivation. The area covered with industrial residues depends on industrial production amount and application rates.

Typical application rates and parameters for transportation and spreading systems of industrial residues in the sugarcane field are comprised in the CanaSoft model and can be customized depending on the system under study.

3.3.3.4 Pesticides and Other Agrochemicals

Agrochemicals used in sugarcane production system may vary a lot depending on the pests' infestation level and several other soil and climate factors.

Concentration and type of specific active substances, market prices, and typical application rates of all the commercial pesticides regulated for sugarcane in Brazil are comprised in the CanaSoft model. However, application rates can be easily customized in the agrochemicals data bank depending on the characteristics of the system under study.

3.3.4 Modeling Sugarcane Production Cost

The farm operating costs in the CanaSoft model consist of several parcels, which will be covered in the subtopics of this section. All of them are calculated considering one hectare of operation and then weighted according to each operation fraction in a representative hectare.

3.3.4.1 Machinery

The cost related to the agricultural machinery considers the uniform recovery of initial investment, accounting for physical depreciation of equipment, plus interest on invested capital. Salvage values are considered in the end of working lifespan: for agricultural machinery and transport trucks, 20 % over initial investment; and for agricultural implements and infield wagon, no salvage values are considered.

3.3.4.2 Diesel and Lubricants

Diesel costs are calculated with the amount of fuel used per hectare and the fuel price in the agricultural market. Costs related to lubricating oil used are generally defined as a fraction of expenditures on diesel. Default values in CanaSoft model are 3 % for lubricating use in trucks and general agricultural machines, and 10 % for lubricating use in harvesters and loaders.

3.3.4.3 Labor

Labor costs are calculated by multiplying the labor use in each operation by the respective manpower cost per hour.

3.3.4.4 Maintenance and Taxes

Maintenance costs include the expenditures on repairs and maintenance of agricultural fleet. It accounts for expenditures on replacement of pieces and garage labor, and on payments for maintenance performed by third parties. It is normally presented as a fraction of machinery costs. Default values in CanaSoft are 10 % for transport trucks and infield wagons, 33 % for general agricultural machinery, and 69 % for sugarcane harvesters, and can be changed if required in a specific assessment.

Other costs related to mechanize operations include expenditures on insurance and taxes; 2 % over machinery costs are considered in CanaSoft.

3.3.4.5 Inputs

Similar to diesel, input costs are calculated with the amount of material used per hectare and the market price of each input.

3.3.4.6 Land Cost and Taxes

Land cost and taxes are included in biomass production cost. The opportunity cost of land is used in CanaSoft model, and it reflects the profit that the landowner could gain if the land was rented for other crop production. Alternatively, if the sugarcane producer is not the owner of the land, it accounts for the effective payment of a land rent. Taxes include tax on production and administrative fees.

3.3.5 Modeling Emissions in the Sugarcane Production System

In addition to inputs and machinery use, it is of fundamental importance to calculate their emissions to air, water, and soil in the sugarcane production system. Coupled, this information will be useful to generate the Life Cycle Datasets of biomass production scenarios assessed on CanaSoft model.

3.3.5.1 Emissions from Diesel Use in Agricultural Machinery

Airborne emissions from diesel combustion in agricultural machinery are calculated based on emission factors by Nemecek and Kagi (2007). The emission factors for hydrocarbons, carbon monoxide (fossil), and nitrogen oxides are estimated as an average from different operations (Nemecek and Kagi 2007). The emission factors for diesel combustion in agricultural machinery engines are presented in Table 3.6.

Table 3.6 Emission factors for diesel combustion in agricultural machinery

Emission	Emission factor	Unit
Carbon dioxide (fossil)	3.12×10^{3}	$g\ kg_{diesel}^{-1}$
Methane (fossil)	1.29×10^{-1}	$g\ kg_{diesel}^{-1}$
Dinitrogen monoxide	1.20×10^{-1}	$g\ kg_{diesel}^{-1}$
Ammonia	2.00×10^{-2}	$g\ kg_{diesel}^{-1}$
Sulfur dioxide	1.01	$g\ kg_{diesel}^{-1}$
Benzene	7.30×10^{-3}	$g\ kg_{diesel}^{-1}$
Cadmium	1.00×10^{-5}	$g\ kg_{diesel}^{-1}$
Chromium	5.00×10^{-5}	$g\ kg_{diesel}^{-1}$
Copper	1.70×10^{-3}	$g\ kg_{diesel}^{-1}$
Nickel	7.00×10^{-5}	$g\ kg_{diesel}^{-1}$
Selenium	1.00×10^{-5}	$g\ kg_{diesel}^{-1}$
Zinc	1.00×10^{-3}	$g\ kg_{diesel}^{-1}$
Benzo(a)pyrene	3.00×10^{-5}	$g\ kg_{diesel}^{-1}$
Polycyclic aromatic hydrocarbons	3.29×10^{-3}	$g\ kg_{diesel}^{-1}$
Benz(a)anthracene[a]	8.00×10^{-5}	$g\ kg_{diesel}^{-1}$
Benzo(b)fluor-anthracene[a]	5.00×10^{-5}	$g\ kg_{diesel}^{-1}$
Chrysene[a]	2.00×10^{-4}	$g\ kg_{diesel}^{-1}$
Dibenzo(a,h)-antracene[a]	1.00×10^{-5}	$g\ kg_{diesel}^{-1}$
Fluoranthracene[a]	4.50×10^{-4}	$g\ kg_{diesel}^{-1}$
Phenanthrene[a]	2.50×10^{-3}	$g\ kg_{diesel}^{-1}$
Hydrocarbons	3.00	$g\ kg_{diesel}^{-1}$
Carbon monoxide (fossil)	5.40	$g\ kg_{diesel}^{-1}$
Nitrogen oxides	4.25×10^{1}	$g\ kg_{diesel}^{-1}$

[a]Emissions grouped into polycyclic aromatic hydrocarbons

3.3.5.2 Emissions from Chemical Fertilizers Use

Emissions from chemical fertilizers are calculated based on the available nutrients (NPK). It includes airborne emissions of nitrogen and carbon dioxide, heavy metal emissions to agricultural soil, and nutrient emissions to water.

Emissions of ammonia to the air
Nitrogen contained in chemical fertilizers can be easily converted into ammonia (NH_3) and released to the air (Nemecek and Kagi 2007). Volatilization from ammonia and urea application in sugarcane fields can reach 50 % of total applied nitrogen, depending on climate conditions and agricultural practices (Trivelin and Franco 2011; Costa et al. 2003). Default figure of 10 and 15 % are considered by IPCC (2006) and Nemecek and Kagi (2007), respectively. In CanaSoft model, it is considered that 30 % of the total nitrogen applied in the form of ammonia and urea is emitted as ammonia under average Brazilian conditions of sugarcane production.

Other nitrogen fertilizers such as ammonium nitrate, ammonium sulfate, and multi-nutrient fertilizers containing N have lower volatilization in Brazilians acid soils (Trivelin and Franco 2011). For these fertilizers, emissions factor from Nemecek and Kagi (2007) are considered, according to Table 3.7.

Values from Table 3.7 are used to calculate the primary emissions of ammonia from chemical fertilizers. Part of the N emitted as NH_3 is subsequently converted to N_2O, and final values of ammonia emissions discount the converted portion.

Nutrients leaching to groundwater
Sugarcane root system usually explore the soil deeply enough to consider that nutrients leaching to groundwater in average Brazilian sugarcane production conditions of soil and clime are not quantitatively relevant (Trivelin and Franco 2011). Only for ammonia and urea, it is considered that part of the applied nitrogen can be leached as nitrate (NO_3^-), and emission factor adopted in CanaSoft model is 5 % of the applied nitrogen (Trivelin and Franco 2011). Since Brazilian soils have acid pHs, there is no evidence of potassium, phosphorous, or nitrogen (applied as

Table 3.7 Ammonia emission from mineral fertilizers

Fertilizer	Emission factor for NH_3 (% N emitted in the form of NH_3)
Ammonia	30
Urea	30
Ammonium nitrate	2
Ammonium sulfate	8
Monoammonium phosphate	4
Diammonium phosphate	4
Potassium nitrate	4

nitrates) leaching to groundwater (Trivelin and Franco 2011). Part of the N emitted as NO_3^- is subsequently converted to N_2O, and calculations of final values for nitrogen leaching must discount the converted portion.

Emissions of nitrous oxide to the air

Direct and indirect emissions of nitrous oxide (N_2O) are included in the model, based on Nemecek and Kagi (2007). For direct emissions, it is considered an emission factor equal to 1 % of the total applied nitrogen. Indirect emissions (nitrogen is firstly emitted as NH_3 or NO_3^- and then converted to N_2O) accounts for 1 % of the nitrogen volatilized as ammonia plus 0.75 % of the nitrogen leached as nitrate (IPCC 2006).

Emissions of nitrogen oxides to the air

These emissions are estimated based on nitrous oxide emissions. The value of 21 % of N_2O emissions is used, without discounting any quantity from N_2O emissions.

Carbon dioxide to the air

As carbon present in urea has a fossil origin, all carbon content in urea is considered as carbon dioxide emission to air (IPCC 2006). It is considered also that all carbon content in limestone is converted into carbon dioxide emitted to air. It was used the emission factor from IPCC (2006), assuming a carbon content of 0.13 kg of carbon per kilogram of limestone. It corresponds to 0.48 kg CO_2 per kg limestone.

Heavy metals to soil

The inputs of heavy metal contained in mineral fertilizers, limestone, and gypsum are assumed to be emissions to soil. The heavy metal content in each mineral fertilizer is described in Table 3.8. It considers the weighted average for Brazilian products and imported fertilizers composition, based on Rodella (2011) and Gabe

Table 3.8 Heavy metal content of chemical fertilizers

	Metal content (mg kg^{-1})					
	Cd	Cu	Zn	Pb	Ni	Cr
Ammonia	0.0	4.0	21.8	1.0	3.1	1.4
Urea	0.0	3.9	28.9	0.7	1.3	1.3
Ammonium sulfate	0.1	4.0	30.0	1.1	1.8	2.0
Ammonium nitrate	0.0	4.8	34.2	1.3	8.9	2.7
Monoammonium phosphate	13.2	33.6	182.5	16.1	25.4	142.9
Diammonium phosphate	23.1	53.2	338.1	22.2	45.2	265.3
Single superphosphate	3.4	26.8	172.4	87.3	45.5	31.0
Triple superphosphate	52.0	45.0	299.0	3.5	44.0	261.0
Potassium chloride	0.1	4.4	40.2	4.8	1.8	1.7
Potassium sulfate	0.1	2.0	32.0	3.3	0.8	2.0
Potassium nitrate	0.0	2.7	30.9	3.5	3.3	39.0
Limestone	0.0	6.0	7.0	0.0	0.0	9.9
Gypsum	0.8	10.0	5.0	9.9	4.9	9.9

and Rodella (1999) for Brazilian fertilizers and Nemecek and Kagi (2007) for imported.

3.3.5.3 Returned Industrial Residues (Vinasse, Filter Cake, and Ash)

Nitrous oxide (N_2O) and nitrogen oxides (NO_x) emissions from returned industrial residues—vinasse and filter cake—are also considered, using the same direct and indirect emission factors for chemical fertilizers. Ammonia emissions to air from vinasse are considered equal to 30 %; the same emission factor is used to calculate nitrate emissions of vinasse and filter cake (Nemecek and Kagi 2007). Nitrogen present in industrial residues is calculated by process simulation and values are feed to CanaSoft model. It is important to observe that no emissions were considered for ashes application in sugarcane field.

3.3.5.4 Straw Burning in the Field

The emission factors from trash burning before harvesting are based on GREET (2010). These values are given in Table 3.9. The emission factors are given in kilograms per dry Mg of sugarcane burned straw. Calculation of straw burning emissions takes into account the amount of straw produced and the percentage of the sugarcane area with preharvesting burning in a representative hectare.

3.3.5.5 Unburned Trash

Emissions of unburned trash are estimated considering emission factors for agricultural residues from Nemecek and Kagi (2007), numerically the same from filter cake emission factors for N_2O, NO_x, and NO_3^- emissions. The nitrogen content of unburned trash is 0.47 % (dry mass) (Franco 2008).

Table 3.9 Emission factors from the straw burning in the field

Substance	Emission factor (in kg per dry Mg of straw)
Volatile organic compounds (VOC)	7
Carbon monoxide (biogenic)	92
Nitrogen oxides	2.5
Particulates, <10 μm and >2.5 μm	7.8
Particulates, <2.5 μm	3.9
Sulfur dioxide	0.4
Dinitrogen monoxide	0.070
Methane (biogenic)	2.70

Table 3.10 Soil emission factors for tire degradation in agricultural machinery

Emission	Emission factor	Unit
Zinc	8.96	$g\ kg_{tire}^{-1}$
Lead	1.456	$g\ kg_{tire}^{-1}$
Cadmium	0.336	$g\ kg_{tire}^{-1}$

3.3.5.6 Root System

Sugarcane root system is renewed each year by re-growth of ratoon. Emissions of root system are estimated using the root's nitrogen content and the amount of root system, calculated with a root-stalks ratio of 0.2 (Smith et al. 2005). The root nitrogen content considered is 0.514 % (Franco 2008), and the same unburned straw emission factors are used.

3.3.5.7 Agrochemicals

In CanaSoft model, it is considered that 1.5 % of the agrochemicals applied in sugarcane are emitted to surface water, via runoff, according to Renouf et al. (2010). This study represents the production of sugarcane at Australia, which is a different environment from what is being considering for Brazilian conditions. However, it was used due to lack of specific data for pesticide emissions from sugarcane production systems in Brazil. Other consideration is that the remaining amount of pesticides (98.5 %) is assumed to be emitted to agricultural soil. The degradation or the absorption of the pesticides is not being considered, either their leaching potential or emissions to air.

3.3.5.8 Agricultural Machinery Tire Degradation

Emissions to soil due to tire degradation by agricultural machinery use are based on Nemecek and Kagi (2007). It is considered the tire/machinery weight relation of 0.0275 and the emission factors are presented in Table 3.10.

3.4 Use of CanaSoft Model

3.4.1 Scenario Input

Analysis of sustainability aspects is performed for a given sugarcane production scenario. Table 3.11 summarizes some of the main parameters that characterize a sugarcane production scenario. This information is used as input in the CanaSoft model for calculations of technical, economic, environmental, and social impacts of sugarcane production system.

3.4.2 Model Outputs

Taking into consideration the sugarcane production system characterized by the parameters presented in Table 3.11, the CanaSoft processes this information and provides economic, environmental, and social outputs. Examples of these outputs are provided in this section.

Table 3.11 Example of main inputs characterizing a sugarcane production scenario

Parameter	Unit	Value
Sugarcane spatial density	(ha cane) (ha total)$^{-1}$	0.247
Size of the mill (crushing capacity)	Mg year^{-1}	4
Sugarcane yield	Mg ha^{-1}	88.9
Crop rotation	–	no
Fraction of planting in reforming area	%	100
Fraction of mechanized planting	%	100
Type of tilling for planting	–	Conventional
Fraction of year-and-half cane	%	0
Number of cuts per cycle	Cuts/cycle	5
Fraction of mechanized harvesting	%	100
Fraction of bunt cane in mechanized harvesting	%	0
Fraction of straw recovery	%	0
Fraction of cane transported with "rodotrem"	%	100
Vinasse disposal rate	m^3 ha^{-1} year^{-1}	170
Vinasse disposal: directly by trucks	%	6
Vinasse disposal: transported by trucks and applied with aspersion system	%	31
Vinasse disposal: transported by open channels and applied with aspersion system	%	63
Bagasse ashes disposal rate	kg(db) ha^{-1} year^{-1}	4000
Filter cake mud disposal rate	kg(db) ha^{-1} year^{-1}	5000
Cost of land	US$ ha^{-1} year^{-1}	418
Diesel cost	US$ L^{-1}	1.06

3.4.2.1 Economic Outputs

Table 3.12 presents a default breakdown of sugarcane production costs provided by the CanaSoft model. Other tables can be formulated based on user needs.

Table 3.12 Breakdown of sugarcane production costs

Production cost	Value	Unit	Fraction of total sugarcane cost (%)
Agricultural operations	**781.73**	**US$ ha^{-1} year^{-1}**	**38.4**
Mechanized operations	715.98	US$ ha^{-1} year^{-1}	35.2
Machinery	*218.80*	*US$ ha^{-1} year^{-1}*	*10.8*
Implements	*89.66*	*US$ ha^{-1} year^{-1}*	*4.4*
Diesel	*157.31*	*US$ ha^{-1} year^{-1}*	*7.7*
Labor	*89.39*	*US$ ha^{-1} year^{-1}*	*4.4*
Maintenance	*146.14*	*US$ ha^{-1} year^{-1}*	*7.2*
Lubricants	*10.85*	*US$ ha^{-1} year^{-1}*	*0.5*
Shelter and insurance	*3.83*	*US$ ha^{-1} year^{-1}*	*0.2*
Manual operations	65.76	US$ ha^{-1} year^{-1}	3.2
Labor	*65.76*	*US$ ha^{-1} year^{-1}*	*3.2*
Inputs	**498.47**	**US$ ha^{-1} year^{-1}**	**24.5**
Seedlings	*126.77*	*US$ ha^{-1} year^{-1}*	*6.2*
Pesticides	*90.71*	*US$ ha^{-1} year^{-1}*	*4.5*
Fertilizers (NPK)	*264.73*	*US$ ha^{-1} year^{-1}*	*13.0*
Limestone and gypsum	*16.26*	*US$ ha^{-1} year^{-1}*	*0.8*
Industrial residues	–	*US$ ha^{-1} year^{-1}*	*0.0*
Transport	**245.93**	**US$ ha^{-1} year^{-1}**	**12.1**
Transport of cane and straw	197.61	US$ ha^{-1} year^{-1}	9.7
Transport of vinasse	18.72	US$ ha^{-1} year^{-1}	0.9
Transport of inputs	29.60	US$ ha^{-1} year^{-1}	1.5
Trucks	*54.30*	*US$ ha^{-1} year^{-1}*	*2.7*
Infield wagon	*18.36*	*US$ ha^{-1} year^{-1}*	*0.9*
Diesel	*103.83*	*US$ ha^{-1} year^{-1}*	*5.1*
Labor	*23.01*	*US$ ha^{-1} year^{-1}*	*1.1*
Tires and lubricants	*37.71*	*US$ ha^{-1} year^{-1}*	*1.9*
Maintenance	*7.27*	*US$ ha^{-1} year^{-1}*	*0.4*
Shelter and insurance	*1.45*	*US$ ha^{-1} year^{-1}*	*0.1*
Vinasse spreading	34.48	US$ ha^{-1} year^{-1}	1.7
Irrigation	–	US$ ha^{-1} year^{-1}	0.0

(continued)

Table 3.12 (continued)

Production cost	Value	Unit	Fraction of total sugarcane cost (%)
Cost of land	417.57	US$ ha^{-1} $year^{-1}$	20.5
Taxes	57.08	US$ ha^{-1} $year^{-1}$	2.8
Total sugarcane cost	**2035.26**	**US$ ha^{-1} $year^{-1}$**	**100.0**
Total sugarcane cost	**26.78**	**US$ Mg^{-1}**	–

3.4.2.2 Environmental Outputs

Table 3.13 presents a summarized version of the sugarcane production life cycle dataset provided by the CanaSoft model. These data are exported to life cycle assessment software for performing a comprehensive environmental impact assessment. Reference flow of figures in this table is per hectare of sugarcane per year. Names of unit process for technosphere inputs reflect names form ecoinvent

Table 3.13 Summarized life cycle dataset for the sugarcane production system

Name	Unit	Value
Products		
Sugarcane	kg ha^{-1} $year^{-1}$	76,010.0
Straw	kg ha^{-1} $year^{-1}$	–
Inputs from nature		
Occupation, arable, non-irrigated	ha ha^{-1} $year^{-1}$	1.0
Inputs from technosphere		
Vinasse	m^3 ha^{-1} $year^{-1}$	40.2
Filter cake	kg ha^{-1} $year^{-1}$	570.1
Ashes	kg ha^{-1} $year^{-1}$	456.1
Urea, as N	kg ha^{-1} $year^{-1}$	101.2
Single superphosphate, as P_2O_5	kg ha^{-1} $year^{-1}$	15.5
Potassium chloride, as K_2O	kg ha^{-1} $year^{-1}$	108.5
Limestone	kg ha^{-1} $year^{-1}$	400
Gypsum	kg ha^{-1} $year^{-1}$	200
Pesticide unspecified	kg ha^{-1} $year^{-1}$	0.31
Glyphosate	kg ha^{-1} $year^{-1}$	0.389
Diuron	kg ha^{-1} $year^{-1}$	0.098
Carbofuran	kg ha^{-1} $year^{-1}$	0.42
Harvester	kg ha^{-1} $year^{-1}$	4.433
Tractor	kg ha^{-1} $year^{-1}$	5.455
Agricultural machinery, general	kg ha^{-1} $year^{-1}$	11.1
Diesel	kg ha^{-1} $year^{-1}$	125.9
Sugarcane transport	tkm ha^{-1} $year^{-1}$	2800
Vinasse transport	tkm ha^{-1} $year^{-1}$	215

(continued)

Table 3.13 (continued)

Name	Unit	Value
Inputs transport	tkm ha^{-1} $year^{-1}$	560
Vinasse transport channel	m ha^{-1} $year^{-1}$	0.048
Vinasse pumping and storage system operation	m^3 ha^{-1} $year^{-1}$	25.3
Vinasse aspersion system operation	m^3 ha^{-1} $year^{-1}$	37.8
Emissions to air		
Dinitrogen monoxide	kg ha^{-1} $year^{-1}$	4.24
Ammonia	kg ha^{-1} $year^{-1}$	40.0
Nitrogen oxides	kg ha^{-1} $year^{-1}$	0.89
Carbon dioxide, fossil		335.1
Carbon dioxide, fossil	kg ha^{-1} $year^{-1}$	392.9
Methane, fossil	g ha^{-1} $year^{-1}$	16.2
Dinitrogen monoxide	g ha^{-1} $year^{-1}$	15.1
Ammonia	g ha^{-1} $year^{-1}$	2.5
Sulfur dioxide	g ha^{-1} $year^{-1}$	127.2
Benzene	mg ha^{-1} $year^{-1}$	919.4
Cadmium	mg ha^{-1} $year^{-1}$	1.3
Chromium	mg ha^{-1} $year^{-1}$	6.3
Copper	mg ha^{-1} $year^{-1}$	214.1
Nickel	mg ha^{-1} $year^{-1}$	8.8
Lead	mg ha^{-1} $year^{-1}$	0.0
Selenium	mg ha^{-1} $year^{-1}$	1.3
Zinc	mg ha^{-1} $year^{-1}$	125.9
Benzo(a)pyrene	mg ha^{-1} $year^{-1}$	3.8
PAH, polycyclic aromatic hydrocarbons	mg ha^{-1} $year^{-1}$	414.3
Hydrocarbons, unspecified	g ha^{-1} $year^{-1}$	377.8
Carbon monoxide, fossil	g ha^{-1} $year^{-1}$	680.1
Nitrogen oxides	kg ha^{-1} $year^{-1}$	5.4
Emissions to water		
Nitrate	kg ha^{-1} $year^{-1}$	163.5
Carbofuran	g ha^{-1} $year^{-1}$	6.30
Diuron	g ha^{-1} $year^{-1}$	1.46
Fipronil	g ha^{-1} $year^{-1}$	0.60
Glyphosate	g ha^{-1} $year^{-1}$	5.83
Hexazinone	g ha^{-1} $year^{-1}$	0.43
Pesticides, unspecified	g ha^{-1} $year^{-1}$	2.10
Tebuthiuron	g ha^{-1} $year^{-1}$	1.50
Emissions to soil		
Cadmium	g ha^{-1} $year^{-1}$	0.47
Copper	g ha^{-1} $year^{-1}$	8.40

(continued)

Table 3.13 (continued)

Name	Unit	Value
Zinc	$g\ ha^{-1}\ year^{-1}$	32.63
Lead	$g\ ha^{-1}\ year^{-1}$	10.55
Nickel	$g\ ha^{-1}\ year^{-1}$	5.53
Chromium	$g\ ha^{-1}\ year^{-1}$	9.23
Carbofuran	$g\ ha^{-1}\ year^{-1}$	413.70
Diuron	$g\ ha^{-1}\ year^{-1}$	96.14
Fipronil	$g\ ha^{-1}\ year^{-1}$	39.40
Glyphosate	$g\ ha^{-1}\ year^{-1}$	382.97
Hexazinone	$g\ ha^{-1}\ year^{-1}$	27.97
Imazapic	$g\ ha^{-1}\ year^{-1}$	137.90
Tebuthiuron	$g\ ha^{-1}\ year^{-1}$	98.50
Zinc	$g\ ha^{-1}\ year^{-1}$	23.0
Lead	$g\ ha^{-1}\ year^{-1}$	3.7
Cadmium	$g\ ha^{-1}\ year^{-1}$	0.864

and its modifications at the Virtual Sugarcane Biorefinery (VSB) life cycle inventories data bank.

3.4.2.3 Social Outputs

Besides the technical description of the sugarcane production systems, the CanaSoft model can provide other useful information for the social assessment of sugarcane. In this regard, CanaSoft model can estimate the total working hours according to the different functions that are necessary for the production of one Mg of sugarcane. For instance, considering the parameters presented in Table 3.11, are necessary 0.34 h of manual labor and 0.18 h machine operators per Mg of sugarcane delivered at industry. Other social information can be organized in different ways based on user needs.

3.5 Adaptation of CanaSoft for Other Biomass

It needs to be emphasized that besides CanaSoft was constructed to sugarcane production systems, it can be used for evaluation of other biomass by changing some parameters. In this section, we describe some of the main changes to be done in the CanaSoft in order assess other biomass production systems, providing some examples.

3.5.1 Energy Cane

Energy cane is a fiber-rich species of cane, obtained by traditional breeding methods of the ancestral species of sugarcane (Matsuoka et al. 2014), with biomass yields expressively higher than those from sugarcane. Energy cane has been envisioned as a promising alternative for second-generation ethanol production, and its production system is very similar to sugarcane's.

Besides higher biomass yield, the number of cuts in a production cycle also increases, since more abundant and vigorous root system of energy cane leads to no decrease on productivities like observed in conventional sugarcane crops (Matsuoka et al. 2014).

Since the energy cane production system is not consolidated yet, to include it on CanaSoft model it was considered a production system with the same operations for conventional sugarcane, using the same equipment adjusted for a biomass with higher productivity. Planting and harvesting of energy cane are considered to be totally mechanized. The same operational efficiencies for planting and the same amount of seedlings used for sugarcane were considered, but a reduction in the operating speed of the harvester (due to the larger amount of biomass for energy cane) was assumed. Besides the fact that the plant more efficiently exploits the soil resources (Matsuoka et al. 2014), application rates of mineral fertilizers were increased to replace nutrient extracted, considering the same extraction rates related to conventional sugarcane. In energy cane harvesting, all produced straw is collected along with the stalks, using the integral harvest system, since straw removal on the field is considered to be impracticable due to thinner and lighter stalks been more difficult to separate by blowing.

All other operations are considered to be equal to those performed for sugarcane, and economic, environmental, and social results are calculated following the same rules described in this chapter for sugarcane.

In the study from Milanez et al. (2015), the CanaSoft model was used to evaluate economic aspects of energy cane production in medium- and long-term scenarios of first- and second-generation ethanol production. Jonker et al. (2015) also used CanaSoft to calculate biomass cost to assess ethanol production costs for different horizon times and different biomasses, including energy cane.

3.5.2 Sweet Sorghum

Sweet sorghum (*Sorghum bicolor* L.) production has been proposed as a promising alternative to extend the industrial operation season of the sugarcane mills since its industrialization requires minor adjustments in equipment used for the sugarcane processing (May 2012a, b).

Sweet sorghum may be produced in the sugarcane replanting areas. The sugarcane crop cycle is about 5 years long, meaning that every year around of 20 % of the area is plowed out after harvest and replanted about 6 months later. This fraction of the area of the mill is available for sweet sorghum production, without interfering in the sugarcane production cycle. Alternatively, additional areas can be used to complement sorghum production during sugarcane off-season. Another advantage is that agricultural machinery and inputs are similar to those used for sugarcane production.

To include sweet sorghum production in CanaSoft model, a set of agricultural operations must be added between soil preparation for sugarcane, which occurs after ratoon removal, and sugarcane planting. Operations for sweet sorghum production include seeding, application of fertilizers and agrochemicals, harvesting, and transportation to the mill. Most of the machinery and inputs used in sweet sorghum production system are already included on CanaSoft data banks for sugarcane. Only a grain seeder and specific agrochemicals had to be added and became part of CanaSoft data banks.

Even though higher yields are achieved if smaller rows spacing is practiced, sorghum is often planted flowing the same sugarcane rows spacing. Previously to sorghum seeding, soil preparation is planned to sugarcane planting. Spacing is also technically limited by the harvester equipment. Sugarcane harvester is employed to collect sweet sorghum, due to the lack of commercial harvesting equipment specifically designed for this biomass. Stalks, leaves, and grain are gathered and transported to the mill by the same trucks used for sugarcane transport. Bulk density of sorghum biomass is smaller than that of sugarcane stalks, and transport capacity of the trucks is properly adjusted.

Excluding adjustments on harvesting and transport operations, no modifications are necessary to other operations, which are similar to the ones practiced on sugarcane production. Besides information about machinery and operations, input application rates and biomass yields must be informed. All the calculations and outputs are quite similar to that modeled for sugarcane.

In the study from Cavalett et al. (2012), different alternatives for integration of sweet sorghum processing during the off-season of sugarcane were assessed and the CanaSoft model was used to evaluate production cost of different sweet sorghum varieties. Cavalett et al. (2013) also used sweet sorghum as a harvest extension alternative, and biomass production costs and environmental impacts were calculated using CanaSoft model. Jonker et al. (2015) used the CanaSoft to calculate biomass cost to evaluate the ethanol production costs in Brazil up to 2030 for different biomass crops, including sweet sorghum.

3.5.3 Corn

Similar to sweet sorghum, corn (*Zea mays*) may be produced in the sugarcane replanting areas, as a rotation crop without interfering in sugarcane cycle. Most of the Brazilian production area, however, is comprised of the area devoted to corn production as the main crop, or by the so-called *safrinha* corn, generally planted as a secondary crop in rotation with soybean.

CanaSoft model can be used to evaluate all these alternative production systems for corn, by using the information on agricultural data banks and including the operations commonly used for corn production. When corn is assessed as a crop rotation with sugarcane, production area is limited by the availability of sugarcane reforming areas. For other production models, there are no limitations of areas, but transport distances need to be adequately adjusted. Machinery and inputs exclusively used in corn production, without correspondence in sugarcane production system, were already included in CanaSoft data banks.

In the study from Milanez et al. (2014), the CanaSoft model was used to evaluate *safrinha* corn production system, using typical information of corn production in rotation with soybean in Mato Grosso state. Results were used to assess the integration of corn processing during the off-season of sugarcane using sugarcane bagasse as energy feedstock for the corn ethanol process.

References

Braunbeck OA, Magalhães PSG (2010) Avaliação tecnológica da mecanização da cana-de-açúcar. In: Cortez LAB (ed) BIOETANOL DE CANA-DE-AÇÚCAR: P&D Para Produtividade E Sustentabilidade. Blucher-FAPESP, São Paulo, pp 413–424

Cardoso TF, Cavalett O, Chagas MF, Moraes ER, Carvalho JLN, Franco HCJ, Galdos MV, Scarpari F, Braunbeck OA, Cortez LAB, Bonomi A (2013) Technical and economic assessment of trash recovery in the sugarcane bioenergy production system. Scientia Agricola 70:353–360

Cavalett O, Cunha, MP, Chagas MF, Junqueira TL, Dias MOS, Pavanello LG, Leal MRLV, Rossell CEV, Bonomi A (2012) An exploratory economic analysis of sugarcane harvest extension using sweet sorghum in the Brazilian sugarcane industry. In: XXVIII International Society of Sugar Cane Technologists (ISSCT) Congress, 2013, São Paulo. XXVIII ISSCT Proceedings. São Paulo: Society of Sugar Cane Technologists, 2013. vol. 1, pp 225–225

Cavalett O, Chagas MF, Cardoso TF, Franco HCJ, Junqueira TL, Pavanello LG, Jesus CDF, Moraes BS, Bonomi A (2013) Development of an agricultural model for biorefineries sustainability optimization. In: 21st European Biomass Conference and Exhibition, 2013, Copenhagen. 21st European Biomass Conference and Exhibition Proceedings, 2013. v. 1. p. 1826-1834

Costa MCG, Vitti GC, Cantarella H (2003) Volatilização de N–NH_3 de fontes nitrogenadas em cana-de-açúcar colhida sem despalha a fogo. Rev Bras Ciênc Solo 27:631–637

DNIT (2009) Departamento Nacional de Infraestrutura de Transporte. INSTRUÇÃO DE SERVIÇO/DG N° 09, DE 21 DE JULHO DE 2009

Don A, Osborne BA, Hastings A, Skiba U, Carter MS, Drewer J, Flessa H, Freibauer A, Jones MB, Lanigan GJ, Mander U, Monti A, Valentine J, Walter T, Zenone T (2012) Land-use change to bioenergy production in Europe: implications for the greenhouse gas balance and soil carbon. Global Change Biology- Bioenergy 4:372–391

Ecoinvent centre (2007) Ecoinvent data v2.2. Ecoinvent reports No.1-25, Swiss Centre for Life Cycle Inventories, Dübendorf

Figueiredo Filho (2011) Os desafios da mecanização em cana-de-açúcar. In: II Simpósio Paulista de Mecanização em Cana-de-açúcar. UNESP. Jaboticabal

Franco HCJ (2008) Eficiência agronômica da adubação nitrogenada de cana planta. Tese (Doutorado). Escola Superior de Agricultura Luiz de Queiros. Universidade de São Paulo. Piracicaba

Gabe U, Rodella AA (1999) Trace elements in Brazilian agricultural limestones and mineral fertilizers. Commun Soil Sci Plant Anal 30(5–6):605–620

Goes T, Marra R, Araújo M De, Alves E, de Souza MO (2011) Sugarcane in Brazil: Current technologic stage and perspectives. Revista de Politica Agricola 1:52–65

GREET (2010) Greenhouse Gases, Regulated Emissions, and Energy Use in Transportation, version 1.8d, Argonne National Laboratory. Argonne, Illinois, USA

IPCC (2006) Intergovernmental Panel on Climate Change. Guidelines for national greenhouse gas inventories, vol. 4—Agriculture, forestry and other land use, Intergovernmental Panel on Climate Change

Jonker JGG, Van Der Hilst F, Junginger HM, Cavalett O, Chagas MF, Faaij APC (2015) Outlook for ethanol production costs in Brazil up to 2030, for different biomass crops and industrial technologies. Appl Energy 147:593–610

Leal MRLV, Walter A, Seabra JEA (2012) Sugarcane as an energy source. Biomass Conversion and Biorefinery. doi: 10.1007/s13399-012-0055-1, 2012

Leal MRLV, Galdos MV, Scarpare FV, Seabra JEA, Walter A (2013) Sugarcane straw availability, quality, recovery and energy use: A literature review. Biomass Bioenergy 53:11–19

Matsuoka S, Kennedy AJ, Santos EGD, Tomazela AL, Rubio LCS (2014) Energy cane: its concept, development, characteristics, and prospects. Advances in Botany, vol. 2014, Article ID 597275, 13 pages. doi:10.1155/2014/597275

May A (2012a) Seminário Temático sobre Sorgo Sacarino. Documentos 137 - Embrapa Milho e Sorgo, Anais do I Seminário Temático sobre Sorgo Sacarino. Sete Lagoas, MG. 83 p

May A (2012b) Sistema Embrapa de Produção Agroindustrial de Sorgo Sacarino para Bioetanol Sistema BRS1G – Tecnologia Qualidade Embrapa. Documentos 139 - Embrapa Milho e Sorgo. Sete Lagoas, MG. 120p

McKendry P (2002) Energy production from biomass (part 1): overview of biomass. Bioresource Technol 83:37–46

Milanez AY, Nyko D, Valente MS, Xavier CEO, Kulay LA, Donke ACG, Matsuura MISF, Ramos NP, Morandi MAB, Bonomi A, Capitani DHD, Chagas MF, Cavalett O, Gouveia VLR (2014) A produção de etanol pela integração do milho-safrinha às usinas de cana-de-açúcar : avaliação ambiental, econômica e sugestões de política. Revista do BNDES 41:147–207

Milanez AY, Nyko D, Valente MS, Sousa LC, Bonomi A, Jesus CDF, Watanabe MDB, Chagas MF, Rezende MCAF, Cavalett O, Junqueira TL, Gouveia VLR (2015) De promessa a realidade: como o etanol celulósico pode revolucionar a indústria da cana-de-açúcar - uma avaliação do potencial competitivo e sugestões de política pública. BNDES Setorial 41:237–294

National Supply Company (Companhia Nacional de Abastecimento - CONAB). Cana-de-açúcar – Safra 2014/15 – Segundo Levantamento – accessed in 2015 http://www.conab.gov.br/cana-de-açúcar

Nemecek T, Kagi T (2007) Life Cycle Inventories of Agricultural Production Systems – Data v2.0, Zurich and Dubendorf: 2007

Renewable Fuels Association (2015) Ethanol Industry Outlook. 36 pp. Available on line in http://www.ethanolrfa.org/pages/annual-industry-outlook

Renouf MA, Wegener MK, Pagan RJ (2010) Life cycle assessment of Australian sugarcane production with a focus on sugarcane growing. Int J Life Cycle Assess 15:927–937

Rodella AA (2011) Personal Communication. Retired professor from Escola Superior de Agricultura Luiz de Queiroz (ESALQ/USP)

Smith DM, Inman-Bamber NG, Thoburn PJ (2005) Growth and function of the sugarcane root system. Field Crops Res 92:169–183

Souza A, Grandis A, Leite DC, Buckeridge M (2013) Sugarcane as a bioenergy source: history, performance, and perspectives for second-generation bioethanol. Bioenergy Res 7:24–35

Trivelin PCO, Franco HCJ (2011) Balanço de nitrogênio de fertilizantes e a sustentabilidade de agrossistemas: o estado da arte e necessidade de pesquisas. Congresso Brasileiro de Ciência do Solo 32

Chapter 4
Biorefinery Alternatives

Edvaldo R. Morais, Tassia L. Junqueira, Isabelle L.M. Sampaio, Marina O.S. Dias, Mylene C.A.F. Rezende, Charles D.F. de Jesus, Bruno C. Klein, Edgardo O. Gómez, Paulo E. Mantelatto, R Maciel Filho and Antonio Bonomi

4.1 Biorefinery Definition

The need for substitutes for oil-based fuels, energy and chemical products is known and arises for different reasons, such as greenhouse gas (GHG) emissions and the possibility of a future oil shortage. In this scenery, the development of biorefineries is of utmost importance. A biorefinery is a facility where biofuels, bioenergy, biochemical, biomaterials, food and feed are coproduced from biomass (Cherubini et al. 2009). According to the International Energy Agency (IEA) Bioenergy Task 42: "Biorefinery is the sustainable processing of biomass into a spectrum of marketable products and energy" (IEA 2008).

A biorefinery would be the analogue of a petroleum refinery, where crude oil is processed and a variety of products is obtained from each fraction, targeting the complete use of feedstock to produce multiple products. In the same way, the objective of a biorefinery is to transform the different biological constituents of biomass (sugar, starch, lipids, cellulose, hemicellulose and lignin) into desired products through a combination of technologies and processes. Figure 4.1 is a

E.R. Morais · T.L. Junqueira · I.L.M. Sampaio · M.C.A.F. Rezende · C.D.F. de Jesus
B.C. Klein · P.E. Mantelatto · R. Maciel Filho · A. Bonomi (✉)
Brazilian Bioethanol Science and Technology Laboratory (CTBE), Brazilian Center of Research in Energy and Materials (CNPEM), Campinas, SP, Brazil
e-mail: antonio.bonomi@bioetanol.org.br

M.O.S. Dias
Institute of Science and Technology, Federal University of São Paulo (ICT/UNIFESP), São José dos Campos, SP, Brazil

E.O. Gómez
Federal University of Grande Dourados (UFGD), Dourados, Mato Grosso do Sul, Brazil

R. Maciel Filho
School of Chemical Engineering, University of Campinas (UNICAMP), Campinas SP, Brazil

A. Bonomi et al. (eds.), *Virtual Biorefinery*, Green Energy and Technology,
DOI 10.1007/978-3-319-26045-7_4

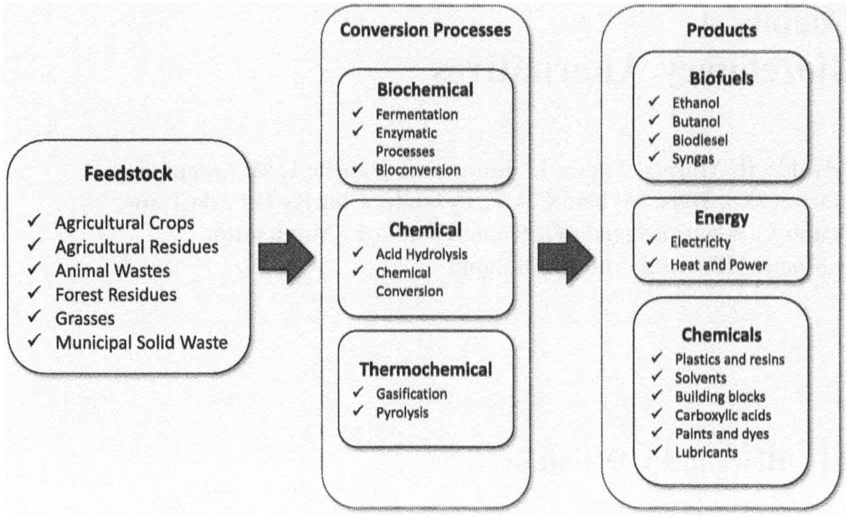

Fig. 4.1 Simple three-step biomass–process–product procedure (adapted from Fernando et al. 2006)

simplified representation of a biorefinery, in which biomass feedstocks are used to obtain a myriad of products such as fuel, power and chemicals, using biological, thermochemical and chemical conversion processes.

4.2 Biorefinery Concepts

A simplified way to classify biorefineries is to arrange them into different categories known as Phases I, II and III according to feedstock, process used and final products (Cherubini et al. 2009; Kamm and Kamm 2004a, b).

Phase I biorefineries have fixed processing capability (and consequently a fixed production capacity for products/coproducts), focusing on one feedstock (normally grains). An example of this biorefinery is the biodiesel production from sunflower. The oil is extracted from the seeds and then transesterified to fatty acids methyl esters (biodiesel) and glycerol (as by-product) using an alcohol catalyst (Pfaltzgraff and Clark 2014).

Phase II biorefineries have more flexibility, being able to produce more than one product (e.g., energy and fuels), depending upon a product demand and prices, using the same facility and one feedstock. Corn wet milling technology for ethanol, high-fructose corn syrup, corn oil and starch production could be classified as a Phase II biorefinery (Fernando et al. 2006). Mills producing ethanol, sugar and electricity could be another example of Phase II biorefineries. In these facilities, sugarcane juice can be diverted to sugar or ethanol production according to market demands (Dias et al. 2015).

Phase III biorefineries allow a combination of different technologies to produce a wide range of products (e.g., biomass pretreatment, biological and/or chemical transformation) from several feedstocks. The innumerous constituents of feedstock can be treated separately to produce a great number of high-valued chemical derivatives. This biorefinery phase can be further divided into subcategories, namely whole-crop, green and lignocellulose feedstock biorefineries (Fernando et al. 2006).

Whole-crop Biorefinery

Whole-crop biorefineries (Fig. 4.2) focus on the use of cereals for nonalimentary purposes. Examples of feedstock for these facilities include wheat, corn and rye.

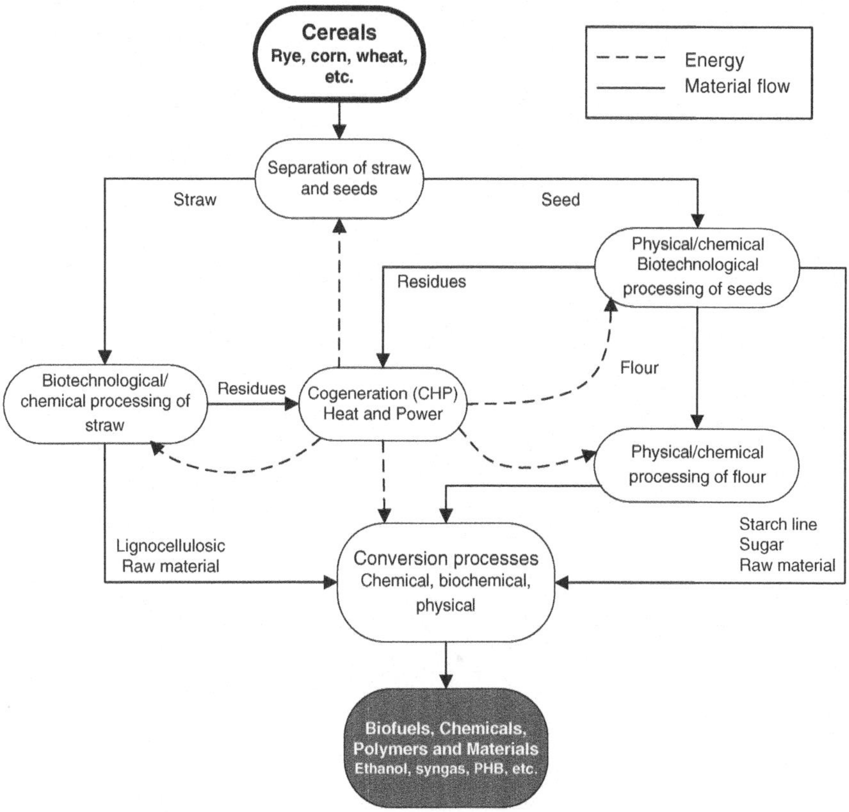

Fig. 4.2 Whole-crop biorefinery block diagram (adapted from Kamm et al. 2006)

The straw and seed are processed separately; the straw can be processed in a lignocellulosic feedstock regimen that will be discussed in further sections (Kamm and Kamm 2004a, b). Cereal can be used to produce starch or meal, and a great number of products through chemical and biological conversion. Some examples of final products of whole-crop biorefineries are as follows: biofuels, bioplastics, organic acids and solvents.

Green Biorefinery

Green biorefineries (Fig. 4.3) operate with "wet" biomass (or green biomass), such as sugarcane or grass. The first step is the wet pressing; products from this operation are the press juice (a liquid mixture containing sugars, proteins and other soluble solids) and press cake (composed mostly of fibers). Lactic acid, amino acids, ethanol and proteins can be obtained from the green juice; levulinic acid, green feed pellets and other raw chemicals are some of the possible usages for the press cake (Kamm and Kamm 2004a, b).

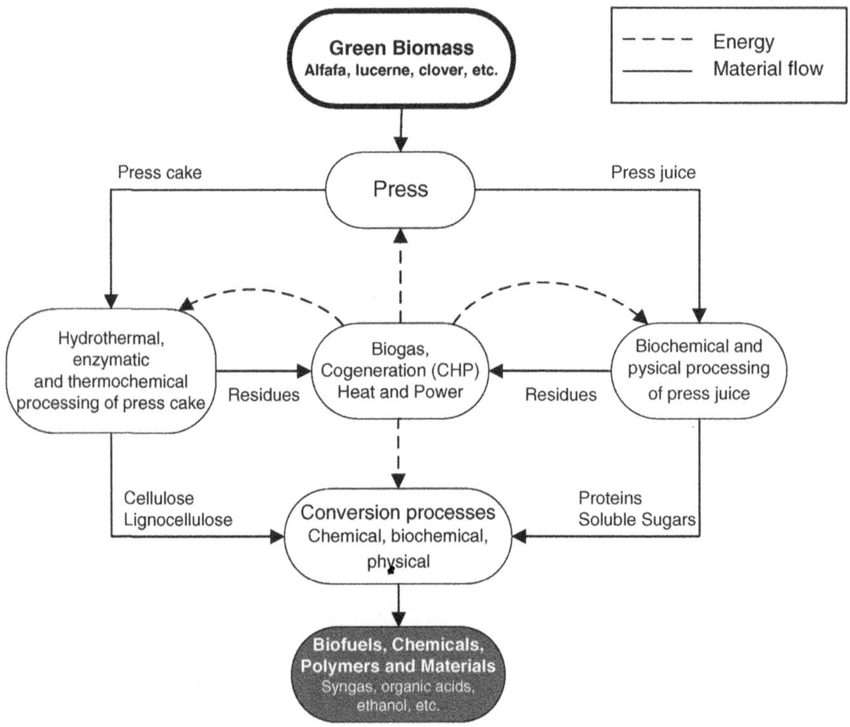

Fig. 4.3 Green biorefinery block diagram (adapted from Kamm et al. 2006)

Lignocellulosic Feedstock Biorefineries

Lignocellulosic feedstock (LCF) biorefineries (Fig. 4.4) operate with materials such as straw, reed, wood, paper and municipal waste.

Lignocellulosic feedstock is basically composed of three types of components: hemicellulose (a sugar polymer mainly made of pentoses), cellulose (glucose polymer) and lignin (phenol polymer) (Kamm and Kamm 2004a, b). The first operation of the LCF biorefinery is the separation of the feedstock into these three chemical fractions.

Cellulose undergoes hydrolysis operation, through chemical or biological ways. From this step, glucose is obtained. This six-carbon sugar can be applied to the production of ethanol, butanol and organic acids such as acid lactic.

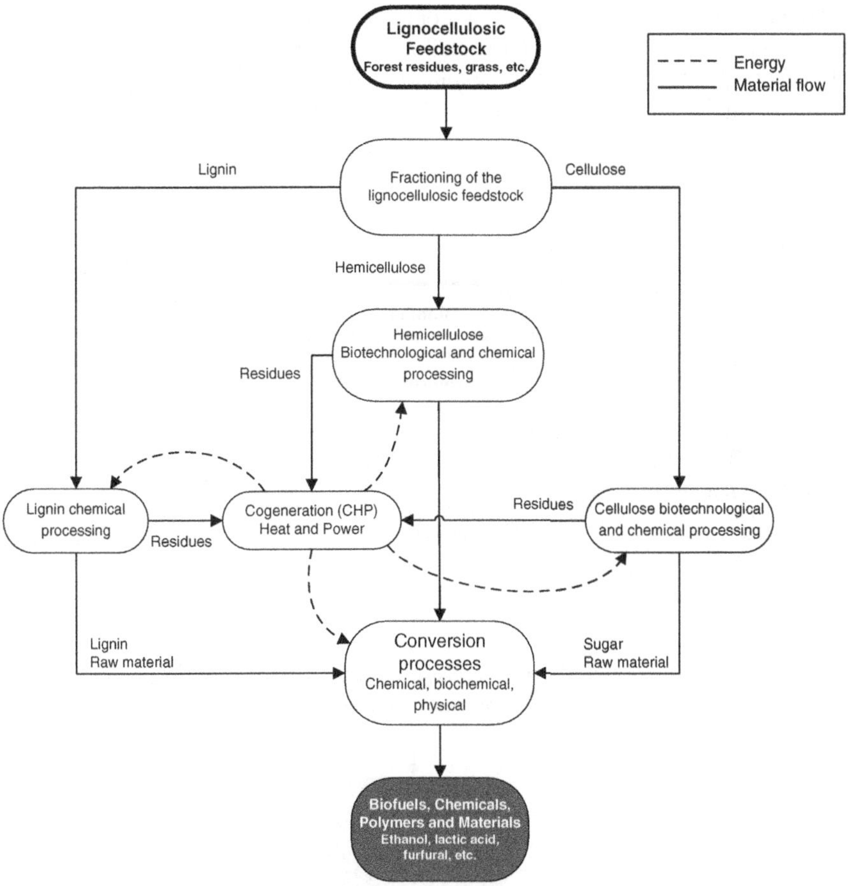

Fig. 4.4 Lignocellulosic feedstock biorefinery block diagram (adapted from Kamm et al. 2006)

Hemicellulose hydrolysis generates five-carbon (pentose) sugars. Fermentation of pentose can produce biofuels, such as ethanol and butanol. Other important chemicals can also be produced such as organic acids (lactic acid, ferulic acid), xylitol and furfural.

Lignin can be used as a solid fuel, gasified or applied for the production of adhesives. However, it is important to note that this destination is not the most attractive, because the decomposition of lignin could generate considerable amounts of monoaromatic hydrocarbons, which could be more interesting if they were produced and separated in energy-efficient ways (Kamm and Kamm 2004a, b).

The Two-Platform Concept

In this model of biorefinery (Fig. 4.5), the feedstock is separated and then processed through two different routes: a biochemical route (sugar platform) and a thermochemical route (or syngas platform). Sugar platform breaks down biomass into different types of sugars. These sugar components are converted to several chemicals and fuels through fermentation or other biological processing. Thermochemical platform converts biomass to hydrogen and carbon monoxide or pyrolysis oil.

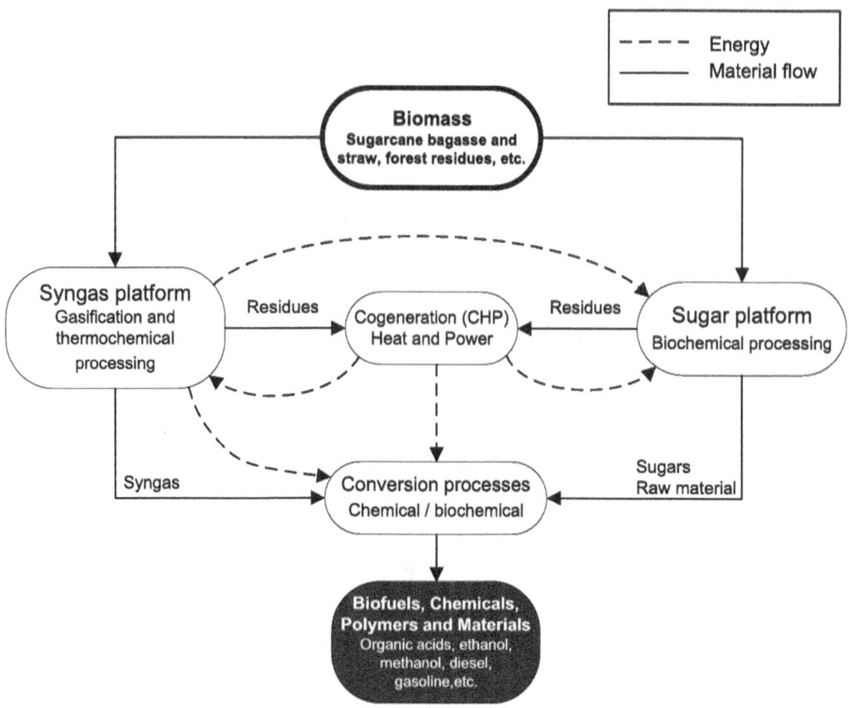

Fig. 4.5 Two-platform biorefinery block diagram (adapted from Kamm et al. 2006)

According to Cherubini (2009), these classification criteria are not homogeneous, and some of these systems can be classified in more than one category. For instance, a system where wood is gasified to syngas can be categorized both as a lignocellulosic feedstock biorefinery (since lignocellulose is used) and as a two-platform biorefinery concept (because syngas is produced).

Several technological processes can be applied to convert biomass feedstock into sought-after products. Such processes are classified according to the type of technology employed in the conversion route: biochemical, thermochemical and chemical. In what follows, routes focused on biochemical and thermochemical processes will be described in more detail. Routes based on chemical conversion processes will not be discussed in this book since no studies using the Virtual Sugarcane Biorefinery—VSB framework for this route were carried out until this moment.

4.2.1 Biochemical Routes

Biochemical routes are pathways that include processes where microorganisms or enzymes are used for the conversion of biomass into different components. Biochemical processes mostly occur at mild conditions (lower temperature and pressure), and examples of these processes include anaerobic digestion, aerobic and anaerobic fermentation and enzymatic conversion (Cherubini et al. 2009).

Biochemical routes in biorefineries can produce biofuels, chemical building blocks and bioplastics, among other biomaterials.

4.2.1.1 Biofuels

Biofuels can be either solid, liquid or gaseous fuels, obtained from renewable feedstock (Demirbas 2009). According to the production process, biofuels can be classified as follows:

- **First-Generation Biofuels**: First-generation biofuels are mainly obtained from food crops, such as sugarcane, sugar beet and corn. They might also be derived from animal fat and vegetable oils. Some significant examples include ethanol production from maize (USA), wheat (Northwest Europe), sugar beet (Northwest Europe), cassava (Thailand), sweet sorghum (China) and sugarcane (Brazil). For biodiesel, some of the main examples include the production from winter oilseed rape (Northwest Europe), soybean (USA and Brazil) and oil palm (Malaysia) (de Vries et al. 2010; Padula et al. 2012). These biofuels are currently produced in industrial scale, and their production processes are at a mature stage.
- **Second-Generation Biofuels**: Second-generation biofuels are the ones obtained from lignocellulosic materials and residues, such as sugarcane bagasse and

straw, forest residues and agricultural wastes. Some of the main advantages of second-generation biofuels over first-generation ones are that there is no competition with food crops and land use change. The concern with land use change has been attracting attention to the cultivation of species that are able to grow in degraded lands, not adequate for food or fiber crop production. Some examples are vegetative grasses such as *Miscanthus* and switchgrass and forest species with a short rotation period, such as *Eucalyptus* (Dutta et al. 2014). Ethanol can also be produced through second-generation technologies, as well as other biofuels such as butanol and isopropanol.

Some of the most relevant biofuels are presented below with a brief process description.

Ethanol
An important biofuel, ethanol is currently produced commercially from renewable sources. Details on ethanol production from sugarcane juice and from lignocellulosic materials can be found in other sections of this book.

Butanol
Butanol is a four-carbon advanced biofuel. It is compatible for ethanol blending and can enhance the ethanol and gasoline mixing. Butanol also has various applications in chemical industry, as a chemical precursor and as a solvent.

This biofuel can be produced through fermentation of sugars with anaerobic bacteria. The most common fermentation process applied is the ABE fermentation (acetone–butanol–ethanol). One example of microorganism employed in ABE fermentations is *Clostridium beijerinckii* (Mariano et al. 2012). Some microorganisms are able to produce butanol from pentose. Using these sugars can be advantageous, because a variety of lignocellulosic materials generate pentoses when they undergo pretreatment.

Biodiesel
The term "biodiesel" is applied to diesel fuel produced with vegetable oils, waste or animal fat as feedstock. Vegetable oils applied for the production of this biofuel include soybean, palm, sunflower, coconut and rapeseed, among others.

These fuels are composed of alkyl esters, displaying long chains. Examples of these esters include methyl and ethyl esters. Alcohols used for the transesterification reaction of the oils are more commonly methanol or ethanol. Catalysts for this reaction can be present in the same phase of reactants (liquid) or on a different one (solid). The first case is referred to as homogeneous catalytic transesterification and the second one as heterogeneous catalytic transesterification. These are both classified as chemical catalysts, but there is a third type available, the biocatalysts (Chouhan and Sarma 2011).

Biodiesel is currently produced in industrial scale. The main producers were Germany, USA, Brazil, Argentina and Indonesia (EIA 2012).

Biogas

Biogas is obtained from biodigestion of organic matter by microorganisms. The composition of the gas produced is variable and dependable on process conditions, feedstock and biological agents involved. Examples of gases that might be present are methane (CH_4), carbon dioxide (CO_2), hydrogen (H_2) and hydrogen sulfide (H_2S), among others.

Feedstocks applied for the production of biogas include organic waste, animal manure, sewage sludge, organic process residues such as vinasse from ethanol-producing plants and energy crops such as maize, sugarcane and grass (IEA 2011).

This gas mixture can be used for various applications such as a fuel for diesel substitution in agricultural operation, as heating fuel in the chemical plants, domestic refrigeration and electricity production (Cortez et al. 2008). A recent trend in biogas usage is the injection of purified biogas (biomethane) into natural gas grids (Moraes et al. 2015).

4.2.1.2 Building Blocks

The components known as "building blocks" are substances used for the manufacturing of an extensive array of chemical derivatives.

A great number of building blocks can be produced from biomass. These compounds are mainly used not only for the production of plastics and polymers, but also for a great number of fine and specialty chemicals. It is possible to say that bioderived chemicals could substitute almost all chemical compounds currently derived from fossil resources, but the cost of production of these renewable substituents is still often times higher, this being an obstacle for their production (de Jong et al. 2012).

In 2004, the US Department of Energy (DOE) released a report containing a list of the most promising biobased building blocks (Werpy and Petersen 2004). Table 4.1 shows the original list and a newer version of this selection, from 2010's article "Technology development for the production of biobased products from biorefinery carbohydrates—the US Department of Energy's 'Top 10' revisited" (Bozell and Petersen 2010).

Technologies and processes are always evolving, and thus, a further revision was made by Choi et al. (2015). This article presented a new look on the previously selected building blocks, highlighting the fact that in these 11 years, since the publication of DOE's primary report, many of these chemical compounds started to be industrially produced from biobased feedstock.

The production process and applicability of some of the most important building blocks obtained from renewable sources are described below.

Table 4.1 Top building blocks from Werpy and Petersen (2004) and Bozell (2010)

Werpy and Petersen (2004)	Bozell (2010)
Succinic, fumaric and malic acids	Ethanol
2,5-Furan dicarboxylic acid	Furans
3-Hydroxypropionic acid	Glycerol and derivatives
Aspartic acid	Biohydrocarbons
Glucaric acid	Lactic acid
Glutamic acid	Succinic acid
Itaconic acid	Hydroxypropionic acid/aldehyde
Levulinic acid	Levulinic acid
3-Hydroxybutyrolactone	Sorbitol
Glycerol	Xylitol
Sorbitol	–
Xylitol/arabinitol	–

Succinic Acid

Succinic acid is used in food, agricultural and pharmaceutical industries and is a potential chemical precursor for the production of a great number of derivatives with higher value, such as 1,4-butanediol (BDO), tetrahydrofuran (THF), g-butyrolactone (GBL), succinimide and the polymer polybutylene succinate (PBS) (Wang et al. 2014).

Current commercial production of succinic acid derives from oil-based sources; however, the production of this acid from renewable sources is gaining crescent attention, with companies such as BioAmber, Reverdia and Myriant preparing for commercialization (Choi et al. 2015).

Different microorganisms can be applied to obtain succinic acid from sugars through fermentation: recombinant *Escherichia coli* (Wang et al. 2014), *Actinobacillus succinogenes* (Luque et al. 2009) and *Corynebacterium glutamicum* (Zhou et al. 2015), among others.

Ethanol

Ethanol is a platform chemical for a myriad of compounds, such as butanol, butadiene and polypropylene. Further details on the production of this compound can be found in other sections of this book.

Lactic Acid

Lactic acid is an important additive in food, textile, leather and pharmaceutical industries. It is also a chemical precursor to the production of polylactate polymers such as PLA, a biodegradable and biocompatible material (Vijayakumar et al. 2008).

This carboxylic acid can be obtained through chemical and biological routes. The chemical pathway uses feedstock of petrochemical origin (lactonitrile); the lactic acid produced is a racemic mixture of D-L-lactic acid. For the case of biological routes of production, fermentation processes using lactic acid bacteria (e.g.,

Lactobacillus delbrueckii) or fungi (e.g., *Rhizopus oryzae*) can be employed. The final product obtained is pure L (+) or D (−) lactic acid. The substrates used for fermentation are carbohydrates such as glucose, sucrose, lactose and maltose, from different sources such as molasses, whey or sugar beet (Ghaffar et al. 2014).

4.2.1.3 Bioplastics

Bioplastics are polymers manufactured using renewable materials. Some examples of bioplastics include aliphatic polyesters (PHA, PHB, PLA), biobased polyethylene (PE) and plastics obtained from starch.

The most common application for bioplastics is for dischargeable packages, such as food packing, but there are other usages. PLA and PHA, for example, for their biodegradability and biocompatibility have been used for surgical implants.

Other than the application of the pure biobased polymers, it is possible to make blends of petroleum-derived and biobased plastics, achieving some desirable characteristics that these materials when pure may not display.

Poly-3-hydroxybutyrate (PHB)

PHB is one of the most common biopolymers from the category known as "PHAs" (polyhydroxyalkanoates). PHAs can be produced by bacterial fermentation using renewable carbon sources. One of the main advantages PHB has shown is that it possesses some important characteristics found in petroleum-based plastics, such as elastomeric and thermoplastic properties. This polymer is also biodegradable and has been used in various medical applications, such as bone plates and nails (Tripathi et al. 2013).

Studies report the production of PHB applying different bacteria species. Some examples include *Alcaligenes sp* (Tripathi et al. 2013) and *Bacillus megaterium* (Gouda et al. 2001). Carbon sources for the production of PHB can vary from sugarcane molasses to corn steep liquor (Gouda et al. 2001), depending on process conditions and microorganisms applied. Other examples include glycerol, ethanol and palm oil (Mahishi et al. 2003).

One of the main factors that negatively influences the use of PHB for substituting petroleum-based plastics is the production cost. Particularly for the downstream process, the costs are still very high. Optimization of fermentation and gene cloning are some of the strategies currently studied to make PHB a more competitive choice (Tripathi et al. 2013).

Biobased Polyethylene (Bio-PE)

Currently, the most common method for production of polyethylene is through polymerization of ethylene, produced by steam cracking of petroleum feedstock. Environmental concerns and the possibility of a future shortage of petroleum have encouraged the search for renewable substitutes for polyethylene production.

Bio-PE manufacturing involves the dehydration of ethanol to ethylene and the subsequent polymerization as in the usual process. Ethanol used as feedstock in this process can be obtained from fermentation of corn syrup, sugarcane juice and sugars from sugar beet, among others.

Braskem, a Brazilian company, produces polyethylene from sugarcane-derived ethanol. The annual production reported by the company is of 200,000 t, and three types of polyethylene are manufactured: high-density polyethylene (HDPE), linear low-density polyethylene (LLDPE) and low-density polyethylene (LDPE). Applications of the Bio-PE include bags, films and different types of packaging (Braskem 2015).

Poly Lactic Acid (PLA)
PLA is a polymer obtained from renewable sources. It is also a biodegradable material. This plastic is produced through a ring opening polymerization of lactide or by polycondensation of lactic acid. The lactide used in the ring opening polymerization is a cyclic dimer of lactic acid (Reddy et al. 2013).

PLA is also a biocompatible material, having a myriad of applications on medical industry, such as orthopedic implants, screws, sutures and controlled drug release (Vijayakumar et al. 2008). Other applications of this polymer include food packages, films and containers.

4.2.2 Thermochemical Routes

Thermochemical routes are based on the concept of integral sugarcane utilization incorporating thermochemical biomass processing technologies to convert lignocellulosic fractions to synthesis gas (syngas), rich in carbon monoxide and hydrogen.

Considering the thermochemical route, the syngas platform can be identified as previously illustrated in Fig. 4.5.

The syngas platform is based on thermochemical biomass conversion processes and conditioning gas technologies to generate syngas to be converted into biofuels, biobased chemicals and heat and power.

4.2.2.1 Building Blocks

According to de Jong et al. (2006), pyrolysis oil and syngas (mainly CO and H_2) are identified as of the main building blocks of the thermochemical route.

Pyrolysis Oil
Pyrolysis process is the biomass thermal depolymerization at 250–550 °C, in the absence or little added oxygen (oxidative pyrolysis) (Kersten and Garcia Perez 2013).

Under this condition, the spectrum of products from biomass pyrolysis depends invariably on the process temperature, pressure and residence time of the pyrolysis vapors released by the reactor. The main product or fraction obtained is named bio-oil or pyrolysis oil, which is a complex mixture of organic compounds that may be upgraded with different technologies to finally obtain value-added compounds. Pyrolysis oil can be also considered a building block since it derives compounds as phenols, organic acids, furfural, HMF, anhydrosugars as levoglucosan, and pyrolytic lignin. Pyrolysis biorefinery presents an important advantage that is the possibility of decentralized production of the oil in regions where abundant biomass is available. Currently, an important issue to obtain value-added compounds is the cost-effective fractionation of the pyrolysis oil (de Jong et al. 2012).

According to Karatzos et al. (2014), pyrolysis oils can be used in stationary heat and power applications or they can be upgraded to drop-in biofuels such as diesel, gasoline and biojet fuel grade hydrocarbons.

Syngas

Syngas (mainly CO and H_2) can be considered a building block since it is the precursor to a wide range of liquid biofuels, such as ethanol, methanol, mixed alcohols, hydrogen, and gasoline, diesel and Fischer–Tropsch Synthetic Paraffinic Kerosene (FT-SPK) synthesis, biobased chemicals, such as formaldehyde acetic acid and ammonia (NH_3), and heat and power. Figure 4.6 illustrates the range of

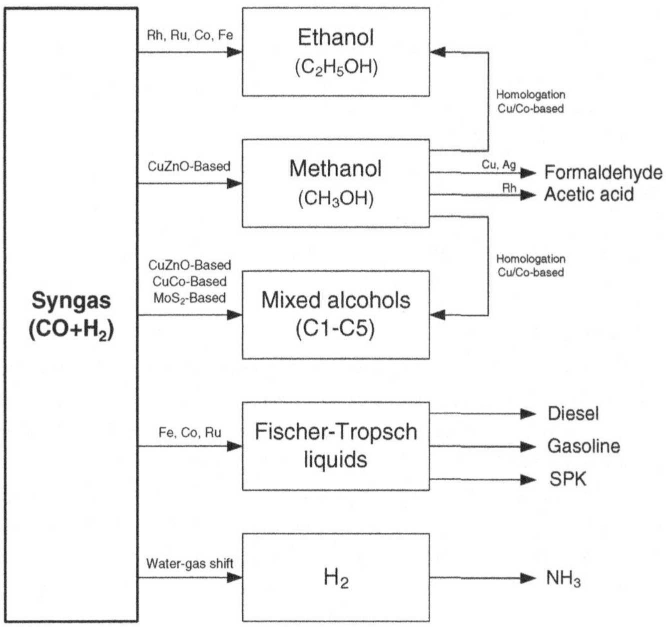

Fig. 4.6 Range of biofuels and biobased chemicals obtained by catalytic conversion of biomass-derived syngas (adapted from Subramani and Gangwal 2008)

biofuels and biobased chemicals that can be obtained by catalytic conversion of biomass-derived syngas.

Depending on the process and the desired product, specific catalysts are employed. Several studies have been developed to assess the catalyst performance in each process and the effect of promoters and additives to improve the conversion and selectivity to the desired product in order to make the catalytic conversion of syngas commercially attractive. Besides the catalyst, other important parameters may be controlled to improve the catalytic synthesis, as reactor type, pressure and temperature conditions and H_2/CO ratio (Subramani and Gangwal 2008; Spivey and Egbebi 2007).

According to Werpy and Petersen (2004), methanol, hydrogen, mixed alcohols and Fischer–Tropsch liquids can in turn be considered as building blocks, since other value-added products are derived from them. A short discussion on methanol and hydrogen acting as building blocks is presented below.

Methanol

Figure 4.6 shows that methanol obtained from catalytic conversion of syngas with CuZnO based is also considered a building block (Subramani and Gangwal 2008), since through methanol it is possible to obtain ethanol, gasoline and olefins, formaldehyde, acetic acid, dimethyl ether (DME), methyl tert-butyl ether (MTBE) and others. According to Spath and Dayton (2003), in a global context, the largest consumption of methanol is in formaldehyde production (35 % of worldwide methanol), followed by MTBE (25 %) and acetic acid (9 %).

Hydrogen

Hydrogen obtained from syngas by water–gas-shift reaction is used as a building block of ammonia production, being ammonia the largest consumer of H_2, according to Spath and Dayton (2003).

4.2.2.2 Biofuels

This section briefly describes biofuels as ethanol and mixed alcohols from syngas, and gasoline, diesel and FT-SPK liquids.

Ethanol and Other Alcohols

As shown in Fig. 4.6, syngas can be catalytically converted to ethanol directly using Rh-based catalysts. However, cost and availability of the Rh-based catalysts and low yield of ethanol are limitations of the direct conversion of syngas to ethanol (Subramani and Gangwal 2008). New types of catalysts with high selectivity are necessary to make feasible this route (Spivey and Egbebi 2007). The mixed alcohol (C_1–C_5 alcohols) route coupled with methanol homologation using CuZnO based is a more interesting route to improve ethanol yield from syngas (Subramani and Gangwal 2008).

In addition to the catalytic route, an alternative route comprising thermochemical–biological processing, named hybrid processing, considers syngas fermentation to produce ethanol, as can be found in Brown (2006).

Fischer–Tropsch Products

In Fischer–Tropsch processes, syngas can be catalytically converted into liquid hydrocarbon mixtures (FT liquids) that can be upgraded to gasoline, diesel and biojet fuels (Karatzos et al. 2014).

According to Spath and Dayton (2003), low temperatures in FT synthesis favor high molecular mass linear waxes, while high temperatures produce gasoline and low molecular weight olefins. Upgrading of waxes in diesel is performed by hydroprocessing. Olefins can be converted to gasoline by hydrogenation (Karatzos et al. 2014).

As reported by Cremonez et al. (2015), some studies have demonstrated the possibility to convert biomass into FT-SPK; however, technical environmental issues such as catalyst production and residue treatment need to be solved.

4.2.2.3 Biobased Chemicals

Among possible biobased chemicals to be obtained from syngas, acetic acid and ammonia can be cited as the most representative ones. These compounds are briefly described below.

Acetic Acid

As shown in Fig. 4.6, acetic acid can be obtained by homogeneous catalytic synthesis through methanol carbonylation with CO in the liquid phase with Co, Rh or Ni catalysts promoted by iodine (Spath and Dayton 2003).

Acetic acid is an important chemical used as raw material for vinyl acetate monomer and acetic anhydride synthesis, and it is used as a solvent for purified terephthalic acid production, as reported by Yoneda et al. (2001).

Although acetic acid is commercially produced via methanol carbonylation, some advances in direct conversion of syngas to acetic acid have been shown (Subramani and Gangwal 2008).

Ammonia

In the world, one of the most produced inorganic chemicals is ammonia. In 2009, a total global production of 109 Mt of ammonia was registered and the global production capacity will increase to nearly 239 Mt by 2020, as reported by Andersson and Lundgren (2014). According to these authors, the economic challenge in the NH_3 production via synthesis of nitrogen and hydrogen in the Haber–Bosch process is the hydrogen production.

As reported by Spath and Dayton (2003), ammonia synthesis reaction is an exothermic chemical reaction and carried out in catalytic converters at 15–35 MPa and 430–480 °C promoted by Fe catalysts. H_2:N_2 ratio is near 2 at high space velocities and near 3 at low space velocities as equilibrium becomes more dominant.

Two recent publications reported an environmental and economical assessment (Gilbert et al. 2014) for the production of ammonia from biomass gasification and a techno-economic analysis for the production of ammonia from integrated biomass gasification (Andersson and Lundgren 2014).

4.3 Sugarcane Biorefinery

4.3.1 First-Generation Sugarcane Biorefinery

First-generation sugarcane processing takes place in autonomous distilleries or sugar mills with annexed distilleries; in the latter, hereafter referred simply as annexed distilleries, a fraction of the sugarcane juice is diverted for sugar production and the remaining fraction, along with the molasses (impure solution of sugars that remains after sucrose crystallization), is used for ethanol production. In the most common scenario, annexed distilleries operate using half of the sugarcane juice for sugar production and the other half, plus molasses from sugar production process, is used for bioethanol production. Anhydrous and hydrous ethanol may be produced, being the former used as gasoline additive and the latter as fuel substitute for gasoline. The few plants, which produce only sugar, sell the molasses for other industrial uses. The flexibility of annexed plants to produce more ethanol or more sugar, depending upon the market demands, is part of the reason for the success of bioethanol production in Brazil. However, the range of operation of an installed plant is somehow limited to the existing design restrictions (Dias et al. 2015; Junqueira 2011).

The sugarcane processing facility is energetically self-sufficient, with all the thermal and electrical energy required for the production process being produced in its combined heat and power system, using bagasse as fuel (Chandel 2014). If sugarcane straw is recovered from the field, it may also be used as a fuel to increase the energy production.

A simplified diagram of the sugar, ethanol and electricity production process from sugarcane is illustrated in Fig. 4.7.

This arrangement is quite common and for decades represents the production process. Advances were achieved, and more energetically efficient solutions have been adopted; new equipment with better design and improved performance is available today, but the main operations are essentially the same. Research is still important to develop and optimize the process, but several comprehensive and in-depth handbooks are available to technicians that work in this field (Camargo 1990; Hugot 1986; Chen 1993; Rein 2007).

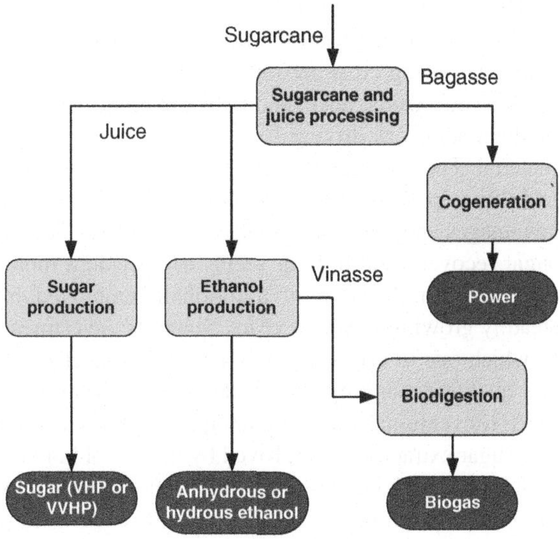

Fig. 4.7 Simplified block diagram for the production of ethanol, sugar and electricity

The first unit operations in the processing of sugarcane, in either autonomous or annexed distilleries, are comprised basically by reception and cleaning of the sugarcane, cane preparation and juice extraction, in which bagasse and juice are separated.

Once the sugarcane is received from the field, it is identified and weighted while loaded in the trucks and is sampled, because the payment is based on its sugar content; it is unloaded by hilo unloaders into feeding tables, a piece of equipment that provides a uniform and continuous cane layer to the following operations. It is passed through a cleaning system for removal of dirt and other impurities entrained during the process of harvesting and transportation. Over the years, changes in the way of harvest of the sugarcane required improvements in cleaning technology. The most common cleaning approach was to wash the cane, but the current trend is to use dry cleaning systems, in order to reduce water consumption and also reduce losses of sugars. This is in line with the increase of the mechanized harvest where the cane comes to the factory already chopped, what prohibits the cleaning with water due to the losses of sugar (Dias 2008; Chavez-Rodríguez 2010).

Cleaned cane is transported in conveyor belts to a preparation system. The cane goes through a set of cane levelers, cutting knives and hammer shredders; the former provides a more uniform material for the latter and prevents choking. The role of this equipment is to open the juice-containing cells of the cane in order to facilitate the juice extraction. At the end of these operations, the final material, referred as *prepared cane*, passes by a magnet to remove tramp iron that may be trapped inside the cane mat (Jenekar et al. 2005). The open-cell index is used to indicate how well prepared was the cane; values above 85 % are good enough for mills, but when the mill uses a diffuser, at least 90 % is required.

The prepared cane follows to the extraction station where the juice, rich in sugars, is separated from the fibrous part, called bagasse. This separation could be performed using crushing mills or diffusers. In the former, mechanical pressure is used to squeeze the juice, and in the latter, besides direct lixiviation, a concentration gradient serves as driving force to separate the sugar-rich stream from the fibrous material. The diffusion process is more sensitive to the quality of the prepared cane, and this step needs more attention when diffusers are used. Both approaches have advantages and drawbacks in specific contexts, but diffusion usually reports a better sugar recovery and milling seems to provide a more flexible operation. In Brazil, the crushing mills are more traditionally employed, but the use of diffuser has been steadily growing in recent years. Since the investment is not so different, the choice of which one to use is not only based on objective aspects, but rather in subjective and qualitative arguments (Modesto et al. 2009).

A conventional milling tandem is composed by five to seven three-roll mills. The sugar extraction is improved by the use of hot brix-free water, imbibition water, which is added in the last roller mill of the equipment. The resulting bagasse presents approximately 50 % moisture and is ready to be burnt in a boiler to produce steam that is used in the industrial process and to produce electrical energy in the cogeneration system. Sugarcane juice is filtered in a rotary screen to remove solid particles (mostly fibers); the fibers obtained in this screen return to the mills for further recovery of sugars, while the juice is sent to juice treatment. Adoption of new technologies is in course in the extraction area, for example, crushing mills driven by electrical motors, new mechanical setups and the use of advanced control schemes.

Following extraction, juice receives a chemical treatment to remove other impurities. The treatment of the juice is slightly different depending if it will be used to produce sugar or ethanol, but the difference remains mainly in the amount of chemicals used. This process consists of juice initial heating, addition of phosphoric acid and lime and a second heating operation, up to 105 °C. Hot juice is flashed to remove dissolved air, and after addition of a flocculant polymer, impurities are removed by a settler, where mud and clarified juice are obtained. A filter is used to recover some of the sugars carried along with the mud, and the separated solids are recycled to the process prior to the second heating operation; bagasse fines (bagacillo) and wash water are used in the filter to improve recovery of sugars. The clarified juice passes through screens to remove possible remaining rough solid particles. Advances in this area comprise the use of better chemicals and more efficient separation technologies, for example, belt press filters substituting the widespread rotary vacuum filters, centrifuge decanters and clarifiers with better design (trayless and low turbulence clarifiers).

In order to concentrate the clarified juice, for sugar or ethanol production, the "old" multiple-effect arrangement with Roberts-type evaporator is typically found in the Brazilian distilleries. This equipment is reliable, and several modifications were tested and present advantages in some extent, for example, different feeding positions, use of vapor recompression and design of tube bundle with singular characteristics. Other types of evaporators are currently in use to concentrate the juice, for instance, climbing and falling film evaporators, plate-falling film evaporators and several company-designed equipment. This operation is very important for the energetic balance of the plant (Rein 2007; Ensinas et al. 2007, 2009; Ensinas 2008; Dias et al. 2010). For this reason, there are many commercial solutions and the choice of the equipment should cover aspects as diverse as investment cost, after-sales support, evaporation rate, residence time, facility to perform CIP (cleaning in place) and available space.

When sugar is produced, the syrup from the evaporator goes to the "boiling house," where the sucrose present in the syrup is separated from the liquor as sugar crystals. In the most usual arrangement, the process is performed in equipment that operates in a similar way as the evaporator, called vacuum pans and crystallizers, usually operated in fed-batch mode and under vacuum, in order to use steam from the evaporators and prevent sucrose degradation. The mixture of sugar crystals and molasses (liquid part) inside the equipment is called massecuite. When the amount of material reaches the limit of the vacuum pan (at the end of a batch), the massecuite is transferred to crystallizers and, after an appropriate residence time, it is sent to centrifuges that separate the crystals and the molasses (Chandel et al. 2014). It is possible to exhaust more the molasses (recuperating more sugar), repeating the process one or two more times. The sugar produced, with up to 2 % of moisture content, is ready for storage after a drying step.

There are several commercial technologies available for the crystallization area; some of them propose a different equipment design or other operational approaches, for example, vertical crystallizers, continuous crystallization—where the syrup is fed continuously and sugar crystallization controlled by cooling. Some chemicals may be used to adjust the massecuite viscosity or to destroy some contaminants, as dextran. Due to the several qualities of sugar (e.g., VHP, VVHP, QLC, LP) and industrial process peculiarities, the main challenges are handled with operational maneuvers and a control system properly set. The research in this area focuses on CFD modeling of the flow pattern inside the equipment, assessment of the effect of impurities in the sugar quality and a better phenomenological understanding about the sugar crystallization, in order to provide information to model, simulate and control the process.

Concentrated juice from the evaporators is sent to the fermentation area and fed into fermenters. For annexed distilleries, concentrated juice is mixed with an appropriate volume of final molasses. The two main ethanol fermentation approaches found in the Brazilian distilleries are the fed-batch with yeast recycle (Melle–Boinot process) and the continuous multistage process—also with yeast recycle (Camargo 1990; Andrietta 1994). In the fermentation process, sugar is exothermically converted to ethanol and carbon dioxide by yeasts. The temperature is controlled and the cell recycling system adjusted to provide a stable operation. The fed-batch processing time goes from 8 to 12 h, and the sugar conversion to ethanol in well-operated processes may reach values around 90 % of the theoretical value.

There is a wide variety of process possibilities and sources of disturbance for this operation, for example, fed-batch or continuous mode, with or without molasses, open and closed fermenters, different yeasts, level of contamination and design of equipment. This explains the widespread dispersion of productivities and conversions found in the industrial plants, but at the same time indicates that all these topics are research targets. Fermentation has been receiving attention because, despite the present high conversion, there is still room to improvements. New and nontoxic chemicals are available, such as antifoam, viscosity reducer, surfactants and antibiotics, in order to improve the industrial operation.

The produced ethanol is recovered from the wine in the distillation area. With the aim of producing hydrous ethanol (92.6–93.8 % w/w), it is necessary a set of distillation columns, A–A1–D, and rectification columns, B–B1, with the columns of each set stacked one at the top of the other. In the former set, wine is fed in column A1 and the three main output streams are the volatile impurities, eliminated in the top of column D, vinasse, the bottom product from column A and phlegm, an ethanol-rich stream that follows to the rectification columns. From the bottom of column B1 is withdrawn a nearly pure water stream, named phlegmasse, hydrated ethanol is produced on top of column B, and a side stream, containing most of the higher alcohols, is also withdrawn from column B (Dias 2008; Junqueira 2010). In order to dehydrate the azeotropic solution from column B to produce anhydrous ethanol (min. 99.3 % w/w), three main technologies are commercially available: azeotropic distillation with cyclohexane, extractive distillation with ethylene glycol and adsorption with molecular sieves. From a technological point of view, the distillation sector is very mature and presents ethanol recovery efficiencies of, at least, 99.0 %.

Even with the use of condensers and heat exchangers to promote heat recovery from the streams, this area of the process demands large amount of process steam. The steam consumption (at 2.5 bar, saturated) for distillation and azeotropic dehydration with cyclohexane ranges from 3 to 3.5 and 1.5–2 kg L^{-1} of ethanol, respectively (BNDES and CGEE 2008). So, the main appeal of several improvements and competing approaches commercially available focuses on the reduction of steam demand.

For decades, the only destination of the vinasse, concentrated or not, was its application in the field as fertilizer, but anaerobic digestion to produce biogas may be a suitable alternative to improve their environmental sustainability and take advantage of its energetic potential. However, unsuccessful empirical approaches and poor understanding about fundamental aspects of such biological process added difficulties, up to now, for the acceptance of anaerobic digestion in the sugarcane mills. The main vinasse treatment plant in Brazil treats vinasse in a 5000-m^3 UASB reactor and uses the biogas to dry part of the yeast withdrawn from fermentation (Moraes 2015). A more recent attempt processes vinasse in codigestion with other residues, aiming electrical energy generation (Geo Energética 2013).

Despite the energetic advantages and renewed research interest in vinasse biodigestion, several challenges are related to biogas production in biorefineries, such as the variable composition of the raw material, high investment, lack of government policies, lack of viability studies for biorefinery size situations and uncertainties concerning not-yet-proven technologies.

The sector that provides all necessary thermal and electric energy for the industrial site is the combined heat and power (CHP) system. Traditionally, since the mid-1990s, the sugarcane mills are able to produce their own energy using only their biomass residues in cogeneration systems (Leal 2010). Bagasse from the extraction area and straw, when available, is burnt in one or more bagasse-fired boilers that generate and provide steam to backpressure turbogenerators, to steam-driven equipment and to unit operations of the process. Unfortunately, the initially installed CHP technology was designed to get rid of the surplus bagasse and not to efficiently generate electricity, but in the last decade, the number of sugarcane processing plants interested in selling surplus electrical energy to the national electric grid has been growing steadily. For this reason, relevant investment is required, from government and sugarcane mills, to develop infrastructure, technology and a regulated environment for using sugarcane residues, bagasse and straw as fuel in advanced cogeneration systems.

Several characteristics and operational choices of the distillery affect the CHP technology to be adopted. The first aspect to be addressed is if the solution is going to be implanted in a new or existing facility. New projects may use more recent technologies and equipment, and the options are not limited to other area constraints. Nowadays, it is possible to find, from national manufactures, boilers operating at 100 bar/530 °C steam conditions, with monodrum design, steam generating capacities up to 400 t h^{-1} and efficiencies up to 89 % (CGEE 2009). Up to now, higher pressure and temperature have not being used by the mills. Despite of this availability, the current standard for greenfield projects, which uses only backpressure turbogenerators, is steam conditions of 65 bar/480° C. More advanced configurations for the CHP system are still object of research and development, such as the technology BIG/GT—biomass integrated gasifier to gas turbines (Walter 2010). Points that eventually will be addressed in this area are the maximum amount of straw that could be incorporated to the bagasse without affecting efficiency and new and more appropriate boiler design, such as bubbling fluidized bed boilers.

4.3.2 Second-Generation Sugarcane Biorefinery

4.3.2.1 Biochemical Route: State of the Art

The focus of the second-generation biochemical route is to produce fermentable sugars from lignocellulosic feedstock. This route is comprised of pretreatment, required to decrease biomass recalcitrance and improve accessibility of enzymes for the subsequent step, enzymatic hydrolysis, in which cellulose is converted to glucose. In the last step, sugars released during pretreatment and hydrolysis must be fermented to ethanol—while C6 sugars released in enzymatic hydrolysis can be easily fermented to ethanol using the same yeasts employed in first-generation ethanol production, C5 sugars, derived from the hemicellulosic fraction of the lignocellulosic feedstock, can only be properly fermented using different microorganisms.

A variety of biomass pretreatment technologies has been investigated; their main purpose is to modify or remove hemicellulose and/or lignin from the lignocellulosic matrix, further exposing cellulose fibers. Some are based on biological reactions, physical transformations, such as milling pretreatment (Silva et al. 2010), or chemical reactions and a combination of these. Among them, pretreatment based on steam explosion is one of the most investigated methods for sugarcane bagasse (Leibbrandt et al. 2011). In this process, the lignocellulosic feedstock is submitted to steam treatment for a few minutes, followed by quick depressurization and cooling, during which hemicellulose is converted to pentoses (C5 sugars) through reactions catalyzed by acids released during pretreatment (autohydrolysis) or added to the reactor (Menon and Rao 2012; Pourbafrani et al. 2014). In the case of sugarcane bagasse, xylose is the main C5 sugar produced. Alvira et al. (2010) provide a throughout review of pretreatment technologies, describing the main features of biological, physical, chemical and physicochemical pretreatments such as steam explosion, liquid hot water, ammonia fiber expansion, wet oxidation, alkali, acid and ionic liquid pretreatments, among others. More recently, Singh et al. (2015) provided a review of steam explosion, acid and alkaline pretreatments, highlighting their characteristics regarding applicability, costs and environmental impacts. The main characteristics of several pretreatment methods are summarized in Table 4.2.

Pretreatment comprises a significant fraction of cellulosic ethanol production cost; an important factor that is often disregarded during development of pretreatment processes at laboratory scale is their energy consumption. Mafe et al. (2015) developed a model to estimate the energy demand of dilute acid pretreatment, verifying that solid loading has a major impact on the process energy

Table 4.2 Main characteristics of selected pretreatment technologies (adapted from Limayem and Ricke 2012)

Pretreatment	Key characteristics	References
Dilute acid	High C5 sugar yield; generates toxic inhibitors; high recovery; equipment costs	Tao et al. (2011)
Liquid hot water	High hemicellulose solubility; no chemicals; high energy usage	Yu et al. (2013)
Lime	High sugar yield; inexpensive reagent; long reaction time	Rabelo et al. (2011a)
Catalyzed steam explosion	Short reaction time; high sugar yields; possible hemicellulose degradation	Behera et al. (2014)
Organosolv	High solvent costs; high lignin removal; formation of toxic inhibitors	Behera et al. (2014)
Ammonia fiber expansion (AFEX)	High NH_3 loading; NH_3 recycle required; high operating pressure	Dale et al. (2010) and Bals et al. (2011)

consumption—low solid loading, in addition to increasing water consumption, requires large amounts of energy since ethanol must be concentrated to be used as a fuel. Second-generation ethanol production from sugarcane bagasse is particularly influenced by the process energy consumption, since the feedstock (bagasse) is also used as fuel for the production of the energy required in the process. The higher the energy consumption of pretreatment, hydrolysis and subsequent concentration and fermentation steps, the lower the amount of feedstock that will be available for second-generation ethanol (Dias et al. 2013).

Hemicellulose removal from the lignocellulosic feedstock significantly increases enzymatic hydrolysis yields (Wyman et al. 2005). In addition to removing hemicellulose, lignin removal and modification of the lignocellulosic fibers contribute to increased enzymatic digestibility of cellulose (Aita et al. 2011). Besides promoting alteration of the lignocellulosic matrix leading to increased cellulose accessibility, an adequate pretreatment should improve production of fermentable sugars in enzymatic hydrolysis, with minimal cellulose degradation and formation of fermentation inhibitory compounds (furfural from xylose and hydroxymethylfurfural from glucose, besides phenolics and acetic acid), as well as reduced energy consumption and production costs (Gupta and Verma 2015; Limayem and Ricke 2012; Singh et al. 2015). If solvents are required, they must be highly and easily recyclable, stable and nontoxic for the subsequent process stages (Sathitsuksanoh et al. 2013)—considering the expected size of the ethanol production unit, even if minor amounts of solvents are lost in the process, the consequent environmental impact can render the process unfeasible (Dias et al. 2012a).

One of the main issues regarding the development of pretreatment processes is the fact that its performance depends on the feedstock (Limayem and Ricke 2012) and on the process configuration (Galbe and Zacchi 2012), so an optimal pretreatment for sugarcane bagasse in a conventional plant is not necessarily optimal for other process conditions and feedstock, such as sugarcane straw.

Pretreatment can solubilize hemicellulose and/or lignin, depending on its characteristics. Generally, a solid–liquid separation is used to separate the pretreated lignocellulose before enzymatic hydrolysis. The liquid stream contains either soluble lignin or pentoses and can be diverted for pentose conversion or lignin recovery. Pretreatment characteristics may limit the use of those streams as feedstock or fuel, e.g., sulfur-impregnated lignin may hinder its use as fuel in boilers due to high sulfur oxide emissions, if no recovery technologies are employed (Maas et al. 2008). Pentoses derived from hemicellulose hydrolysis can be used as feedstock for ethanol fermentation, biogas or production of chemicals (Dias et al. 2013; Mariano et al. 2013; Rabelo et al. 2011b).

C5 fermentation to ethanol can be carried out using different microorganisms naturally able to metabolize it, such as *Scheffersomyces stipitis* (*Pichia stipitis*) and *Candida shehatea*, since *Saccharomyces cerevisiae* is not capable of utilizing pentoses. While *S. stipitis* is considered the most efficient microorganism, issues regarding its glucose utilization make *S. stipitis/S. cerevisiae* or *S. stipitis/Z. mobilis* coculture a more interesting option; given the fact that each strain operates under different environmental conditions, coculture remains a challenge (Karagöz and

Özkan 2014). Engineered microorganisms have been investigated as well, focusing on promoting their ability to ferment both pentoses and hexoses, maximizing their ethanol productivity. Gírio et al. (2010) provide a detailed review regarding hemicellulose conversion to ethanol, including detailed information about the different microorganisms and technologies being investigated. Future application of engineered microorganisms in sugarcane biorefineries, however, is currently uncertain as the need for a sterile environment and relatively long fermentation time, among other technical risks, may hinder their application.

After separation, pretreatment solids containing most of the cellulose must undergo hydrolysis, or saccharification, in order to produce fermentable C6 sugars to be subsequently used in alcoholic fermentation. This process is based on the use of enzymes or acids that convert cellulose to glucose. Due to several drawbacks of acid hydrolysis, such as cellulose and hemicellulose degradation and requirement of corrosion-resistant equipment, the more specific enzymatic hydrolysis has been intensively investigated.

During enzymatic hydrolysis, pretreated lignocellulose is diluted to promote an efficient contact between enzymes and feedstock, and an enzyme mixture (usually comprised of endoglucanase, exoglucanase and β-glucosidase) is used to attack cellulose crystalline structure, remove cellobiose and hydrolyze it to produce glucose. One of the main drawbacks of enzymatic hydrolysis is cellulose enzyme inhibition in the presence of cellobiose and glucose in high concentrations (Baeyens et al. 2015).

For enzymatic hydrolysis, an alternative configuration of separate hydrolysis and fermentation (SHF) is the simultaneous hydrolysis (saccharification) and fermentation (SSF), in which enzymes are not inhibited by their reaction products (dimeric and monomeric sugars), since they are consumed by yeasts before their concentration in the reactor becomes too high (Weusthuis et al. 2011). If fermentation of pentoses, released during pretreatment, is carried out in the same vessel, the process is known as simultaneous saccharification and cofermentation (SSCF). The integrated processes present advantages when compared with the separate hydrolysis and fermentation, such as lower enzyme requirement, shorter reaction time and cost reduction due to reduced equipment, but also face potential decrease on yields due to suboptimal operating conditions (enzymes' optimal temperatures are slightly higher than those of yeast) (Baeyens et al. 2015).

Most studies available in the literature focus on maximizing cellulose conversion during enzymatic hydrolysis; this, however, often comes at the expense of large reactors, high enzyme concentration and low solid loading (Giese et al. 2013; Silva et al. 2011), which lead to increased water and energy use as well as equipment size and investment, thus increasing production costs (Caspeta et al. 2014). In the case of second-generation ethanol production, unreacted solids may be used as fuels to provide energy for the process, so it may not necessarily be advantageous to achieve extremely high yields. For sugarcane plants, since sugarcane bagasse is also fuel for production of steam and electricity, high energy consumption during hydrolysis decreases the amount of feedstock available for second-generation ethanol. A comparison between hydrolysis at 5–20 wt% in the integrated first- and second-generation ethanol production process from sugarcane showed that the lowest overall ethanol production was obtained for hydrolysis with low solid loadings, in spite of its higher hydrolysis yield (Dias et al. 2013).

New commercial enzyme cocktails have been developed to improve the ability of cellulolytic enzymes to work at high solid loading, including the use of improved β-glucosidases and the introduction of a new class of enzymes, known as lytic polysaccharide monooxygenases (LPMOs). The LPMOs are not present in previous *exo*- and *endo*cellulases' cocktails, and work as cleavers of cellulose chains and act synergistically with the hydrolytic enzymes (Rodríguez-Zúñiga et al. 2015). The use of enzymatic cocktails containing LPMOs on hydrolysis of lignocellulosic materials is still being investigated; their reaction mechanism has not been fully resolved (Westereng et al. 2013), but represents a potential breakthrough in enzymatic hydrolysis development (Agger et al. 2014).

Glucose released in cellulose hydrolysis can be readily fermented to ethanol using *Saccharomyces cerevisiae*, the same yeast used in first-generation ethanol production. For sugarcane processing, integrating first- and second-generation ethanol production can take advantage of mixed sucrose and glucose (from cellulose hydrolysis) fermentation, which could decrease impacts of fermentation inhibitors generated during lignocellulose pretreatment. Andrade et al. (2013) showed that high yields (98–100 % of the theoretical) could be obtained for the fermentation of sugarcane bagasse hydrolysate (pretreated with hydrogen peroxide) in a mixture with sugarcane molasses, even in the presence of inhibitors generated during pretreatment like acetic acid. The authors showed that lower yields (approximately 75 %) were obtained for the fermentation of pure hydrolysate due to the large amount of fermentation inhibitors, showing that integrating first− and second−generation allows a more efficient fermentation process. This integration, however, must be carefully evaluated if engineered microorganisms able to ferment both pentoses and hexoses are employed for fermentation of glucose and pentose liquors obtained from lignocellulose.

After fermentation, ethanol produced in the integrated process is purified in the set of rectification and distillation columns and dehydrated using the same processes already described for first-generation ethanol production.

4.3.2.2 Thermochemical Routes: State of the Art

Thermochemical routes consist of operations from biomass handling and feeding systems, thermochemical conversion processes, gas cleaning and conditioning and conversion technologies.

Among the thermochemical conversion processes, the most important are combustion, pyrolysis, torrefaction and gasification. Pyrolysis and torrefaction can be considered as processes of biomass energy densification and can be employed as biomass pretreatment before gasification processes. In this case, torrefaction process generates a torrefied material and pyrolysis process can be controlled to increase the production of biochar or bio-oil.

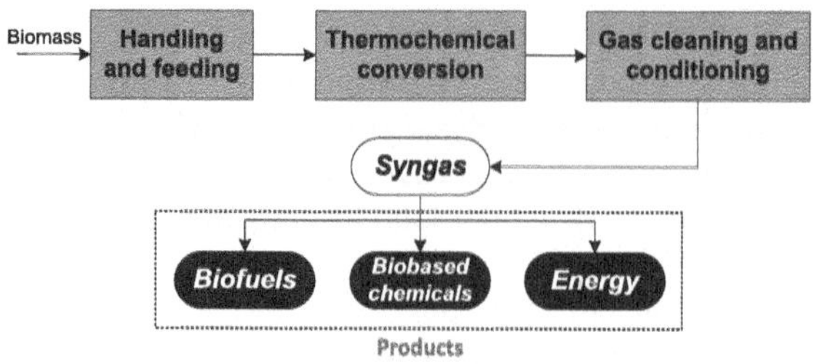

Fig. 4.8 Products obtained from biomass thermochemical conversion

Figure 4.8 depicts a simplified thermochemical route and the unit operations involved.

A brief description of the state of the art of some processes of thermochemical routes is presented.

Biomass Handling and Feeding Systems

Biomass handling and feeding consist of upstream operations in gasification processes. According to Basu (2013), biomass handling and feeding represent one of the most common problems in gasification plants. In order to provide a good operation of the gasifier, this author shows that the biomass handling system is split into steps of receiving, storage and screening, feed preparation, conveying and feeding. Biomass feed preparation is an important step to eliminate the presence of foreign materials, for size reduction and classification, and to adequate the level of moisture. Typically, gasifiers require a moisture content of less than 10–15 % attained in suitable driers, being the temperature of hot gas used for drying in the range 50–60 °C.

Pyrolysis

Pyrolysis is a thermochemical process performed in the absence of oxygen at temperatures between 250 and 550 °C (Kersten and Garcia-Perez 2013). The pyrolysis process produces different phases depending basically on parameters such as temperature, pressure, reaction time and heating rate.

Fast pyrolysis is carried out at temperatures around 500 °C, atmospheric pressure and reaction times of up to 2 s. To perform this process, biomass size should be around 1 mm with 10–15 % of moisture (Vaz Júnior 2013). Fast pyrolysis process provides high yields of liquids, up to 75 % by weight, which can be used directly in different applications or as an efficient energy carrier (Bridgwater 2012a), thus justifying the recent interest in this process.

The bio-oil from pyrolysis usually undergoes an upgrading process in order to reduce or remove one or more of its undesirable characteristics or properties (Bridgwater 2012b). With this purpose, catalytic upgrading of pyrolysis bio-oil is performed or, alternatively, it has been studied the use of catalysts during pyrolysis process ("in situ catalytic pyrolysis") in order to change bio-oil characteristics, as reported by Compton et al. (2011). Conventional transport fuels as diesel, gasoline, kerosene, methane and liquefied petroleum gas can be obtained by bio-oil upgrading through deoxygenation and refining (Bridgwater 2012a).

Alternatively, anhydrosugars present in bio-oil, particularly levoglucosan, can be fermented to ethanol (hybrid processing), as reported by Brown (2006).

In slow pyrolysis, biomass is heated to temperatures around 400 °C in the absence of oxygen producing approximately equal masses of syngas, bio-oil and biochar. The scale-up of slow pyrolyzers is a difficult task since heat transfer into biomass is slow, requiring a long residence time in the reaction chamber. Among biochar applications, it can be used as fuel, as a filter material to remove contaminants from effluent streams and as a soil amendment, depending on its quality (Laird et al. 2011).

Torrefaction

Torrefaction takes place in an inert atmosphere as pyrolysis process, but the temperature requirement is between 200 and 300° C, lower than the temperature requirement for pyrolysis process (Bergman and Kiel 2005).

As reported by Koppejan et al. (2012), there are torrefaction technologies processing feedstock with both small and large particles. In this way, the choice of torrefaction technology depends on the characteristics of the feedstock or if the feedstock needs to be preprocessed before torrefaction for particle size reduction. The moisture content of biomass feeding the torrefaction reactor should be less than 15 %. The above-referenced authors provide a description about important reactor technologies that can be employed in torrefaction, such as rotating drum, moving compact bed, belt driers and microwave reactor.

According to Pelaez-Samaniego et al. (2014), one of the most promising uses of torrefaction is energy densification, via fuel pellet production, in addition to uses as pretreatment method for fast pyrolysis and gasification.

Gasification

Gasification is a chemical process that converts biomass into useful gases and chemicals requiring a medium for reaction, which can be a gas or supercritical water. As gaseous medium, air, oxygen, subcritical steam or a mixture of them can be used (Basu 2010).

As reported by Hofbauer et al. (2007), gasification is a two-step, endothermic process. In the first step, pyrolysis, the volatile components of the biomass are vaporized at temperatures below 600° C according to a set of complex reactions. In the second step, the char, which is not vaporized, is gasified through reactions with oxygen, steam, carbon monoxide and hydrogen. The heat required for the endothermic gasification reactions can be generated by combustion of part of the biomass, char, or gases, depending on the reactor technology.

Gasification technologies can be classified as direct or indirect. In direct gasification, part of the biomass is burnt to provide the required heat to the initial pyrolysis step, working in substoichiometric conditions with an air factor of 0.2–0.4. In indirect gasification, being the heat provided in indirect ways, this gasification is considered as a variant of the pyrolysis process (Lora and Venturini 2012).

The review of Puig-Arnavat et al. (2010) presents the following types of reactor design that can be used for gasification: fixed bed (updraft, downdraft and cross-draft), fluidized bed (bubbling, circulating and twin-bed), entrained flow bed and stage gasification with physical separation of pyrolysis, oxidation and/or reduction zones.

Theoretically, syngas from gasification is composed of equimolar amounts of hydrogen (H_2) and carbon monoxide (CO) obtained by water–gas-shift reaction (WGS). The gas from gasification of lignocellulosic biomass is named producer gas and it is composed of impurities as dust, tar, halogen and alkali compounds with inorganic impurities, being hydrogen sulfide (H_2S), ammonium (NH_3), hydrogen chloride (HCl), methane (CH_4) and other light hydrocarbon (C_2H_6) contaminants of catalysts in downstream processes. In this way, specific operations are required to remove such contaminants (Obando et al. 2010).

Gas Cleaning and Conditioning

Biomass composition and the type of gasification process impact the quality of the resulting gas. A gas cleaning is required to remove contaminants and provide a syngas within specifications for downstream processes and gas utilization, while a gas conditioning system is required to eliminate undesired main gas compounds and adjust the H_2/CO ratio (Hofbauer et al. 2007).

Stevens (2001) presents some technologies that can be used to remove undesirable components from the producer gas leaving the gasifier. For particulate removal, cyclonic filters, barrier filters, electrostatic filters and wet scrubbers are included. For alkali removal, the producer gas is cooled in order to condensate the alkali vapors into solid particulates. These solids can be removed by various filtration systems, such as those mentioned before. Tar can be removed through physical, catalytic or thermal reduction methods. For removal of nitrogen compounds, catalytic destruction of ammonia and wet scrubbing are possible. For sulfur removal, wet scrubbers can be employed.

Richardson et al. (2012) showed a promising approach to integrate cleaning into gasifier operations such as catalytic tar cracking/reforming, CO_2 elimination, H_2 separation and elimination of particles and other contaminants in a process intensification concept, reducing the cost of producing clean syngas in the required specifications for different applications.

Integrated to the gas cleanup, gas conditioning aims at the adjustment of the H_2/CO ratio and at the CO_2 removal. In relation to the first, to increase the H_2/CO ratio, the water–gas-shift reaction is used, and in relation to the second, since CO_2 is undesired in most processes, it can be removed from syngas by chemical and physical absorption with a washing liquid or by adsorption with solid adsorbents (Hofbauer et al. 2007).

Syngas Utilization

Syngas from gas cleaning and conditioning can be catalytically converted into a wide range of liquid fuels and chemicals, such as gasoline and diesel by Fischer–Tropsch (FT) synthesis, methanol, dimethyl ether (DME), NH_3, mixed alcohols, ethanol and others.

This section briefly describes catalytic synthesis of syngas to produce ethanol, Fischer–Tropsch (FT) synthesis to produce gasoline and diesel and an advanced cogeneration system to be used in sugarcane biorefineries.

Catalytic Synthesis of Ethanol Syngas can be directly converted into ethanol using Rh-based catalysts. The main reactions involved are basically hydrogenation reactions of CO or CO_2, and side reactions, such as water gas shift and methanation reactions (Spivey and Egbebi 2007).

In relation to the operation conditions, since alcohol synthesis reactions are exothermic reactions, an adequate temperature control is important for the reaction selectivity and catalyst life (He and Zhang 2008). According to Subramani and Gangwal (2008), higher pressures will thermodynamically favor alcohol-forming and methanation reactions. The limit value on pressure for the synthesis reactor is determined by the pressure employed in the gasifier associated with the availability of the subsequent compression of the syngas. The types of reactor commonly employed for alcohol synthesis are fixed bed, fluidized bed and slurry bubble column reactors. High H_2/CO ratio favors lower alcohol and hydrocarbon production; on the other side, ratio values lower than 1 could lead to catalyst deactivation, due to coke formation (Subramani and Gangwal 2008).

Fischer–Tropsch Synthesis In Fischer–Tropsch synthesis, a strongly exothermic reaction of CO and H_2 conversion to hydrocarbons and water with either a cobalt- or iron-based catalyst is carried out (De Deugd 2004), as follows:

$$n\mathrm{CO} + 2n\mathrm{H_2} \rightarrow -(\mathrm{CH_2})_n- + n\mathrm{H_2O} \quad (4.1)$$

In the FT synthesis, the formation of hydrocarbons follows the classical Anderson–Schulz–Flory (ASF) product distribution (Ramos et al. 2011). The ASF model is expressed by Eq. 4.2 (Spath and Dayton 2003):

$$Wn = n(1-\alpha)^2\alpha^{n-1} \quad (4.2)$$

where Wn is the weight percentage of a product containing n carbon atoms and α is the chain growth probability. The chain growth probability depends on both catalyst and process conditions (De Deugd 2004).

According to Spath and Dayton (2003), if diesel is the desirable product, a slurry reactor at low temperature (200–240 °C) with a cobalt catalyst is the best FT synthesis configuration. For gasoline, a fixed bed reactor at high temperature (300–350 °C) with an iron catalyst is the best choice. Specific H_2/CO ratio is required for FT synthesis, depending on the catalyst type. For cobalt catalyst, H_2/CO ratio is 2.15, and for iron catalyst, it is 1.7, since for iron catalyst, WGS reaction is performed in addition to the FT reaction. In relation to the operating pressure of FT reactors, typically the range of 10–40 bar is employed. After FT synthesis, the upgrading of FT liquids by hydrotreating, hydrocracking and hydroisomerization is performed, in addition to product separation.

Advanced Cogeneration System An attractive option of cogeneration system in Brazilian sugarcane industry is the biomass integrated gasification/gas turbine combined cycle (BIG/GTCC). Biomass gasification allows power energy production with the use of gas turbine as prime drivers.

A relevant question to be observed is that gas turbines, usually designed to operate with natural gas [heating value of 35–40 MJ N^{-1} m^{-3} (Williams and Larson 1991)], should be adapted to operate with syngas [heating value of 3–19 MJ N^{-1} m^{-3}, depending on the gasifying medium (Lora and Venturini 2012)]. Operating either in the same maximum allowed temperature, or close to this value, the flue gas flow from syngas should be higher, generating higher mechanical stress on the gas turbine expander. This is only one of the technical reasons why this technology is not yet commercially applied.

4.3.3 Other Feedstocks for Biorefinery

Sugarcane biorefineries operate 6–8 months per year according to sugarcane harvesting (season) period: from September to March, in the North–Northeast region and from April to November in Center–South region, where most facilities are located (CGEE 2009). More prolonged periods of operation are desirable for a better use of existing industrial capacity, reducing contribution of investment on production costs. As sugarcane cannot be stored for more than a few days, if manually harvested, or several hours, in the case of mechanical harvesting (chopped sugarcane), operation in the off-season period must rely on alternative feedstock (BNDES and CGEE 2008).

Besides, processing a complementary feedstock in the same industrial facility allows increasing ethanol and, in some cases, electricity production, thus increasing annual revenues. Nowadays, three alternative feedstocks are being considered in Brazil: sweet sorghum, corn and energy cane.

Sweet sorghum is a short-cycle culture (4 months) and can be utilized in rotation with other annual crops and potentially with sugarcane. For instance, sweet sorghum could be planted as a rotation culture or in land where sugarcane yields are limited due to marginal soils (Swayze 2009). Nevertheless, for the inclusion of sweet sorghum in the sugarcane chain, there are still some challenges to overcome, such as definition of a strategy for planting and adaptation of the machinery for an efficient harvesting (Novacana 2013).

Sweet sorghum, similar to sugarcane, contains readily fermentable sugars—such as sucrose and reducing sugars—and also generates bagasse that can be used as fuel in the cogeneration system to supply steam and electricity to the process. Thus, it may be processed using the existing industrial infrastructure of a sugarcane biorefinery (Durães 2011). However, sweet sorghum is not a suitable feedstock for sugar production, since it presents a higher proportion of reducing sugars, aconitic acid and starch, when compared to sugarcane, which hinders sucrose crystallization (Cutz and Santana 2014).

Corn, commonly used for ethanol production in the USA, is another potential alternative for integration with sugarcane in Brazil as its availability increased due to the corn–soybean rotation in the Center–West region (Milanez et al. 2014). Similar to sugarcane, corn is a C4 plant from the grass family, with annual production cycle (BNDES and CGEE 2008). However, for its processing to ethanol, an additional hydrolysis step to convert starch into fermentable sugars is required, which is associated with retrofitting costs. In order to adapt to Brazilian reality, instead of using natural gas and electricity, bagasse is stored to provide energy to the process (Informa Economics FNP 2013; Milanez et al. 2014). Corn ethanol production includes dried distillers grains with solubles (DDGSs) and fructose as coproducts, which heavily influence overall economic and environmental efficiency of this process (IEA 2011).

More recently, energy cane, a cane selected to have a large increment of fibers than sugars, has been pointed out as a promising alternative feedstock. The high biomass productivity of energy cane reduces the need for land, requiring about half of the area to produce the same dry mass of sugarcane. Besides, this feedstock can be available all year-round, since fiber content does not vary during the year (Matsuoka et al. 2014). Juice can be extracted from energy cane, probably with lower efficiency due to the higher fiber content, and destined to ethanol production, while bagasse can be used to increase electricity generation or as feedstock for 2G ethanol production.

The 2G ethanol production arises as another possibility for year-round operation, since lignocellulosic materials (such as sugarcane bagasse and straw) can be stored for off-season operation. Additional steps (e.g., pretreatment and hydrolysis) are required to make available fermentable sugars from lignocellulosic material; thus, significant investment on equipment is required for biochemical conversion of biomass to ethanol. Other sections of the plant such as juice concentration, fermentation, distillation and cogeneration system can be shared between 1G and 2G processes (Dias et al. 2012a). Figure 4.9 presents a simplified diagram of these operating extension possibilities.

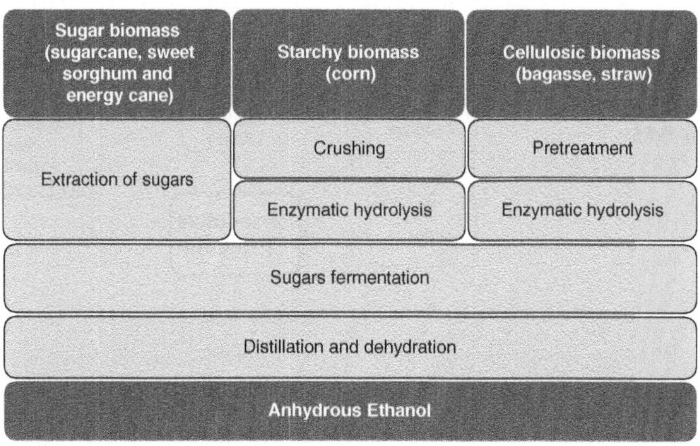

Fig. 4.9 Simplified block flow diagram for ethanol production using different feedstocks (adapted from BNDES and CGEE 2008)

4.4 Process Modeling and Simulation

Process modeling and simulation is an important tool for process design. Its applications include the process synthesis definition through the construction of the flowsheet, the rigorous project of processing equipment and the design of process control system (Pantelides 2004).

4.4.1 Historical Background

The simulation of any chemical process begins with the representation of the process by means of mathematical models. These models are solved with the aid of computer algorithms, which are bundled in a process simulation program and are used to determine the performance and behavior of a desired chemical process under a given set of conditions. The simulation program may be used for steady-state or dynamic simulations and may be a tailor-made simulator to simulate a particular process or may be general to simulate more complex process configurations (Fig. 4.10).

According to Westerberg et al. (1979), the first fifteen years of history in the flowsheeting programs can be broken roughly into three distinct periods. The first development period occurred during 1955–1959. In this period, the computers were not adequate (size and speed), and high-level computer languages such as FORTRAN were not universally available. The successful programs had a limited scope, usually to perform the design calculations for single units.

The first process simulation system was developed in the late 1950s. In 1958, the Flexible Flowsheet was presented by M.W. Kelloggs Corp (Kesler and Kessler 1958) apud by Westerberg et al. (1979). This program introduced the sequential

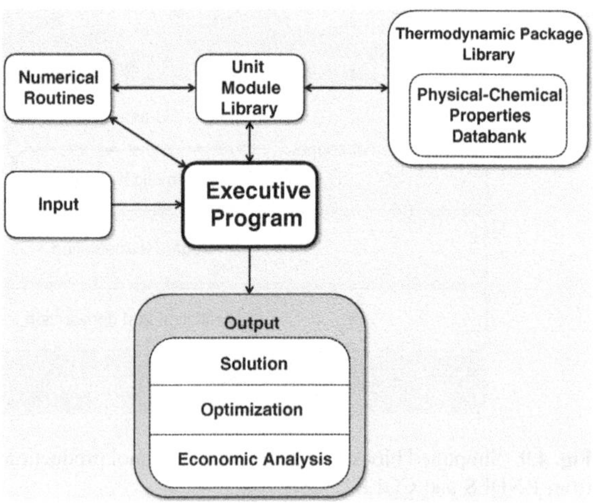

Fig. 4.10 General simulation program based on sequential modular approach (adapted from Raman 1985)

modular approach, where each unit module or building block or process unit is solved separately in sequential form. The main advantage of this approach is that each process unit had a specific and appropriate resolution algorithm.

The second development period is between 1960 and 1964. During this period, the development of process modeling environments, based on sequential modular approach, received a special attention. Those systems were developed in some high-level languages, such as FORTRAN, providing rigorous physical properties' correlations and were made robust enough so that the engineer could usually get his simulation to work (Westerberg et al. 1979). Throughout the next five years (1965–1969—the third development period), powerful large-scale computers had become available, and by 1970s, the flowsheet programs were seen as a useful and cost-effective alternative.

Several companies, research institutes and universities developed their own process modeling environment, e.g., PACER (1963), PROCESS (1966) and FLOWTRAN (1970). According to Westerberg (1998), more than 200 different packages were developed in this period. However, a great effort was required to these in-house developments as well as to their source code maintenance.

Despite the large number of process simulators, it was Monsanto's FLOWTRAN that set the stage for the definition of the new technology. Based on FORTRAN, it articulated a paradigm that would become the basis for subsequent commercial programs and the platform out of which new designs would emerge with rapid succession (Stephanopoulos and Reklaitis 2011).

Besides the sequential modular approach, two other different approaches were developed in this period: simultaneous modular approach and equation-oriented approach.

4.4.2 Sequential Modular Approach

In this approach, each unit module is solved separately in sequential form. The unit models are connected by data sets that represent the streams of material and energy flowing between the process units. Thus, the outputs of one unit module are the inputs of the next one. An executive program is responsible for supervising the information flow between the unit modules. When there is a recycle stream, the simulation convergence is achieved iteratively with the aid of some strategies such as the stream tearing technique. In this technique, different initial guesses are chosen for the tear stream and information is passed from unit to unit until new values of the variables of the tear stream are computed. The computation stops when both the units and the tear streams satisfy some convergence criteria, usually the closure of the material and heat balances. Sequential modular approach was the first computational strategy used for flowsheet simulation, but it remained as the dominant technology for steady-state simulations (Dimian 2003).

The sequential modular approach presents several advantages such as easiness to add and remove unit operations, since individual modules may be readily incorporated in the process flowsheet. Its main disadvantage is related to the solution of complex systems with several recycle streams, because establishing the computation or precedence order may be inefficient, unless an optimal solution sequence is implemented into the executive program (Raman 1985).

4.4.3 Simultaneous Modular Approach

The simultaneous modular approach is similar in principle to sequential modular. The main difference is that once the first estimate of the input and output values for each operation unit is obtained, linear relationships between them are constructed, resulting in a linear equation model. These interconnection equations are solved simultaneously with the whole system model to obtain a new set of inlet stream values. If two successive iterates of assumed stream values converge within a present tolerance, the simulation is completed. Otherwise, the simulation reinitializes.

In the simultaneous modular approach, the calculations alternate between sequential modular and solution of linear equations, which reduces the computation time (Raman 1985). The development of this approach was motivated by ASPEN (Advanced System for Process Engineering) project, developed by the Massachusetts Institute of Technology (MIT) and US Department of Energy (DOE) in the late 1970s.

4.4.4 Equation-Oriented Approach

The basis of this approach is that the entire process can be represented by a large sparse system of nonlinear algebraic equations (NAEs) in steady-state simulation and differential-algebraic equations (DAEs) in dynamic simulation, which are solved simultaneously for all unknowns (Dimian 2003). The set of equations consist of model equations, interconnection equations, design specifications, physical properties and thermodynamic correlations. A medium-sized industrial plant, for instance, may result in a system that contains several thousand equations, which can be solved separately through suitable decomposition techniques (Raman 1985).

Among its advantages are the better handling of systems with many recycle streams, without using tear stream strategies. This approach also presents a more flexible environment for the specification of parameters and variables (e.g., inputs, outputs and internal variables). However, the equation-oriented approach demands greater programming effort and has difficulties dealing with large DAE systems as well as convergence and debugging (Dimian 2003).

4.4.5 *Interfacing Standard*

Since the last decade, the chemical engineering community has been working on a uniform standard for interfacing the current process simulators. This interface allows computer-aided process engineering (CAPE) applications to interoperate. One of the recent initiatives is the CAPE-OPEN consortium. This consortium, funded by the European Union, has undertaken the task of defining a common application framework for CAPE. The CAPE-OPEN Laboratories Network (CO-LAN) is the international organization for the management of CAPE-OPEN standards, which define the rules to be implemented in flowsheeting software.

The CAPE-OPEN standard is the de facto standard for interfacing process modeling software components for use in the design and operation of chemical processes. It is based on universally recognized software technologies, such as COM/DCOM[1] and CORBA[2] (CO-LAN 2010). Recently, Microsoft® had designed and developed a new specification for software interoperability, called .NET, which represents a great step forward into interoperable and easy-to-develop "cross-platform" applications (Fermeglia and Parenzan 2007).

Numerous industrial and academic projects have adopted CAPE-OPEN as their preferred means of interoperating (e.g., ANSYS, AspenTech, COMSOL, DECHEMA, Imperial College, and Process Systems Enterprise—PSE) (CO-LAN 2010). Barrett Jr and Yang (2005) described with details in their paper, the development of a chemical process modeling environment based on CAPE-OPEN interface standards using the Microsoft .NET framework.

4.4.6 *Impacts of Computer Hardware Development*

As the mathematical models become more detailed and rigorous, it is inevitable that their mathematical solution requires increasingly greater use of computing resources. Thus, it is essential to take advantage of existing computational advances, in terms of both hardware and architecture.

4.4.6.1 Parallel Computation

The first works covering the hardware architecture advances for process modeling and simulation made use of vector supercomputers; however, its availability was somewhat limited. Vector supercomputers appeared in the early 1970s and were the dominant projects for supercomputers from 1970 to 1990. The fast decline in the price–performance ratio of conventional microprocessor projects led to the death of

[1]COM/DCOM—Component Object Model/Distributed Component Object Model.

[2]CORBA—Common Object Request Broker Architecture.

the vector supercomputer after 1990. From this point, most of the researches focused in the use of multiple processors communicating through shared memory (e.g., OpenMP[3]) or message passing protocols (e.g., PVM[4] or MPI[5]). The use of parallel computing concept for process modeling and simulation remained for a long time restricted to academic research applications, and the use of this technology for process modeling was not incorporated in the major commercially available process simulation environments. Fluid dynamics and computational chemistry, unlike process modeling and simulation, went through significant improvements with the use of parallel computing (Pantelides 2004).

One of the reasons that may have contributed to this fact was the fast increase in processing computation of single-core processors since the early 1990s, which allowed the solution of even complex simulations in few minutes and delayed the use of parallel strategies in this field (Pantelides 2004). Currently, all process simulation environments are able to run on modern personal computers.

4.4.6.2 Computer Memory Improvement

The size of mathematical systems has increased substantially in last decades, both due to the incorporation of significant level of details in the unit operation models and by the number of processes involved in these units. Consequently, the demand for computer memory and computational speed represents the main requisites for the adequate use of process modeling and simulation tools. This issue becomes even more apparent as new developments on process simulators attempt to improve their usability (e.g., graphical user interface), providing better diagnostic tools for the numerical solution and improving the interface/communication between simulation softwares (e.g., middleware such as CORBA and COM/DCOM, and frameworks such as .NET). All these features increase computer memory demands (Pantelides 2004). Nevertheless, the computer memory, cost reduction and the use of 64-bit architecture help to overcome such obstacles.

4.4.7 Current Process Simulators

In recent decades, the market for process simulators underwent severe transformations, and relatively few systems survived during this period. Nowadays, most of the current process simulation environments adopted a new development paradigm following an object-oriented approach using languages such as C++ and Java. The substitution of procedural to object-oriented programming paradigm brought new

[3]OpenMP—Open Multiprocessing.

[4]PVM—Parallel Virtual Machine.

[5]MPI—Message Passing Interface.

possibilities to the development of process simulators and will certainly contribute immensely to the process engineering community (Martín 2014). Another important change is that now the current simulators are classified, depending on their licenses, as commercial or open-source application. Nevertheless, some process simulators are not open source, but distributed free of charge.

Well-established process simulation environments such as AspenPlus, ChemCAD, Aspen HYSYS, PRO/II, SuperPro Design and gPROMS are some of the current commercial simulators that are still under constant development. Currently, AspenPlus is the industry market-leading process simulation environment. Based on FLOWTRAN (1970) process simulator, which eventually becomes the kernel upon which it was built, it is based on simultaneous modular approach and used as process modeling tool for conceptual design, optimization and performance monitoring. Featuring the world's largest database of pure component and phase equilibrium data for conventional chemicals, electrolytes, solids and polymers, it has been used with success in a wide range of research and industrial applications.

Some new process simulators have been developed following the open-source concept. In this software development model, the source code is available to the general public for use and/or modification from its original design. Open-source code is meant to be a collaborative effort, where programmers improve upon the source code and, ideally, share the changes within the community (OSI 2015). Some examples include OpenModelica and DWSIM.

OpenModelica is an open-source Modelica-based modeling and simulation environment intended for industrial and academic usage (OpenModelica 2015). It is an object-oriented language for modeling large, complex, and heterogeneous physical systems, e.g., systems containing mechanical, electrical, electronic, hydraulic, thermal, control, electric power or process-oriented subcomponents (Modelica 2014).

DWSIM is a multiplatform CAPE-OPEN compliant chemical process simulator based on the sequential modular approach. It features a graphical user interface, advanced thermodynamic calculations, reaction support and petroleum characterization/hypothetical component generation tools. DWSIM is able to simulate steady-state, vapor–liquid, vapor–liquid–liquid, solid–liquid and aqueous electrolyte equilibrium processes (DWSIM 2015).

An example of free-of-charge distributed process simulator is EMSO (Environment for Modeling, Simulation and Optimization) simulator. EMSO is a multiplatform CAPE-OPEN compliant equation-oriented process simulator with graphical user interface for modeling complex, dynamic or steady-state processes (EMSO 2015). Prebuilt models are available at the EMSO Modeling Library (EML). New models can be written for EMSO using its own object-oriented modeling language or be developed externally in C, C++ or FORTRAN and then be embedded into the process simulation environment.

4.4.8 *Biorefinery Applications*

Several works about biorefinery process simulations have been developed in recent years. In 2002, the National Renewable Energy Laboratory (NREL) published a techno-economic report analyzing a 2G ethanol production process from corn stover. In this report, a complete flowsheet simulation was built in AspenPlus® process simulation environment (Aden et al. 2002). An update of this study was made by Humbird et al. (2011). Other biorefinery-related studies, using a process simulation environment, were published by NREL such as ethanol production via thermochemical route (Dutta and Philips 2009) and hydrogen production from biomass (Spath et al. 2005).

In 2011, the Brazilian Bioethanol Science and Technology Laboratory (CTBE) released the report "The Virtual Sugarcane Biorefinery (VSB)," assessing the technical, economic and environmental impacts of different scenarios of first- and second-generation ethanol production with the VSB framework (CTBE 2011).

Many scientific papers covering process synthesis and analysis as well as process simulation in the context of biorefineries were also published. Cardona and Sánchez (2007) performed a techno-economical evaluation of fuel ethanol processes using AspenPlus® as process design tool. In 2012, Macrelli et al. (2012) carried out a complete techno-economic evaluation of 1G2G ethanol production from sugarcane bagasse and straw using the same fore-mentioned simulation tool.

This book presents examples of applications of process simulation in the context of sugarcane biorefineries. These studies include first- and second-generation ethanol production (Chaps. 7 and 8), integration of butanol production to an autonomous distillery (Chap. 7) and integration of the thermochemical route for obtention of ethanol and superior alcohols to an annexed plant (Chap. 7), among others.

4.5 Process Design: Sugarcane Biorefinery

4.5.1 *Input Data*

4.5.1.1 Feedstock Composition

Input data for the simulation performed with the Virtual Sugarcane Biorefinery (VSB) are based on the industrial reality for sugarcane mills. Sugarcane received at the industrial site is comprised of stalks (where most of the sucrose is present), mineral impurities (dirt carried along during harvest) and vegetable impurities (sugarcane straw—tops and leaves). When sugarcane straw is considered to be collected from the fields, it is received as a separate stream since baling system was assumed. Stalk composition varies with time, soil, climate and other aspects, so average figures (14 % sucrose, 13 % fibers) were considered (CTBE 2012).

Fibers in the sugarcane are mainly composed of cellulose, hemicellulose and lignin. For the sugarcane plant, approximately 10 % of the hemicellulose is comprised of acetate, so a division between acetate and xylan was considered when defining sugarcane hemicellulose content.

Vegetable impurities carried along with sugarcane stalks comprise 4.8 % (wet basis) of the sugarcane plant (stalks and vegetable impurities) delivered at the industrial site. Their moisture content is of 52.1 %, higher than that of baled sugarcane straw (15 %) (Hassuani et al. 2005). The total amount of straw produced at the field is assumed to be equal to 140 dry kg TC^{-1} (TC—tonne of sugarcane stalks) (Hassuani et al. 2005); when straw is assumed to be recovered and used in the industrial plant, 50 % is left in the field for agricultural purposes. Therefore, for each tonne of sugarcane stalks, 24.15 kg of straw (dry basis) is carried along as vegetable impurities and 45.85 kg is baled and transported to the industrial site. Straw composition was based on data provided by Oliveira et al. (2014).

The amount of soil dragged along with stalks during harvest is assumed to be equal to 1.2 % of the amount of stalks (Finguerut and Paes 2014). Soil is present in the baled straw in the proportion of 4 %, based on the amount of baled straw (wet basis) (Cardoso et al. 2013).

Data for the optimized sugarcane distillery processing 4 million tonnes of sugarcane per year and recovering 50 % of sugarcane straw are presented. Mass flow and composition for these streams are shown in Table 4.3.

Table 4.3 Composition and total flow of the feedstock received at the industrial site

Component—composition wt%	Stalks	Mineral impurities (stalks)	Vegetable impurities (stalks)	Sugarcane (total)	Straw (bale)	Mineral impurities (bale)
Reducing sugars	0.60	–	0.11	0.57	0.19	–
Water	70.29	–	52.10	68.63	15.00	–
Phosphate	0.03	–	0.01	0.03	0.02	–
Minerals	0.20	–	–	0.19	–	–
Organic acids	0.56	–	0.26	0.54	0.47	–
Salts	1.31	–	1.20	1.54	2.13	–
Sucrose	14.00	–	2.45	13.29	4.34	–
Fibers	13.00		43.87	11.85	77.85	
Acetyl group	0.33	–	1.30	0.37	2.31	–
Cellulose	5.99	–	18.27	6.51	32.42	–
Lignin	3.21	–	11.60	3.57	20.59	–
Xylan	3.21	–	12.70	1.13	22.53	–
Ash	0.27		–	0.27		
Soil	–	100.00	–	0.37	–	100.00
Total flow (kg h^{-1})	833,333	10,000	42,017	885,350	44,950	1,798

Table 4.4 Sugarcane bagasse composition in the VSB

Component	Fraction (wt%, wet basis)	Fraction (wt%, dry basis)
Cellulose	21.61	43.25
Acetyl group	1.21	2.42
Xylan	11.81	23.63
Lignin	11.72	23.46
Extractives	2.14	4.28
Organic acids	0.09	0.18
Reducing sugars	0.08	0.17
Sucrose	1.96	3.92
Phosphates	0.01	0.01
Ash	1.48	2.96
Salts	0.98	1.96
Minerals	0.12	0.25
Dirt	0.38	0.75
Water	50.04	–

Based on estimates provided by specialists, the VSB assumes that 2/3 of the ash obtained in sugarcane bagasse analyses is inherent to the fiber, and the remaining 1/3 is derived from the sugarcane stalks. Sugarcane impurities are represented by minerals, salts and organic acids, which comprise both sugarcane stalk ash and bagasse ash. Sugarcane bagasse composition also varies along with the harvest, being affected by climate, soil, sugarcane variety, etc. In the VSB, sugarcane bagasse composition was defined based on data provided by Rocha et al. (2010), as shown in Table 4.4.

4.5.1.2 Component Databank

After sugarcane composition is defined, it is necessary to insert its components in the AspenPlus simulator. Since components of the lignocellulosic material (cellulose, hemicellulose and lignin) were not available in the software databank, their properties were obtained from the databank for biofuel components developed by the National Renewable Energy Laboratory (NREL) (Wooley and Putsche 1996); lignin structure, however, was modified to represent sugarcane lignin, with the molecular formula $C_9O_{2.9}H_{8.6}(OCH_3)$, and its enthalpy of formation was determined based on enthalpy of combustion (27,000 kJ kg^{-1}) given by Stanmore (2010), resulting in −25,689 kJ kg^{-1}. Fiber components (cellulose, hemicellulose and lignin) were inserted as solids; no particle distribution for solids was assumed, but a defined molecular weight.

In addition to fiber components, structural components must also be included in the simulation. All the components of the sugarcane plant, such as minerals and salts, were included with characteristics based on those of specific representative chemicals:

- Ash is assumed to be comprised of the following:
  - Minerals: represented by K_2O (Mantelatto 2005; Chen and Chou 1993);
  - Salts: represented by KCl (Mantelatto 2005), since potassium salts are those present in greater proportion (approximately 60 % of the total salts in the ash);
  - Dirt: represented by SiO_2;
- Extractives are assumed to be comprised of the following:
  - Organic acids: represented by aconitic acid, whose concentration is around 3 times higher than that of all the other organic acids (Mantelatto 2005);
  - Sucrose;
  - Glucose;
- Phosphoric acid: represents the phosphates present in the sugarcane plant.

Several components are also required to represent process inputs and reaction products in the simulation. Some of these components are not present in AspenPlus database, and when inserted, their properties must be defined (either by Aspen estimator or provided by the user). Similar to structural sugarcane components, some of them were represented by specific chemicals:

- Flocculant polymer: represented by polyacrylamide.
- HMF (hydroxymethylfurfural): inserted by importing its molecular structure.
- GLUCOLIG (glucose oligomers): inserted as glucose (dextrose), having its molecular weight (MW) and standard enthalpy of formation (DHFORM), recalculated with data from cellulose (Wooley and Putsche 1996).
- XYLOLIG (xylose oligomers): inserted as xylose (d-xylose), having its molecular weight (MW) and standard enthalpy of formation (DHFORM), recalculated with data from xylan (Wooley and Putsche 1996).
- ACETATE (acetyl group): represented by acetic acid, having its standard enthalpy of combustion changed to match the one calculated for acetate (s) + $2O_2 \rightarrow 2H_2O + 2CO_2$.
- LGNSOL (soluble lignin): similar to LIGNIN with same physicochemical properties. Its solubilization is represented by a phase change (solid to mixed).
- CAL-ACON (calcium aconitate): properties collected from gypsum ($CaSO_4{-}2H_2O$) available at NREL databank (Wooley and Putsche 1996), except by its molecular weight, which was calculated.
- YEAST: molecular weight and solid standard enthalpy of formation retrieved from SuperPro Designer® databank. Molar volume was calculated based on its density (1180 kg m^{-3}, Scherrer et al. 1977) and molecular weight.
- ENZYME: the enzyme component inserted as a solid component with user-defined properties. The values for the molar weight and enthalpy of formation and volume of the solid were retrieved from Humbird (2011).

All the components used in the simulation are shown in Table 4.5.

User-defined properties are shown in Table 4.6.

Table 4.5 Components used in the simulation, with their AspenPlus ID, name and formula

Component	AspenPlus ID	Type	Aspen name
Acetic acid	ACET-AC	Conventional	ACETIC-ACID
Calcium oxide	CAO	Conventional	CALCIUM-OXIDE
Lime	$CAOH_2$	Conventional	CALCIUM-HYDROXIDE
Methane	CH_4	Conventional	METHANE
Carbon dioxide	CO_2	Conventional	CARBON-DIOXIDE
Ethanol	ETHANOL	Conventional	ETHANOL
Furfural	FURFURAL	Conventional	FURFURAL
Glucose oligomers	GLUCOLIG	Conventional	DEXTROSE
Glucose	GLUCOSE	Conventional	DEXTROSE
Glycerol	GLYCEROL	Conventional	GLYCEROL
Water	H_2O	Conventional	WATER
Sulfuric acid	H_2SO_4	Conventional	SULFURIC-ACID
Phosphoric acid	H_3PO_4	Conventional	ORTHOPHOSPHORIC-ACID
Hydroxymethylfurfural	HMF	Conventional	5-HYDROXYMETHYLFURFURAL
Isoamyl alcohol	ISOAMIL	Conventional	3-METHYL-1-BUTANOL
Isobutanol	ISOBUTOH	Conventional	ISOBUTANOL
Minerals	MINERALS	Conventional	POTASSIUM-OXIDE
Nitrogen	N_2	Conventional	NITROGEN
Caustic soda	NAOH	Conventional	SODIUM-HYDROXIDE
Ammonium hydroxide	NH_4OH	Conventional	AMMONIUM-HYDROXIDE
Oxygen	O_2	Conventional	OXYGEN
Organic acids	ORG-AC	Conventional	TRANS-ACONITIC-ACID
Salts	SALTS	Conventional	POTASSIUM-CHLORIDE
Sulfur dioxide	SO_2	Conventional	SULFUR-DIOXIDE
Soil	SOIL	Conventional	SILICON-DIOXIDE
Sucrose	SUCROSE	Conventional	SUCROSE
Xylose oligomers	XYLOLIG	Conventional	D-XYLOSE
Xylose/arabinose	XYLOSE	Conventional	D-XYLOSE
Acetyl group	ACETATE	Solid[a]	ACETIC-ACID
Calcium aconitate	CAL-ACON	Solid[a]	*USER DEFINED*
Calcium phosphate	CAL-PHOS	Solid[a]	CALCIUM-PHOSPHATE
Calcium sulfite	CAL-SULF	Solid[a]	CALCIUM-SULFITE
Cellulose	CELLULOS	Solid[a]	*USER DEFINED*
Enzymes	ENZYME	Solid[a]	*USER DEFINED*
Flocculant	FLOCCUL	Solid[a]	POLY(ACRYLAMIDE-STYRENE)
Xylan	XYLAN	Solid[a]	*USER DEFINED*
Lignin	LIGNIN	Solid[a]	*USER DEFINED*
Soluble lignin[b]	LGNSOL	Solid[a]	*USER DEFINED*
Sulfur	SULFUR	Solid[a]	SULFUR
Yeast	YEAST	Solid[a]	*USER DEFINED*

[a]Solid components do not take part in phase equilibrium
[b]Solubilization represented by a phase change (solid to mixed)

Table 4.6 User-defined properties

Component	Molecular weight	Solid standard enthalpy of formation	Standard enthalpy of formation	Pc	Dipole moment	Tb	Tc	Radius of gyration[h]
Aspen code	MW	DHSFRM	DHFORM	PC	MUP	TB	TC	RGYR
Unity	g mol^{-1}	J $kmol^{-1}$	kcal mol^{-1}	Bar	Debye	°C	°C	meter
CAL-ACON	462[a]	−2E + 09[b]						
CELLULOS	162.1436[b]	−9.8E + 08[b]						
XYLAN	132.117[b]	−7.6E + 08[b]						
LIGNIN	194.1968[a]	−3.5E + 08[a]						
LGNSOL	194.1968[a]	−3.5E + 08[a]						
YEAST	30.25542[c]	−9.7E + 07[b]						
NH4OH	35.046[c]		−86.27114	117.06[c]	6.86371[e]	−2.15[c]	132.39[c]	
GLUCOLIG	162.1424		−192.87534					
XYLOLIG	132.11612		−149.41258					
ACETATE		−4.56E + 08						
ORG-AC					1.74[d]			2.61E−10[d]
H_3PO_4					3.46106[g]			
XYLOSE					2.749101[f]			4.75E−10[f]
NAOH					6.86371[g]			
ENZYME	24.0156[i]	−7.494E + 07[i]						

[a]Calculated
[b]Wooley and Putsche (1996)
[c]SuperPro Designer® databank
[d]Aspen databank—value for acetic acid
[e]Same as that for NaOH
[f]Same as those available at Aspen databank for glucose
[g]Colby College (2012)
[h]Parameter required for Hayden O' Connell equation of state
[i]Humbird et al. (2011), "Process Design and Economics for Biochemical Conversion of Lignocellulosic Biomass to Ethanol"

4.5.2 First-Generation Process

4.5.2.1 Process Design

In order to describe the process configuration for first-generation process, an optimized annexed plant is presented as example. A simplified process flow diagram is depicted in Fig. 4.11. This industrial facility processes 4 million tonnes of sugarcane per year, with an effective operation period of 200 days $year^{-1}$.

After reception, when the sugarcane that arrived in trucks was weighted and sampled, the cane follows to the cleaning station. It is assumed that mechanically harvested cane is processed, and for this reason, dry cleaning is used to remove part of the mineral and vegetal impurities. Conveyors transport the cleaned cane throughout a set of levelers and cutting knives and hammer shredders in order to break the sugarcane structure and facilitate the extraction.

Juice is extracted from the prepared cane in a mill tandem, producing bagasse as fibrous residue. Imbibition water is used to raise the sugar extraction, and the usual sugar recovery in the juice is 96 %. The juice follows to the treatment station where it is heated and has chemical reagents added to precipitate impurities (e.g., phosphoric acid, hydrated lime, polyelectrolyte). Reheated to about 105 °C, the juice flashes to atmosphere to remove air and goes to a decanter. The muddy stream is filtered; the liquid part is recycled to the juice before the chemical treatment in an

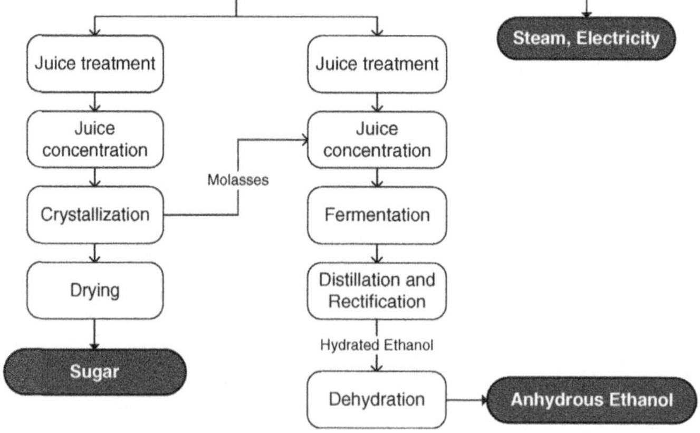

Fig. 4.11 Block flow diagram for an optimized annexed plant

earlier point of the process, and filter cake is separated. In the juice treatment for sugar production, more lime is necessary and more filter cake is obtained.

The clarified juice obtained in the clarifier is concentrated by evaporation in a five-effect evaporator, producing syrup for sugar production. The juice diverted to ethanol production is evaporated in a single-effect evaporator. Steam at 2.5 bar is fed to the first effect, and part of the steam produced in this effect is recuperated and, together with steam from single-effect evaporator, is used to attend the heat requirements of other areas, such as juice treatment and crystallization. For sugar production, it was assumed that crystals were separated using the two-boiling-system approach, where two types of sugar are produced from the syrup: the grade "A" sugar (final product) and the grade "B" sugar (intermediate sugar that is produced and recycled inside the process). The final sugar, VVHP type, is dried in a rotary drier and cooled before storage. Sugar fines carried by the exit airflow are recovered in a scrubber and the diluted solution recycled.

The final, exhausted molasses is mixed with the concentrated juice from the single-effect evaporator, and this stream is sent to fermentation. A fed-batch fermentation process with cell recycle was assumed. During fermentation, the effluent gases are washed in an absorption column to recover ethanol. After fermentation reactions cease, the wine is sent to the centrifuges, where cells are separated from the ethanol solution. Aiming at reducing bacterial contamination, the obtained cells are treated in a separate reactor, by the addition of sulfuric acid and water, and recycled for a new batch. Wine is mixed with the alcoholic solution from the absorption column and sent to purification.

The production of hydrated ethanol is performed using the distillation configuration conventionally employed in Brazilian mills, which consists of a set of 5 columns: A, A1, D, B and B1. Wine is heated up and fed in column A1, vinasse exits in the bottom of column A, and the phlegm follows to the rectification columns (B and B1). Phlegmasse is the bottom stream of column B1, and hydrated ethanol is produced on top of column B. Hydrated ethanol is sent to the molecular sieves, where anhydrous ethanol is produced, and the diluted alcoholic stream from them is recycled to column B.

Straw, which comes from the field in bales, is chopped, sieved and mixed with bagasse and with vegetal impurities separated through dry cleaning and sent to the boilers. In the combined heat and power generation system, the produced steam feeds steam turbogenerators that provide process steam to the production areas and generate electrical energy. The surplus energy is exported to the grid.

4.5.2.2 Process Simulation

Hierarchy blocks were used to organize process flowsheet, since sugarcane processing includes several unit operations. The process flowsheet for an optimized annexed plant is depicted in Fig. 4.12. At a first level, it is possible to identify the main process steps: sugarcane preparation and extraction (PREP-EXT), ethanol production (ETHANOL), sugar production (SUGAR) and cogeneration unit (CHP).

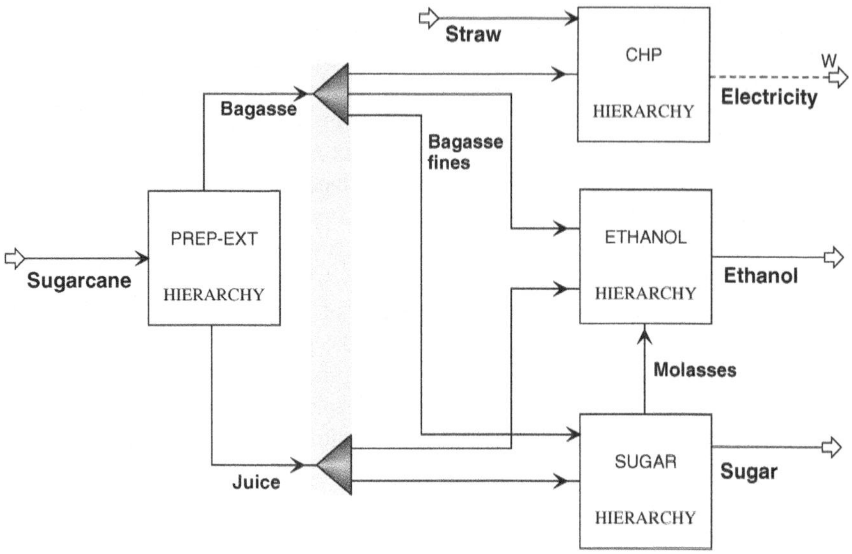

Fig. 4.12 Process flowsheet for an optimized annexed plant

Fig. 4.13 Process flowsheet for fermentation section

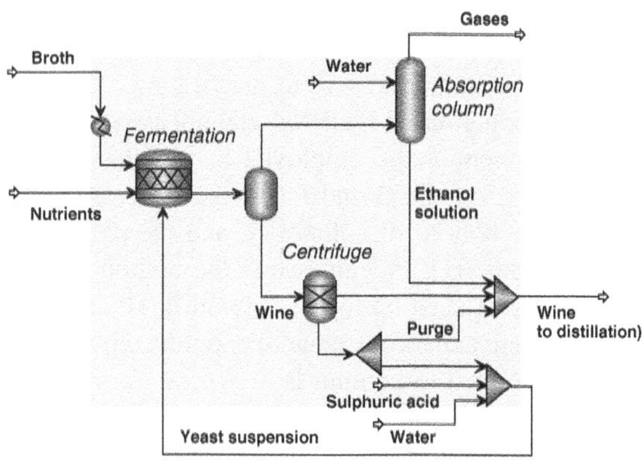

As an example, arrangement of unit operations in the fermentation section, inserted in the ETHANOL hierarchy, is shown in Fig. 4.13.

The main property method used throughout the simulation was NRTL for calculation of activity coefficients of the liquid phase, while ideal behavior was assumed for the vapor phase in most of the simulation. When deviations from the ideal behavior are expected, such as operation under moderate pressures (around 10 bar), or high sugar concentration (e.g., sugar processing and process steps previous to fermentation), thermodynamic properties of the vapor phase were estimated based on Redlich–Kwong equation of state. The Hayden O'Connell (NRTL-HOC) equation was used for vapor-phase calculation when the concentration of acetic acid and other carboxylic acids is significant, such as on fermentation and distillation units.

In the boiler from cogeneration, the method RKS-BM (RKS—Redlich–Kwong–Soave and BM—Boston–Mathias alpha function) was used, since it is adequate for high-temperature gases (Aspentech 2009). For the steam streams, steam tables (STEAMNBS) were used, since they accurately represent pure water and steam for a wide range of pressures and temperatures.

Several operations (mills, filters, settlers and molecular sieves, among others) were represented as component splitters, due to the lack of more adequate blocks in the simulator database. Separation efficiency for each component of the mixture was supplied in such a way that the equipment efficiency and other characteristics of the materials, such as composition, agreed with those found in the literature or in the industry. For simulation of reactors (liming, fermentation and burner), reactions and conversions were defined; models based on vapor–liquid equilibrium were used for flashing, heating/cooling, evaporation, distillation and rectification. Auxiliary equipments such as mixers, splits, valves and pumps were also included in the simulation.

In the simulation of fermentation process, the following reactions were considered: sucrose inversion (Eq. 4.3), production of ethanol (Eq. 4.4), cell growth (Eq. 4.5) and formation of by-products (Eqs. 4.6–4.9). The stoichiometry of reactions was not based on metabolic reactions, which are more complex and not always available, but defined based on mass balance. As a simplification, reducing sugars (fructose and glucose) were represented only by glucose.

$$\text{Sucrose} + \text{H}_2\text{O} \rightarrow 2\text{Glucose} \tag{4.3}$$

$$\text{Glucose} \rightarrow 2\text{Ethanol} + 2\text{CO}_2 \tag{4.4}$$

$$\begin{aligned} &3.2618\text{Glucose} + 2.8469\text{NH}_4\text{OH} + 2.1352\text{CO}_2 \\ &\quad \rightarrow 1.0359\text{Acetic acid} + 6.9454\text{H}_2\text{O} + 19.6342\text{Yeast} \end{aligned} \tag{4.5}$$

$$0.7630\text{Glucose} \rightarrow 0.7630\text{Isobutanol} + 1.5260\text{CO}_2 + 0.7630\text{H}_2\text{O} \tag{4.6}$$

$$0.6107\text{Glucose} + 0.5236\text{H}_2\text{O} \rightarrow 0.5234\text{CO}_2 + 1.047\text{Glycerol} \tag{4.7}$$

$$\text{Glucose} \rightarrow 3\text{Acetic acid} \tag{4.8}$$

$$0.8271\text{Glucose} \rightarrow 0.6617\text{Isoamyl alcohol} + 1.6542\text{CO}_2 + 0.9927\text{H}_2\text{O} \tag{4.9}$$

The combustion reactions for each component present in the lignocellulosic material (cellulose, hemicellulose and lignin, among others) were inserted in the simulation, and conversion was set as 100 %; inefficiencies of the boiler were represented as the loss of a fraction of the hot gases after combustion. Based on the enthalpy of combustion for each component, lower heating value (LHV) of sugarcane bagasse was calculated as 7.5 MJ kg^{-1} (50 % moisture) and 14.9 MJ kg^{-1} for straw (15 % moisture).

Table 4.7 Main parameters of dry cleaning system and sugar extraction

Parameter	Value	References
Bale reception		
Efficiency for dirt removal from bales	65 %	Cardoso (2014)
Electricity demand in sieve and choppers	3.73 kWh $t^{-1}{}_{dry\ straw}$	Delfini (2012)
Dry cleaning system		
Efficiency for straw removal	55 %	Cardoso (2014)
Efficiency for dirt removal	65 %	Cardoso (2014)
Losses of sugarcane stalks	0.5 %	Cardoso (2014)
Electricity demand in conveyors and fans	0.692 kWh $t^{-1}{}_{inlet\ material}$	Delfini (2012)
Electricity demand in sieve and choppers	3.73 kWh $t^{-1}{}_{dry\ straw}$	Delfini (2012)
Sugar extraction[a]		
Amount of imbibition water	280 kg TC^{-1}	Pedra Mill (2006)
Efficiency of sugar extraction in the mills	96 %	Walter et al. (2008)
Bagasse moisture	50 %	Rocha et al. (2010)
Efficiency of dirt and bagasse removal in the screen	65 %	Mantelatto (2010)

[a]Since only physical changes to the sugarcane structure occur in the preparation, only electricity demand is included in the overall consumption of the plant

The main parameters adopted in the simulation per section of the plant are presented from Tables 4.7, 4.8, 4.9, 4.10 and 4.11.

The process descriptions and parameters discussed in this chapter were used for the simulation of first-generation ethanol production considering different scenarios. The results of these studies are presented in Chaps. 7 and 8.

4.5.3 Second-Generation Process: Biochemical Route

4.5.3.1 Process Design

A simplified process flow diagram, illustrated in Fig. 4.14, shows the integration of second-generation process, through biochemical route, to an autonomous distillery. Definition of the process design for 2G, represented by the light gray blocks in Fig. 4.14, is based on the study performed by Milanez et al. (2015).

In the so-called integrated process, surplus bagasse and straw (lignocellulosic material) are pretreated through steam explosion (autohydrolysis), where only steam is injected to achieve the desired reactor temperature, and after held in this condition for few minutes, the pressure is quickly reduced, causing an explosive

Table 4.8 Main parameters of juice treatment and concentration

Parameter	Value	References
Juice treatment		
Temperature—first juice heating	70 °C	Mantelatto (2009)
Phosphate content of the juice after phosphoric acid addition	250 ppm	Mantelatto (2010)
Amount of lime added in liming (ethanol/sugar production)	0.6/1.0 kg CaO TC^{-1}	Mantelatto (2010)
Temperature—second juice heating	105 °C	Copersucar (1989)
Amount of flocculant polymer	2.5 g TC^{-1}	Pedra Mill (2006)
Efficiency of settling of insoluble solids	99.7 %	Mantelatto (2010)
Filter cake moisture content	60–70 %	Mantelatto (2010)
Insoluble solid retention in the filter	65 %	Mantelatto (2010)
Filter cake pol content	1.0 %	Payne (1989)
Amount of wash water related to filter cake	150 %	Mantelatto (2009)
Bagasse fines added in the filter	6 kg TC^{-1}	Mantelatto (2009)
Efficiency of removal of insoluble solids in the clarified juice screen	65 %	Mantelatto (2010)
Juice concentration		
Number of effects in evaporation	5	Mantelatto (2010)
Syrup	65° Brix	Mantelatto (2010)

Table 4.9 Main parameters of sugar production

Parameter	Value	References
Crystallization		
Brix of the "A" sugar	99.0	Ribeiro (2003)
Purity of the "A" sugar	99.6 %	Bazico (2010)
Brix of final molasses	85.0 %	Calculated
Sugar overall recovery (as "A" sugar)	76.5 %	Mantelatto (2010)
Drying		
Moisture content of the dry sugar	0.1 %	Bazico (2010)
Temperature of the outlet sugar	35 °C	Mantelatto (2010)
Sugar dust in the outlet air	0.8 %	Mantelatto (2010)
Sugar recovered from the outlet air (via scrubber) (%)	99.5	Mantelatto (2010)

decompression. Most xylan is solubilized into C5 sugars (pentoses), oligomers and, in a small proportion, degradation products (e.g., furfural). Small fractions of cellulose and lignin are also solubilized.

Table 4.10 Main parameters for ethanol production

Parameter	Value	References
Fermentation and cell treatment		
Fraction of the reactor fed with yeast solution	25 wt%	Pedra Mill (2006)
Fermentation temperature	33 °C	Pedra Mill (2006)
Conversion of sugars to ethanol	89.5 %	Mantelatto (2010)
Efficiency of solid retention in the centrifuges	99 %	Mantelatto (2010)
Ethanol content in the wine (distillation feed)	80 g L^{-1}	Mantelatto (2010)
Ethanol content of the yeast concentrated solution obtained in the centrifuges	6.5 wt%	Mantelatto (2010)
Sulfuric acid addition in yeast treatment (on 100 % basis)	5 kg $m^{-3}{}_{ethanol}$	Rossell (2011)
Distillation		
Vinasse and phlegmasse ethanol content	<200 ppm	Meirelles (2006)
Amount of fusel oil per ethanol produced	0.2 % (v/v)	Garcia (2008)
Hydrated ethanol purity	93 wt%	Mantelatto (2010)
Molecular sieves		
Feed temperature	150 °C	Dias (2008)
Steam pressure	6 bar	Dias (2008)
Ethanol recovered as final product	81.4 %	Mantelatto (2010)
Anhydrous ethanol purity	99.6 wt%	Dias et al. (2012a)

After pretreatment, a diffuser is employed for solid–liquid separation using water to promote soluble solids' recovery, generating C5 liquor and cellulignin. The C5 liquor undergoes an inline pH adjustment with the addition of alkali and then is concentrated through evaporation. Subsequently, this stream is sent to fermentation with genetically modified organisms (GMOs), suitable for C5 conversion. Simultaneously, enzymes are added in order to break the oligomers into fermentable sugars, increasing ethanol production in the reactor. GMO cells are recovered by means of centrifugation and partially recycled to the reactor.

The cellulignin is transferred to enzymatic hydrolysis reactors for conversion of cellulose into glucose. Part of the remaining xylan is also solubilized in this step. A filter is employed to separate the solid residues and C6 liquor. The latter is mixed to sugarcane juice, rich in sucrose (C12 sugar), and concentrated through evaporation prior to fermentation with conventional yeast.

The processing of the alcoholic streams (wine) from C5 and C6/C12 fermentation reactors is performed using the usual 1G technology; that is, these streams are combined and pumped to distillation and rectification in order to obtain hydrous

Table 4.11 Main parameters of cogeneration system

Parameter	Value	References
Fraction of bagasse for start-ups of the plant	5 %	Seabra (2008)
Pressure of the boiler system	65 bar	Dias et al. (2012b)
Boiler efficiency (LHV basis)	87.7 %	Dias et al. (2012b)
Gases outlet temperature	160 °C	Dias et al. (2012b)
Steam temperature	485 °C	Dias et al. (2012b)
Turbine isentropic efficiency	85 %	Dias et al. (2012b)
Generator efficiency	98 %	Dias et al. (2012b)
Condensing pressure	0.11 bar	Seabra (2008)
Electric energy demand of the process (with electric drivers), except dry cleaning system	30 kWh TC^{-1}	Seabra (2008)
2.5-bar steam demand (assuming thermal integration)	80 % of that calculated	Dias et al. (2012a)
6-bar steam demand (for molecular sieves)	To achieve ethanol feed temperature	Calculated
Condensate losses	5 %	Seabra (2008)

ethanol. Then, molecular sieves are employed for dehydration, producing anhydrous ethanol.

The solid residues are sent to cogeneration system, and in order to achieve the energy balance of the integrated plant, part of the lignocellulosic material is also burnt. In this case, steam is produced only to meet process requirement and only backpressure turbines are employed.

Year-round operation for 2G process was considered. For operation in the sugarcane off-season, part of the lignocellulosic material is stored in such a way that the capacity of 2G process is kept the same all year-round. However, process areas shared with 1G process operate with lower flows in the off-season, for example, the fermentation and distillation areas. The same applies for CHP unit, since steam consumption is much lower in the off-season compared to season, when 1G process is also operational. Besides, C6 liquor is fermented independently because juice is not produced in the off-season. As most facilities maintain multiple pieces of equipment for the same function, it is not expected a decrease in efficiencies due to operation outside of the specified flow. In practice, some equipment will be idle in the off-season period, which is useful for the usual maintenance that occurs in this period.

4.5.3.2 Process Simulation

For simulation of second-generation process through biochemical route, representative unit operations, as depicted in Fig. 4.15, were inserted in a hierarchy block in the AspenPlus® process flowsheet.

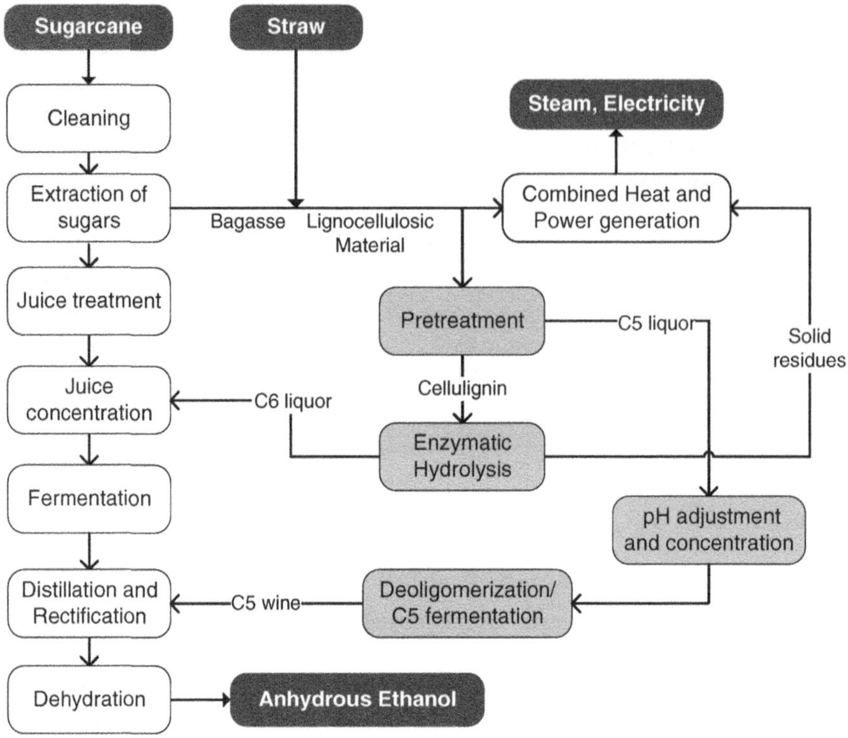

Fig. 4.14 Block flow diagram for integrated 1G2G ethanol production

In the 2G process steps, with exception of fermentation, NRTL for liquid phase and Redlich–Kwong equation of state for vapor phase were considered due to sugars' presence and moderate pressure (around 10 bar). For fermentation units, similar to 1G fermentation process, NRTL-HOC was chosen to account for the nonideality.

Stoichiometric model reactors were used to represent pretreatment and hydrolysis reactors as well as pH adjustment and fermentation of C5 liquor. Solid–liquid separation units (diffuser and filter) as well as centrifuge were simulated considering separation efficiencies.

For pretreatment simulation, reactions were inserted to represent hydrolysis of xylan and cellulose into monomers and oligomers (Eqs. 4.10–4.13) as well as degradation products (Eqs. 4.14 and 4.15). Although degradation products are

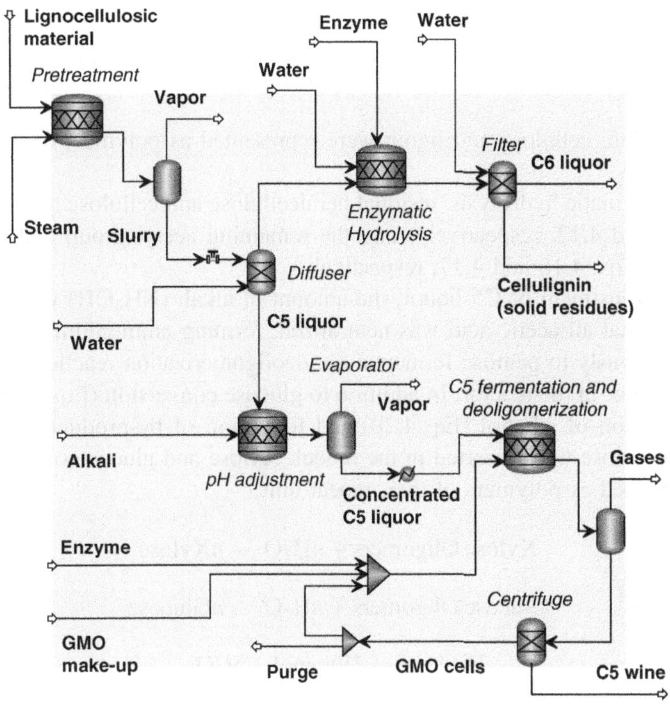

Fig. 4.15 Process flowsheet for second-generation process

formed as a consequence of sugar dehydration, reactions were inserted in such a way that conversion would be calculated based on carbohydrate polymers. In addition, solubilization of acetyl group (Eq. 4.16) and lignin (Eq. 4.17) takes place in the reactor.

$$(\text{Xylan})_n + n\text{H}_2\text{O} \rightarrow n\text{Xylose} \tag{4.10}$$

$$(\text{Xylan})_n + n\text{H}_2\text{O} \rightarrow n\text{Xylose Oligomers} \tag{4.11}$$

$$(\text{Cellulose})_n + n\text{H}_2\text{O} \rightarrow n\text{Glucose} \tag{4.12}$$

$$(\text{Cellulose})_n + n\text{H}_2\text{O} \rightarrow n\text{Glucose Oligomers} \tag{4.13}$$

$$(\text{Xylan})_n \rightarrow n\text{Furfural} + 2n\text{H}_2\text{O} \tag{4.14}$$

$$(\text{Cellulose})_n \rightarrow \text{Hydroxymethylfurfural} + 2n\text{H}_2\text{O} \tag{4.15}$$

$$\text{Acetil group} \rightarrow \text{Acetic acid} \tag{4.16}$$

$$(\text{Lignin})_n \rightarrow n\text{Soluble lignin} \tag{4.17}$$

Note: Xylan, cellulose and lignin were represented as polymers of their repeat unit.

In the enzymatic hydrolysis, residual hemicellulose and cellulose are hydrolyzed (Eqs. 4.10 and 4.12, respectively) and the remaining acetyl group and lignin are solubilized (Eqs. 4.16 and 4.17, respectively).

For pH adjustment of C5 liquor, the amount of alkali (NH_4OH) was defined in such a way that all acetic acid was neutralized, forming ammonium acetate.

Simultaneously to pentose fermentation, deoligomerization reactions (Eqs. 4.18 and 4.19) occur in the reactor. In addition to glucose conversion (Eqs. 4.4 and 4.6–4.9), production of ethanol (Eq. 4.20) and formation of by-products (Eqs. 4.21–4.24) from xylose were inserted in the model. Xylose and glucose oligomers were also represented as polymers of their repeat unit.

$$\text{Xylose Oligomers} + n\text{H}_2\text{O} \rightarrow n\text{Xylose} \tag{4.18}$$

$$\text{Glucose Oligomers} + n\text{H}_2\text{O} \rightarrow n\text{Glucose} \tag{4.19}$$

$$3\text{Xylose} \rightarrow 5\text{Ethanol} + 5\text{CO}_2 \tag{4.20}$$

$$6\text{Xylose} \rightarrow 5\text{Isobutanol} + 10\text{CO}_2 + 5\text{H}_2\text{O} \tag{4.21}$$

$$7\text{Xylose} + 5\text{H}_2\text{O} \rightarrow 5\text{CO}_2 + 10\text{Glycerol} \tag{4.22}$$

$$2\text{Xylose} \rightarrow 5\text{Acetic acid} \tag{4.23}$$

$$3\text{Xylose} \rightarrow 2\text{Isoamyl alcohol} + 5\text{CO}_2 + 3\text{H}_2\text{O} \tag{4.24}$$

A simplified representation of partial recycle of GMO cells was included in the simulation, since no detailing about conditions for treatment was available.

Process parameters and conditions are exemplified for an advanced 2G scenario in Table 4.12. This scenario is based on the same process flowsheet, but the assumptions are more optimistic, including higher yields and recoveries and lower residence times.

Since solid residues from hydrolysis are burnt together with part of bagasse and straw (lignocellulosic material, "LCM") in the cogeneration system, an iterative process is carried out as illustrated in Fig. 4.16. A fraction of LCM is burnt to meet process steam demand, and the remaining LCM is diverted for pretreatment (2G

Table 4.12 Main parameters of advanced second-generation process (Milanez et al. 2015)

Parameter	Value
Steam explosion pretreatment	
Temperature	210 °C
Residence time	5 min
Solid content	Defined by the steam required to achieve reactor temperature
Cellulose solubilization	5.5 %
Xylan conversion to xylose	60 %
Xylan conversion to xylose oligomers	20 %
Xylan degradation to furfural	10 %
Diffuser	
Proportion of water added	180 % of cellulignin fibers
Insoluble solid retention	99.5 %
Soluble solids recovered in the liquor	98.0 %
Cellulignin moisture	50 %
Enzymatic hydrolysis	
Temperature	65 °C
Residence time	36 h
Solid content	25 %
Cellulose conversion to glucose	80 %
Xylan conversion to xylose	80 %
Filter	
Solid retention	99 %
Soluble solids recovered in the liquor	99 %
Solid residues moisture	50 %
C6/C12 fermentation	
Operational conditions	Same as 1G %
C6/C12 conversion to ethanol	90
Deoligomerization and C5 fermentation	
Temperature	33 °C
Residence time	24 h
Xylose oligomer conversion to xylose	90 %
C6 conversion to ethanol	90 %
C5 conversion to ethanol	85 %
GMO cell recycling	95 %
Ethanol content in C5 wine	70 g L^{-1}

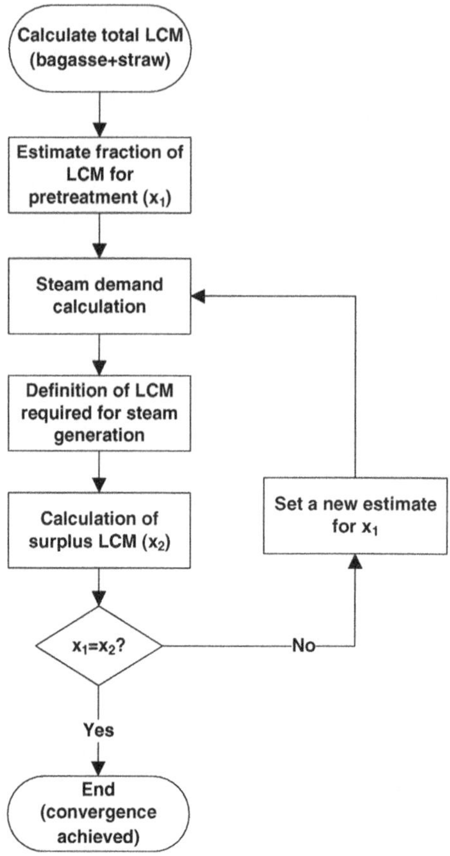

Fig. 4.16 Convergence flowchart for process simulation

process), which, consequently, alters steam demand. The convergence is only achieved when the LCM estimate for pretreatment (x_1) is equal to the surplus LCM (x_2) calculated.

Additional complexity related to year-round operation is observed, since part of the lignocellulosic material is stored for off-season operation. This proportion is defined in such a way that 2G process has the same feed in pretreatment stage all year-round. In the simulation, lignocellulosic material is divided into three streams: The first one is destined for 2G ethanol production in the season (200 days); the second stream is sent to CHP unit along with solid residues to supply energy to the process; and the third one is stored to operate 2G process in the off-season. The iterative process becomes even more complex, since the third parcel must be enough to supply the energy (complementing solid residue combustion) and operate 2G process at full capacity during 130 days (off-season operation).

Simulations of second-generation ethanol production via biochemical route based on the process design and parameters discussed in this chapter and their results are presented in Chaps. 7 and 8.

4.5.4 Second-Generation Process: Thermochemical Route

4.5.4.1 Process Design

Besides the 1G and 2G ethanol production alternatives described, which employ biochemical routes, a 2G thermochemical route to obtain anhydrous ethanol and superior alcohols from sugarcane bagasse and straw was also simulated and assessed in the VSB, evaluating technical, economic, and environmental impacts.

A simplified process flow diagram, illustrated in Fig. 4.17, shows the general flow of processes and equipment in the thermochemical production of mixed alcohols. The process is based on the same process reported by Dutta et al. (2011), with modifications to the operation in the Brazilian context (namely adjusting conversions to sugarcane bagasse and straw as feedstock and considering a plant with energy self-sufficiency).

The thermochemical route is based on the conversion of the lignocellulosic fraction of sugarcane (bagasse and straw) into end products through the incorporation of reactions performed under high temperatures and high pressures. As shown in Fig. 4.17, a considerable number of steps are needed for the process to be carried out.

Initially, since sugarcane lignocellulosic material (LCM) presents high moisture content (above 45 %), the feedstock must be dried to 10 % moisture to improve its conversion in the gasifier (Dutta et al. 2011). Subsequently, the LCM is converted to synthesis gas (syngas) with an indirect steam gasifier, which operates with

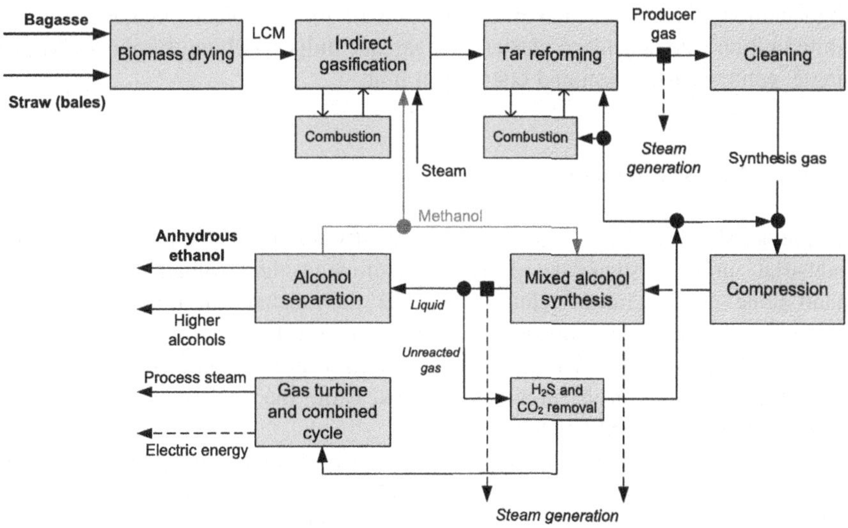

Fig. 4.17 Block flow diagram of the thermochemical conversion of sugarcane lignocellulosic material (LCM) to mixed alcohols

reactions of gasification and combustion occurring in separate vessels. Heat for the gasifier is transferred from the combustor with the flow of synthetic olivine sand (calcined magnesium silicate) (Dutta et al. 2011). The producer gas, i.e., syngas with unadjusted H_2/CO molar ratio, is sent to tar reforming along with other recycle streams to convert tar and char into gases that synthesis reactor is capable of processing. Afterward, the gas stream is cooled down and cleaned in a gas scrubber operating with water to remove particulates, residual tars, ammonia and halides, thus being suitable to be compressed over 200 bar. The pressurized syngas is then mixed with recycle streams and sent to the mixed alcohol synthesis reactor, which operates at high pressure and converts CO, H_2 and methanol (produced in this reactor, separated in the purification area and recycled to the synthesis reactor inlet) into ethanol, propanol and other higher alcohols. After the reaction, the mixture of liquid products and unconverted gases is separated into two streams: The gas fraction passes through acid gas removal and is recycled to one of many possible points in the process (inlet of the synthesis reactor, tar reformer, tar combustor or gas turbine), while the liquid fraction is sent to purification. In this last step, three main products are obtained: hydrated higher alcohols (by-product), anhydrous ethanol (main product) and methanol, which is not commercialized but rather recycled to the inlet of the synthesis reactor, as mentioned earlier, to increase ethanol yields or undergo conversion in the tar reformer.

Besides the described backbone of the process, steam and electric energy generation are vital for the operation of the thermochemical unit and any other integrated plants. The production of steam and electric energy is not concentrated in a single section, alternatively being decentralized over several sections of the process. In an integrated 1G2G thermochemical biorefinery, the consideration of consuming the totality of the LCM in the gasifier may imply in the removal of the cogeneration (CHP) heat and power section of the 1G plant. Another possible configuration is to consider a cluster of sugarcane mills supplying surplus LCM to a single, central gasification and conversion unit.

4.5.4.2 Process Simulation

As previously described, AspenPlus is the software used in the VSB to model industrial units of sugarcane processing, allowing the evaluation of different stand-alone or integrated technologies. The thermochemical route for ethanol production from syngas obtained through sugarcane LCM gasification was also inserted in the simulator. The methodology used to insert sugarcane processing in AspenPlus is shown in the previous sections, while the modeling of the thermochemical unit is detailed below.

For computer simulation of bioprocesses, several studies in the literature consider biomass as a nonconventional stream taking into account their proximate and ultimate analyses, which is also the case in the modeling of sugarcane LCM used in the thermochemical unit in the VSB. Scenarios of 1G and 2G ethanol production by biochemical routes simulated in the VSB do not directly employ elemental analysis

Table 4.13 Sugarcane bagasse composition from ultimate and proximate analyses (VSB database)

Sugarcane bagasse and straw composition	
Ultimate analysis (wt%, dry basis)	
C	46.69
H	5.72
O	44.03
N	0.32
S	0.05
Cl	0.04
Proximate analysis (wt%, dry basis)	
Fixed carbon	16.79
Volatile matter	80.06
Ash	3.15
Moisture	10.00

in LCM (bagasse + straw) composition, as shown earlier in Tables 4.3 and 4.4, rather relying on the mass fraction of identified compounds.

Biomass and ash were inserted as nonconventional (NC) components, and the global stream class is MIXCINC (with substreams MIXED, CISOLID and NC), commonly used when both conventional and nonconventional solid are present, but no particle size distribution (PSD) is available. In this way, the combined conventional sugarcane bagasse and straw composition considered at the inlet of the gasifier, defined by both ultimate and proximate analyses, is shown in Table 4.13.

Three main property methods were used: PR-BM (Peng–Robinson equation of state with Boston–Mathias alpha function) is the method of choice for calculations involving gases; NRTL (nonrandom two-liquid) model is employed in equipment handling mixtures of alcohols and low quantities of gases; and STEAMNBS model is used in calculations involving pure water (steam generation and cooling). Different Henry component (HC) groups were defined and used in specific equipment to ensure better modeling of gas removal.

Char is considered to be pure carbon (C), while tar formed in the gasifier is represented by naphthalene ($C_{10}H_8$).

The operations in each section of the thermochemical unit are detailed in the next pages.

Biomass Drying

The residual heat from flue gases obtained in the combustors of both gasification and tar reforming sections is used to heat air, which in turn exchanges heat with sugarcane bagasse and straw. Flue gas temperature falls from 440 to 130 °C, while the LCM is heated to over 100 °C. A flash vessel performs the separation of water and biomass. The temperature of the entering flue gas is calculated in order to achieve biomass with 10 % moisture in the bottom product of the flash vessel.

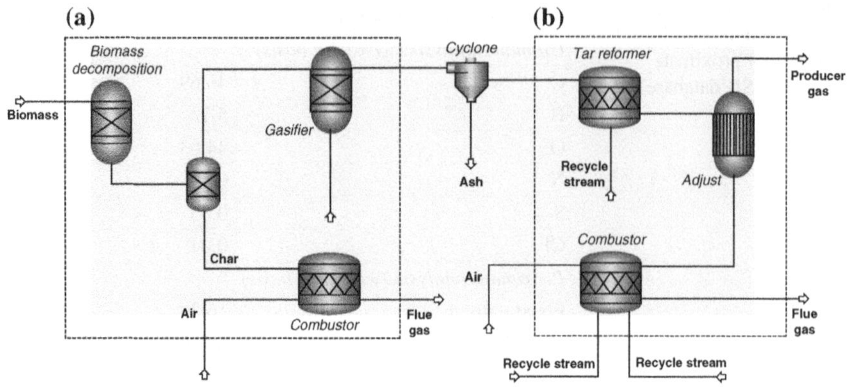

Fig. 4.18 AspenPlus diagram of the gasification (**a**) and tar reforming (**b**) sections of the thermochemical unit

Biomass Gasification

Sugarcane bagasse and straw enter the gasifier at 80 °C and with moisture content of 10 %. The indirect steam gasifier, represented by three interconnected reactors and one separator, is shown in Fig. 4.18.

The operation of the indirect steam gasifier is described as follows:

- Initially, a reactor is used to simulate the decomposition of biomass—a NC component—into conventional and other NC components. The proximate and ultimate analyses of the LCM are used in the calculation of the fractions of the components (H_2, O_2, N_2, Cl_2, S, H_2O, C and ash) leaving the reactor.
- A separator block divides the amount of C going to the gasifier and to the char combustor. The amount of C diverted to the combustor is calculated as to close the energy balance of the gasification section, i.e., the heat given off by both the combustion and gasification equals the heat consumed by the decomposition reactor.
- The gasifier employs predetermined component yields to give producer gas at a defined composition. For sugarcane LCM gasification with steam, the gas composition falls in the range presented in Table 4.14, according to Hofbauer et al. (2007). Steam enters the gasifier at a ratio of 0.4 kg per kg of bone dry biomass, along with part of the methanol produced during alcohol purification.

Table 4.14 Composition of gas from gasifier

Component	Mole fraction (%)
H_2	35–40
CO	25–30
CO_2	20–25
CH_4	9–11

Table 4.15 Operating parameters of gasification section

Gasification section		
	Gasifier	Steam input
Temperature (°C)	869	310
Pressure (bar)	2.3	2.4
	Combustor	Air input
Temperature (°C)	987	200
Pressure (bar)	1.0	1.0

- The char combustor, represented by a stoichiometric reactor, burns char (C) produced in the gasifier with 10 % excess air. The air stream is considered 21 % of O_2 and 79 % of N_2 in molar fraction and enters the reactor at 200 °C.

Table 4.15 presents the operating parameters of the gasification section.

Tar Reforming

After ash removal in a separator acting as a cyclone, the producer gas enters the reforming section in order to increase its H_2/CO ratio through conversion of tar and other compounds. In a similar manner to the gasification section, the reformer is modeled as an array of three reactors (shown in Fig. 4.18).

- The reformer itself employs reforming reactions with predetermined conversions given by Dutta et al. (2011). Part of the unconverted syngas after alcohol synthesis reaction also enters the reformer to increase the output of syngas for compression. The considered reactions are presented in Eqs. 4.25–4.31.

$$CH_4 + H_2O \rightarrow CO + 3H_2 \tag{4.25}$$

$$C_2H_4 + 2H_2O \rightarrow 2CO + 4H_2 \tag{4.26}$$

$$2NH_3 \rightarrow N_2 + 3H_2 \tag{4.27}$$

$$\text{Tar} + 10H_2O \rightarrow 10CO + 14H_2 \tag{4.28}$$

$$CH_4 \rightarrow \text{Char} + 2H_2 \tag{4.29}$$

$$C_2H_6 + 2H_2O \rightarrow 2CO + 5H_2 \tag{4.30}$$

$$CH_3OH + H_2O \rightarrow CO_2 + 3H_2 \tag{4.31}$$

- An adjustment reactor following the reformer calculates water-gas-shift equilibrium in order to adjust CO_2, CO, H_2 and H_2O fractions in the syngas.

Table 4.16 Operating parameters of the tar reforming section

Tar reforming section		
	Tar reformer	
Temperature (°C)	910	
Pressure (bar)	1.8	
	Combustor	Air input
Temperature (°C)	987	200
Pressure (bar)	1.0	1.0
	Adjustment reactor	
Temperature (°C)	910	
Pressure (bar)	1.8	

- The nongaseous fraction exiting the adjustment reactor, mainly consisting of unconverted char and tar, enters a combustor in which it is burned alongside other recycled unconverted syngas streams with 10 % excess air. The energy balance of the three reactors is closed by adjusting the amount of unconverted syngas burnt in the combustor.

Table 4.16 shows the operating parameters of the tar reforming section.

Syngas Cleaning and Compression

Prior to its cleaning step, syngas is cooled down to around 54 °C. A simple arrangement of a flash vessel separating gas and liquid fractions combined with a recycle loop for the liquid stream simulates the operation of a gas scrubber. Part of the contaminants (NH_3, HCl and residual tar) is removed by this equipment.

Afterward, the syngas enters a six-stage centrifugal compressor at 1.6 bar, leaving it at 207.8 bar. The compressor is simulated as six separate sections of compressor/heat exchanger/flash vessel. All compressors work with identical pressure ratios, and the intermediate heat exchangers cool down the syngas to 43 °C.

Mixed Alcohol Synthesis, Acid Gas Removal and Unconverted Gas Recycle

The clean, pressurized syngas is mixed with other recycle streams from the reaction section and sent to the mixed alcohol synthesis reactor, which operates with predetermined reactions (Eqs. 4.32–4.38) and conversions. Methanol recycled from the alcohol purification section to the reactor plays an important part in the achievable ethanol yield of the reactor. Water–gas-shift equilibrium is calculated in a reactor downstream of the synthesis reactor, in a similar manner as described in the reforming section.

$$CO + H_2O \rightarrow H_2 + CO_2 \tag{4.32}$$

$$CO + 2H_2 + CH_3OH \rightarrow C_2H_5OH + H_2O \tag{4.33}$$

$$CO + 2H_2 \rightarrow CH_3OH \tag{4.34}$$

$$CH_3OH + H_2 \rightarrow CH_4 + H_2O \tag{4.35}$$

$$CO + 2H_2 + C_2H_5OH \rightarrow C_3H_7OH + H_2O \tag{4.36}$$

$$CO + 2H_2 + C_3H_7OH \rightarrow C_4H_9OH + H_2O \tag{4.37}$$

$$CO + 2H_2 + C_4H_9OH \rightarrow C_5H_{11}OH + H_2O \tag{4.38}$$

After exchanging heat with the reactor inlet stream, the mixture of unconverted syngas and liquid products is primarily separated using a flash vessel: The gas fraction undergoes acid gas removal, and the liquid fraction is sent to the alcohol purification section.

Removal of H_2S in the form of elemental sulfur (S), of CO_2 and of NH_3 is mainly performed with separators, flash vessels and stoichiometric reactors. Initially, an absorber with dimethyl ether of polyethylene glycol (DEPG) removes nearly all H_2S and part of the CO_2 from the gas stream, recycled to the synthesis reactor and to the tar reformer/fuel combustor. The off-gas from the absorber is sent to the acid gas enrichment section, which utilizes a selective amine to increase H_2S concentration in the acid gas for downstream processes. The gas is then treated with methanol to recover H_2S and recycle it in controlled amounts to the alcohol synthesis reactor. Finally, the acid gas from the methanol absorption system is sent to a sulfur recovery unit.

The unconverted syngas with low H_2S and CO_2 content at different pressures is either recycled to the inlet of the alcohol synthesis reactor, recycled to the tar reformer/combustor or recycled to the gas turbine for electric energy and steam generation in a combined cycle.

Alcohol Purification

Initially, the mixture of alcohols and gases entering the alcohol purification section passes through a flash vessel for additional gas removal, which is sent to the tar reformer. Afterward, the liquid fraction undergoes water removal in a molecular sieve. The crude alcohol is then fractionated in two sequential distillation columns with partial condensers. The first column removes water, propanol and higher alcohols in the bottom stream. The top product containing mainly methanol and ethanol is separated in the second distillation column: Anhydrous ethanol is recovered in the bottom stream, while methanol from the top stream is recycled to various parts of the thermochemical plant. Table 4.17 contains the main parameters of the alcohol purification section.

Table 4.17 Main parameters of alcohol purification section

Molecular sieve	Value
Steam pressure	6 bar
Outlet water content	0.5 wt%
Distillation columns	Value
Higher alcohols purity	80 %
Anhydrous ethanol purity	99.5 wt%

Steam and Electricity Generation

Steam and electricity generation in the thermochemical unit is decentralized, unlike a conventional 1G plant where the CHP section is responsible for the production of utilities.

Several streams with high heating capacity are cooled down both to reheat recycle streams and to generate steam at different pressure levels. Four main steam production points can be identified in the process flow diagram:

- Cooling of producer gas after tar reforming and prior to cleaning: generation of 22-bar steam, which is further expanded to 6 bar and 4 bar in backpressure turbines to provide process steam for the molecular sieves and acid gas removal equipment, respectively, and electric energy.
- Cooling of gas–liquid mixture after alcohol synthesis reaction: generation of 2.5-bar steam, directly used in distillation columns.
- Heat removal from alcohol synthesis reactors: Due to the highly exothermic character of the reactions and to maintain temperatures around 322 °C, water is fed to the shell side of the reactor and saturated 90-bar steam at around 300 °C is produced. This steam is superheated to 530 °C through heat exchange with flue gases from the combustors and then expanded to 2.5 bar in a backpressure turbine to yield process steam and electric energy.
- Combined cycle: A gas turbine, using unconverted syngas as fuel, operates at 1370 °C and emits flue gas at 735 °C. This flue gas exchanges heat with water to produce steam at two pressure levels: 22 bar and 2.5 bar.

Ultimately, all backpressure turbines coupled with electric generators produce work in the form of electric energy after expanding 90-bar and 22-bar steam to 6 bar, 4 bar and 2.5 bar. Another significant portion of the electric energy generated in the plant comes from the gas turbine burning unconverted syngas and the turboexpander placed in the recycle loop of the mixed alcohol synthesis section.

Heat Integration

A vast array of heat exchangers is used to perform heat integration in the plant. Flue gases exiting the combustors have high mass flow rate and temperature (987 °C), which allow them to supply heat to several equipment and streams. For example, streams entering the reformer are heated to temperatures of 927 °C by cooling flue gases. After these exchanges and other operations, residual heat contained in the flue gases is used to dry bagasse and straw from around 45 % moisture to 10 % prior to gasification.

Examples of simulations of second-generation ethanol production via thermochemical route based on the process design and parameters discussed in this chapter, and their results are presented in Chap. 7.

4.6 Industrial Validation Process

The main goal of the validation process of the Virtual Sugarcane Biorefinery is to guarantee that the methodology used for the construction of the AspenPlus simulations is able to correctly represent different process configurations currently found in ethanol-producing facilities. As second-generation ethanol plants are still in the development stage, this process is carried out for first-generation plants only.

For data collection, partnerships were established between CTBE and 2 sugar mills located in the Center–South region. These facilities will be referenced as Mill A and Mill B in the following text.

The selection of the mills took into consideration various aspects such as crushing capacity, steam demand of the plant and the presence of different technologies such as vinasse concentration and reduced water withdrawal.

Mill A

This mill has a crushing capacity of about 4,000,000 t of sugarcane per season, producing raw crystal sugar (VHP and VVHP, using a three-massecuite system), anhydrous and hydrated ethanol and electricity. Ethanol is produced from molasses and secondary juice and sugar exclusively from primary juice. Bagasse is used for steam production, supplying the plant with energy (electricity and steam). The surplus of produced electricity is exported to the grid.

Mill A performs two steam extractions at the multiple-effect evaporators applied for juice concentration. The vapor extracted from the first (V1) and second (V2) evaporators is applied as a heating utility on equipment from the sugar house and the distillery.

As for water collection, the value presented by this mill represents the average value for mill in the Center-South region of Brazil.

Mill B

The crushing capacity of this mill is approximately 2,000,000 t of sugarcane per season, producing raw crystal sugar (using a two-massecuite system), hydrated ethanol and electricity. Ethanol is mainly produced from molasses; about 15 % of the clarified juice is used for ethanol production, and the remainder is destined to the sugar factory.

Like Mill A, this facility performs two steam extractions at the multiple-effect evaporators (V1 and V2). The steam obtained in this operation is applied as a heating utility in various equipments found at the plant.

One of the most remarkable features of Mill B is that this mill was designed to operate with reduced water collection. Mill B uses an evaporative condenser for condensing steam obtained at the last effect of the multiple-effect evaporators and has a system for vinasse concentration; this can be cited as two of the strategies applied to minimize the water collection in this facility.

Methodology Employed for the Construction of the Simulations

The data were provided by the mills in the form of bulletins. Additional data were collected through technical visits and questionnaires, when available.

The main steps of the validation process are described below:

- Collection of information about the plant such as layout, operational conditions and equipment (collected in technical visits);
- Compilation of information about the process through bulletins and files provided by mill;
- Construction on Excel of spreadsheets for calculation of preliminary mass and energy balances;
- Construction of the simulation process flow diagram;
- Interactive adjustment of calculations to obtain rigorous agreement between the values of the process and those from the Excel spreadsheet;
- Introduction of the main values from the Excel spreadsheet in the AspenPlus simulator;
- Adjustment of the AspenPlus simulations to represent accurately the inventory of the plant;
- Calculation of the mass and energy balances using AspenPlus;
- Comparison and interpretation of the results.

Results of the Validation Procedure for First-Generation

In this section, the main results from the process of validation previously described are shown. August was chosen as the month with the most representative data of the average operational conditions for both mills. This month is usually the best one for sugarcane processing, due to characteristics related to amount of rain and sugarcane quality. The months from the beginning of the harvest season (March or April), for example, usually display nonrepresentative data because the operating conditions are still being established.

Tables 4.18 and 4.19 show the amount, on mass basis, of processed sugarcane and the TRS (total reducing sugars) mass processed for sugar and ethanol production for Mills A and B. In Tables 4.20 and 4.21, data regarding the production of sugar and ethanol are presented. In all tables, the last column has information about the average value for the season (percentages) or the total value accumulated for the season (kg). The season period considered for these cases goes from April (or May for Mill A) to August of 2012.

Table 4.18 Sugarcane processed in August and accumulated in season—data from Mill A's bulletin

Destination	Monthly (kt)	Average—monthly (%)	Accumulated—season (kt)	Average—season (%)
Crushed for sugar production	435.53	63.1	1,176.34	59.35
Crushed for ethanol production	254.97	36.9	805.69	40.65
Total crushed	690.50	100	1,982.03	100
Total TRS processed	104.33	100	299.49	100

Table 4.19 Sugarcane processed in August and accumulated in season—data from Mill B's bulletin

Destination	Monthly (kt)	Average—monthly (%)	Accumulated—season (kt)	Average—season (%)
Crushed for sugar production	178.78	66.73	585.01	56.00
Crushed for ethanol production	89.16	33.27	459.81	44.00
Total crushed	267.94	100	1,044.82	100
Total TRS processed	39.84	100	134.47	100

Table 4.20 Sugar and ethanol produced in August and accumulated in season—data from Mill A's bulletin

Destination	Monthly	Accumulated—season
Ethanol 100 % (kL)	29,166.42	82,681.59
Sugar 100 % (t)	45,726.05	114,284.20

Table 4.21 Sugar and ethanol produced in August and accumulated in season—data from Mill B's bulletin

Destination	Monthly	Accumulated—season
Ethanol 100 % (kL)	6,603.39	30,420.79
Sugar 100 % (t)	21,963.05	63,409.50

Using information provided on bulletins and supplementary data collected from the technical visits, it was possible to calculate the general mass balance for each mill. Those calculations were initially performed on an Excel spreadsheet, and later, these results were used as inputs to run the simulation in AspenPlus. In all cases, when no information was available and necessary for the conclusion of the simulation and balances, data from CTBE database—previously found in the literature or estimated by experts—were used. These values have been corrected and analyzed repeatedly, until the calculated values (generated in the simulation) and the compiled values (from the bulletin or process data) were as similar as possible.

Mill A Validation

The next tables provide some examples of the comparisons between the results obtained from the simulation and the data found in the bulletins for Mill A. Table 4.22 presents information about the juice treatment section and Table 4.23 results of the cogeneration section.

In Table 4.23, it is possible to see that the total energy produced that is estimated by the simulation is more than twice higher than the value presented in Mill A's bulletin. One reason assumed for this deviation between the two values is that Mill A had recently changed its cogeneration system during the 2012 harvest, and

Table 4.22 Comparison between the results of brix and pol, from the juice treatment sector, with bulletin data—Mill A

Stream	Brix			Pol		
	Bulletin	Aspen	Deviation (%)	Bulletin	Aspen	Deviation (%)
Clarified juice for sugar	0.1653	0.1653	0.00	0.1462	0.1511	3.32
Clarified juice for alcohol	0.1105	0.1105	0.00	0.0955	0.1010	5.76
Slurry	0.1164	0.1159	−0.42	0.0968	0.0942	−2.73
Filtered juice	0.0789	0.0788	−0.18	0.0648	0.0698	7.77
Filter cake	–	–	–	0.0246	0.0246	−0.11

Table 4.23 Comparison between the results from the cogeneration sector, with bulletin data—Mill A

Stream	Unit	Bulletin	Aspen	Deviation (%)
Energy produced (month)	MW	18,972.80	45,344.50	139.00
Production of 65-bar steam	kg_{steam} $kg^{-1}{}_{bagasse}$	2.229	2.282	2.36
Deaerator steam	kg h^{-1}	5,000	5,000	0.00

therefore, the new operational conditions were still being adjusted. More recent bulletins from the 2013/2014 harvest show similar values for total energy produced as those estimated by the simulation.

The total amount of steam consumed by Mill A is 374.22 kg t^{-1} of sugarcane. The total energy consumed in Mill A reaches 1.38 GJ t^{-1} of sugarcane.

Data related to alcohol production are compared with the results from the simulation in Table 4.24. The ethanol content of the alcohols obtained from the distillation columns in the simulation (INPM) was very similar to that from the data provided by the Mill A. The production of ethanol calculated by the simulation is higher than the production informed on the bulletins; one possible explanation for this is that some losses of the real process might not have been reported by the mill.

Table 4.24 Comparison of the alcohol production data from the simulation and mill's bulletin—Mill A

Stream	Unit	Bulletin	Aspen	Deviation (%)
Hydrated ethanol	m^3 h^{-1}	10.78	10.81	0.24
Anhydrous ethanol	m^3 h^{-1}	28.98	30.79	6.24
Absolute alcohol	m^3 h^{-1}	39.20	41.39	5.58
INPM of hydrated ethanol	°C	93.39	92.70	−0.74
INPM of anhydrous ethanol	°C	99.53	99.53	0.00

Table 4.25 Comparison of the results related to the sugar factory, with bulletin data—Mill A

Stream	Unit	Bulletin	Aspen	Deviation (%)
Sugar	kg h^{-1}	61,618.55	62,844.47	1.99
Sugar 100 %	kg h^{-1}	61,500.54	62,380.25	1.43

The main results of the sugar factory were also compared with data found in bulletins. The results obtained from this comparison show that the sugar production estimated by the simulation is similar to the one reported by the mill, as shown in Table 4.25.

Lastly, the main yields of Mill A are presented in Table 4.26. These values indicate that the results from the simulation are close to the ones informed by the mill. Therefore, these results can be considered satisfactory. Ethanol production for the simulation is higher than the value reported by the mill, possibly because of unreported losses, as previously mentioned in this text.

Mill B Validation

Next tables present some of the results for the simulation constructed with data from Mill B. Comparisons between these results and the data are also shown in these tables. Table 4.27 presents the results for the juice concentration sector and Table 4.28 for the cogeneration sector.

Mill B simulation was implemented using previously constructed simulations as basis, modifying them to match the layout of equipment informed by the mill. Data from the VSB database were used in cases where the mill did not provide the necessary information. Reported data for some sectors of the plant were scarce, especially for the juice treatment sector.

Table 4.27 presents briefly the results from the juice concentration sector of the plant. The comparison shows that for this sector, the results were satisfactory.

Table 4.26 Comparison of yields calculated from the results of the simulation made on AspenPlus, with bulletin data—Mill A

Yield on TRS base	Bulletin (%)	Aspen (%)
Total	89.98	92.04
Sugar factory	92.30	94.11
Alcohol factory	86.03	88.93

Table 4.27 Comparison between the results of brix and pol, from the juice concentration sector, with bulletin data—Mill B

Stream	Brix		
	Bulletin	Aspen	Deviation (%)
Syrup pol	0.5011	0.5013	0.03
Syrup brix	0.5841	0.5840	−0.02

Table 4.28 Comparison between the results of the simulation for the cogeneration sector with bulletin data—Mill B

Stream	Unit	Bulletin	Aspen	Deviation (%)
Energy produced	kW	21,588.02	45,841.10	−112.35
Steam produced	kg	207,140.00	215,844.20	−4.20
Energy production per tonne of sugarcane processed	kW TC^{-1}	59.94	127.29	−112.36

The higher discrepancy observed from the comparison between the simulation results and bulletin data is for the total energy produced in the cogeneration sector, as shown in Table 4.28. Actual values found in the literature (Oliverio et al. 2010) for similar operational conditions as those reported by Mill B (new boilers operating at a pressure of 65 bar, electromechanical drives for the mills) show an electricity production of about 117.40 kW per tonne of sugarcane crushed. This value is closer to the value obtained by the simulation, of 127 kW per tonne of sugarcane processed, and very different from the value reported by the mill (59.94 kW per tonne of sugarcane). It was informed by Mill B that during the 2012/2013 harvest, the cogeneration sector ceased operating several times, due to operational problems.

The main results for the distillation sector of the plant are displayed in Table 4.29. Comparison between the data from Mill B's bulletin and the simulation results shows that the production of ethanol and fusel oil presents similar values. The ethanol content (INPM) of the main product of the distillation sector, hydrated ethanol, is also coherent with the value observed in the real plant.

Main results related to the sugar factory were also compared with data found in bulletins, and the results of this comparison are shown in Table 4.30. Sugar production matches the values reported in bulletins.

A simplified water mass balance was also estimated for Mill B using the entrances and outlets of water found in the simulation. The purpose of this calculation was to determine the water collection of the mill. This value was found to be of about 0.59 m^3 of water per tonne of sugarcane processed. Mill B reported a value of 0.53 m^3 of water per tonne of sugarcane (11 % lower than the value estimated). Considering that no information concerning water treatment or water reuse was provided by the mill, this estimation seems acceptable. Average water collection for Mill B between April and October of 2012 was 0.55 m^3 of water per tonne of

Table 4.29 Comparison of the distillation product data from the simulation and mill's bulletin—Mill B

Stream	Unit	Bulletin	Aspen	Deviation (%)
Hydrated ethanol	L h^{-1}	9,287.90	9,289.00	0.01
INPM of hydrated ethanol	°C	93.10	93.21	0.12
Fusel oil	L h^{-1}	37.23	37.73	1.34

Table 4.30 Comparison of the results related to the sugar house, with bulletin data—Mill B

Stream	Unit	Bulletin	Aspen	Deviation (%)
Sugar	kg h^{-1}	29,686.49	29,668.03	−0.06
Sugar 100 %	kg h^{-1}	29,520.23	29,602.93	0.28

Table 4.31 Comparison of yields calculated from the results of the simulation made on AspenPlus, with bulletin data—Mill B

Yield on TRS base	Bulletin (%)	Aspen (%)
Industrial efficiency—TRS	83.48	83.46
Alcohol factory	86.62	85.43

sugarcane processed and between April and October of 2013, 0.62 m^3 of water per tonne of sugarcane.

The main yields of Mill B are presented in Table 4.31. The results show that the yields estimated are very close to those reported in bulletins. These results can be considered satisfactory.

Final Remarks

Calculations of the mass and energy balances of Mills A and B presented some difficulties, since the bulletins provided by the mills did not contain all information and parameters required for its completion. The parameters that were not found in bulletins had to be estimated, based on data from the database (VSB) and assessed by CTBE specialists. The simulations were based on existing ones in the VSB, which were modified to represent the process diagram of the chosen mills.

Differences between the results of the simulations and the data from the bulletins were expected because of several assumptions made and various parameters estimated. These discrepancies were of small magnitude, with the exception of energy production for both mills. As previously discussed in this text, both mills presented issues on the operation of the cogeneration sector during the time of the collection of the data for the bulletins.

VSB is an important tool for evaluating new routes and technologies and their possible prospects. The industrial validation work is of great relevance to evaluate the capacity of this tool to correctly represent existing facilities and thus used to assess routes that are not yet implemented on industrial scale. The results obtained from the industrial validation indicate that VSB is mature enough and can be successfully used to assess new routes and technologies, given that information concerning the processes that comprise these routes is available (experimental data from laboratory or pilot plant).

References

Aden A, Ruth M, Ibsen K et al (2002) Lignocellulosic biomass to ethanol process design and economics utilizing co-current dilute acid prehydrolysis and enzymatic hydrolysis for corn stover. National Renewable Energy Laboratory—NREL. http://www.nrel.gov/docs/fy02osti/32438.pdf. Accessed May 2015

Agger JW, Isaksen T, Várnai A et al (2014) Discovery of LPMO activity on hemicelluloses shows the importance of oxidative processes in plant cell wall degradation. Proc Natl Acad Sci USA 111(17):6287–6292. doi:10.1073/pnas.1323629111

Aita GA, Salvi DA, Walker MS (2011) Enzyme hydrolysis and ethanol fermentation of dilute ammonia pretreated energy cane. Bioresour Technol 102(6):4444–4448. doi:10.1016/j.biortech.2010.12.095

Alvira P, Tomás-Pejó E, Ballesteros M et al (2010) Pretreatment technologies for an efficient bioethanol production process based on enzymatic hydrolysis: a review. Bioresour Technol 101:4851–4861. doi:10.1016/j.biortech.2009.11.093

Andersson J, Lundgren J (2014) Techno-economic analysis of ammonia production via integrated biomass gasification. Appl Energ 130(1):484–490. doi:10.1016/j.apenergy.2014.02.029

Andrade RR, Rabelo SC, Maugeri Filho F et al (2013) Evaluation of the alcoholic fermentation kinetics of enzymatic hydrolysates from sugarcane bagasse (*Saccharum officinarum* L.). J Chem Technol Biotechnol 88(6):1049–1057. doi:10.1002/jctb.3937

Andrietta SR (1994) Modelagem, simulação e controle de fermentação alcoólica contínua em escala industrial. Ph.D. Thesis, Universidade Estadual de Campinas—UNICAMP

Aspentech (2009) Aspen physical property system—physical property methods and models.

Baeyens J, Kang Q, Appels L et al (2015) Challenges and opportunities in improving the production of bio-ethanol. Prog Energ Combust 47:60–88. doi:10.1016/j.pecs.2014.10.003

Bals BD, Wedding C, Balan V et al (2011) Evaluating the impact of ammonia fiber expansion (AFEX) pretreatment conditions on the cost of ethanol production. Bioresour Technol 102:1277–1283. doi:10.1016/j.biortech.2010.08.058

Barrett WM, Jun Y (2005) Development of a chemical process modeling environment based on CAPE-OPEN interface standards and the Microsoft.NET framework. Comput Chem Eng 30 (2):191–201. doi:10.1016/j.compchemeng.2005.08.017

Basu P (2010) Biomass gasification and pyrolysis: practical design and theory. Elsevier Inc, Amsterdam

Basu P (2013) Biomass gasification, pyrolysis and torrefaction: practical design and theory. Elsevier Inc., USA

Bazico (2010) Açúcar. Bazico Tecnologia Comercial. http://www.bazico.com.br/produto/com_acucar.htm. Accessed May 2011

Behera S, Arora R, Nandhagopal N et al (2014) Importance of chemical pretreatment for bioconversion of lignocellulosic biomass. Renew Sust Energ Rev 36:91–106. doi:10.1016/j.rser.2014.04.047

Bergman PCA, Kiel JHA (2005) Torrefaction for biomass upgrading. In: 14th European biomass conference and exhibition, Paris, 17–21 Oct 2005

BNDES, CGEE (2008) Sugarcane-based bioethanol: energy for sustainable development. Coordination Banco Nacional de Desenvolvimento—BNDES and Centro de Gestão e Estudos Estratégicos—CGEE, Rio de Janeiro

Bozell JJ, Petersen GR (2010) Technology development for the production of biobased products from biorefinery carbohydrates—the US Department of Energy's "Top 10" revisited. Green Chem 12(4):539–554. doi:10.1039/b922014c

Braskem (2015) I'm Green. http://www.braskem.com.br/site.aspx/plastic-green. Accessed Apr 2015

Bridgwater AV (2012a) Review of fast pyrolysis of biomass and product upgrading. Biomass Bioenerg 38:68–94. doi:10.1016/j.biombioe.2011.01.048

Bridgwater AV (2012b) Upgrading biomass fast pyrolysis liquids. Environ Prog Sustain Energy 31(2):261–268. doi:10.1002/ep.11635

Brown RC (2006) Biomass refineries based on hybrid thermochemical-biological processing—an overview. In: Kamm B, Gruber PR, Kamm M (eds) Biorefineries—industrial processes and products, vol 1. Wiley-VCH, Weinheim, pp 227–252

Camargo CA (1990) Manual de recomendações para conservação de energia na indústria do álcool e açúcar. Instituto de Pesquisas Tecnológicas—IPT, São Paulo

Cardona CA, Sánchez O (2007) Fuel ethanol production: process design trends and integration opportunities. Bioresour Technol 98:2415–2456. doi:10.1016/j.biortech.2007.01.002

Cardoso TF (2014) Avaliação socioeconômica e ambiental de sistemas de recolhimento e uso da palha de cana-de-açúcar. Ph.D. Thesis, Universidade Estadual de Campinas—UNICAMP

Cardoso TDF, Cavalett O, Chagas MF et al (2013) Technical and economic assessment of trash recovery in the sugarcane bioenergy. Sci Agric 70(5):353–360

Caspeta L, Caro-Bermúdez MA, Ponce-Noyola T et al (2014) Enzymatic hydrolysis at high-solids loadings for the conversion of agave bagasse to fuel ethanol. Appl Energ 113:277–286. doi:10.1016/j.apenergy.2013.07.036

CGEE (2009) Bioetanol combustível: uma oportunidade para o Brasil. Centro de Gestão e Estudos Estratégicos—CGEE, Brasília

Chandel AK, Junqueira TL, Morais ER et al (2014) Techno-economic analysis of second-generation ethanol in Brazil: competitive, complementary aspects with first-generation ethanol. In: Silva SS, Chandel AK (eds) Biofuels in Brazil: fundamental aspects, recent developments, and future perspectives. Springer Science & Business Media, London, pp 1–29

Chávez-Rodríguez M (2010) Uso de água na produção de etanol de cana-de-açúcar. Dissertation, Universidade Estadual de Campinas—UNICAMP

Chouhan APS, Sarma AK (2011) Modern heterogeneous catalysts for biodiesel production: a comprehensive review. Renew Sustain Energ Rev 15(9):4378–4399. doi:10.1016/j.rser.2011.07.112

Chen JCP, Chou CC (1993) Cane sugar handbook: a manual for cane sugar manufacturers and their chemists. Wiley, New York

Cherubini F, Jungmeier G, Wellisch M et al (2009) Toward a common classification approach for biorefinery systems. Biofuel Bioprod Biorefin 3(5):534–546. doi:10.1002/bbb.172

Choi S, Woo C, Ho J et al (2015) Biorefineries for the production of top building block chemicals and their derivatives. Metab Eng 28:223–239. doi:10.1016/j.ymben.2014.12.007

CO-LAN (2010) COLAN—the CAPE-OPEN laboratories network—2010 Brochure. http://www.colan.org/Communication/Brochure2010.pdf. Accessed Jun 2015

Colby College (2012) Computational chemistry lab. http://www.colby.edu/chemistry/webmo/mointro.html. Accessed May 2015

Compton DL, Jackson MA, Mihalcik DJ et al (2011) Catalytic pyrolysis of oak via pyroprobe and bench scale, packed bed pyrolysis reactors. J Anal Appl Pyrol 90:174–181. doi:10.1016/j.jaap.2010.12.003

COPERSUCAR (1989) Curso Básico sobre Tratamento do Caldo. Centro de Tecnologia Copersucar, Piracicaba

Cortez LAB, Lora EES, Gómez EO (eds) (2008) Biomassa para energia. Editora Unicamp, Campinas

Cremonez PA, Feroldi M, Araújo AV et al (2015) Biofuels in Brazilian aviation: current scenario and prospects. Renew Sust Energ Rev 43:1063–1072. doi:10.1016/j.rser.2014.11.097

CTBE (2012) The virtual sugarcane biorefinery (VSB): 2011 Report. National Laboratory of Science and Technology of Bioethanol (CTBE). http://ctbe.cnpem.br/pesquisa/avaliacao-integrada-biorrefinarias. Accessed Mar 2015

Cutz L, Santana D (2014) Techno-economic analysis of integrating sweet sorghum into sugar mills: the Central American case. Biomass Bioenerg 68:195–214. doi:10.1016/j.biombioe.2014.06.011

Dale BE, Bals BD, Kim S et al (2010) Biofuels done right: land efficient animal feeds enable large environmental and energy benefits. Environ Sci Technol 44(22):8385–8389. doi:10.1021/es1031618

de Deugd RM (2004) Fischer-Tropsch synthesis revisited; efficiency and selectivity benefits from imposing temporal and/or spatial structure in thereactor, Thesis, Delft University of Technology, Netherlands, Ponsen & Looijen B.V., Wageningen, 159p

de Jong E, Van RR, Van TR et al (2006) Biorefineries for the chemical industry—a Dutch point of view. In: Kamm B, Gruber PR, Kamm M (eds) Biorefineries—industrial processes and products, vol 1., Wiley-VCHWeinheim, Germany, pp 85–109

de Jong E, Higson A, Walsh P et al (2012) Bio-based Chemicals: value added products from biorefineries. In: IEA Bioenergy Task 42 Biorefinery. http://www.ieabioenergy.com/wp-content/uploads/2013/10/Task-42-Biobased-Chemicals-value-added-products-from-biorefineries.pdf. Acessed May 2015

de Vries SC, van de Ven GWJ, van Ittersum MK et al (2010) Resource use efficiency and environmental performance of nine major biofuel crops, processed by first-generation conversion techniques. Biomass Bioenerg 34(5):588–601. doi:10.1016/j.biombioe.2010.01.001

Delfini P (2012) Cana crua x Extração. In: 13° Seminário Brasileiro Agroindustrial: a usina em Números. Revista da Sociedade dos Técnicos Açucareiros e Alcooleiros do Brasil—STAB. Ribeirão Preto, 24–25 Oct 2012

Demirbas MF (2009) Biorefineries for biofuel upgrading: a critical review. Appl Energ 86:S151–S161. doi:10.1016/j.apenergy.2009.04.043

Dias MOS (2008) Simulação do processo de produção de etanol a partir do açucar e do bagaço, visando a integração do processo e a maximização da produção de energia e excedentes do bagaço. Dissertation, Universidade Estadual de Campinas—UNICAMP

Dias MOS, Cunha MP, Jesus CDF et al (2010) Simulation of ethanol production from sugarcane in Brazil: economic study of an autonomous distillery. In: European symposium on computer aided chemical engineering—ESCAPE 20, Ischia, June 2010. Computer Aided Chemical Engineering, vol 28. Elsevier B.V., Amsterdam, pp 733–738

Dias MOS, Junqueira TL, Cavalett O (2012a) Integrated versus stand-alone second generation ethanol production from sugarcane bagasse and trash. Bioresour Technol 103(1):152–161. doi:10.1016/j.biortech.2011.09.120

Dias MOS, Junqueira TL, Jesus CDF et al (2012b) Improving second generation ethanol production through optimization of first generation production process from sugarcane. Energy 43(1):246–252. doi:10.1016/j.energy.2012.04.034

Dias MOS, Junqueira TL, Rossell CEV et al (2013) Evaluation of process configurations for second generation integrated with first generation bioethanol production from sugarcane. Fuel Process Technol 109:84–89. doi:10.1016/j.fuproc.2012.09.041

Dias MOS, Maciel Filho R, Mantelatto PE et al (2015) Sugarcane processing for ethanol and sugar in Brazil. Environ Dev 15:35–51. doi:10.1016/j.envdev.2015.03.004

Dimian AC (2003) Integrated design and simulation of chemical processes. In: Gani R (ed) Computer-aided chemical engineering, vol 13. Elsevier Science B.V., Amsterdam

Durães FOM (2011) Sorgo sacarino: desenvolvimento de tecnologia agronômica. Agroenergia em Revista, 3rd edn. Empresa Brasileira de Pesquisa Agropecuária—EMBRAPA, Brasília

Dutta A, Phillips SD (2009) Thermochemical ethanol via direct gasification and mixed alcohol synthesis of lignocellulosic biomass. National Renewable Energy Laboratory—NREL. http://www.nrel.gov/biomass/pdfs/45913.pdf. Accessed Apr 2015

Dutta A, Talmadge M, Hensley J et al (2011) Process design and economics for conversion of lignocellulosic biomass to ethanol: thermochemical pathway by indirect gasification and mixed alcohol synthesis. National Renewable Energy Laboratory—NREL. http://www.nrel.gov/biomass/pdfs/51400.pdf. Accessed May 2015

Dutta K, Daverey A, Lin JG (2014) Evolution retrospective for alternative fuels: first to fourth generation. Renew Energ 69:114–122. doi:10.1016/j.renene.2014.02.044

DWSIM (2015) DWSIM Wiki. http://dwsim.inforside.com.br/wiki/index.php?title=Main_Page. Accessed Jun 2015

EIA (2012) International energy statistics. U.S. Energy Information Administration—EIA. http://www.eia.gov/cfapps/ipdbproject/IEDIndex3.cfm. Accessed May 2015

EMSO (2015) EMSO—environment for modeling, simulation, and optimization. http://vrtech.com.br/rps/emso.html. Accessed Jun 2015

Ensinas AV (2008) Integração termica e otimização termoeconomica aplicadas ao processo industrial de produção de açucar e etanol a partir da cana-de-açucar. Ph.D. thesis, Universidade Estadual de Campinas—UNICAMP

Ensinas AV, Nebra SA, Lozano MA et al (2007) Analysis of process steam demand reduction and electricity generation in sugar and ethanol production from sugarcane. Energ Convers Manage 48(11):2978–2987. doi:10.1016/j.enconman.2007.06.038

Ensinas AV, Modesto M, Nebra SA et al (2009) Reduction of irreversibility generation in sugar and ethanol production from sugarcane. Energy 34(5):680–688. doi:10.1016/j.energy.2008.06.001

Fermeglia M, Parenzan M (2007) Cape open interface revisited in term of class-based framework: an implementation in.NET. Chem Eng Trans 11:235–240

Fernando S, Adhikari S, Chandrapal C et al (2006) Biorefineries: current status, challenges, and future direction. Energ Fuel 20(3):1727–1737. doi:10.1021/ef060097w

Finguerut J, Paes, LAP (2014) Impurezas e qualidade de cana-de-açúcar—levantamento dos níveis de impurezas nas últimas safras. In: Workshop sobre a Limpeza da Cana Crua e Qualidade da Matéria-Prima. Stab, Piracicaba, SP

Galbe M, Zacchi G (2012) Pretreatment: the key to efficient utilization of lignocellulosic materials. Biomass Bioenerg 46:70–78. doi:10.1016/j.biombioe.2012.03.026

Garcia V (2008) Subproduto de destilaria de óleo fúsel: caracterização da composição química e estudo de sua aplicação industrial. Dissertation, Instituto de Tecnologia de Mauá

GEO Energética (2013) Tecnologia inovadora para a produção de biogás revoluciona a agroindústria brasileira. http://www.geoenergetica.com.br/imprensa/16/Tecnologia-inovadora-para-a–producao-de-biogas-revoluciona-a-agroindustria-brasileira. Accessed Jun 2015

Ghaffar T, Irshad M, Anwar Z et al (2014) Recent trends in lactic acid biotechnology: a brief review on production to purification. J Radiat Res Appl Sci 7(2):222–229. doi:10.1016/j.jrras.2014.03.002

Giese EC, Pierozzi M, Dussán KJ et al (2013) Enzymatic saccharification of acid-alkali pretreated sugarcane bagasse using commercial enzyme preparations. J Chem Tech Biotechnol 88(7):1266–1272. doi:10.1002/jctb.3968

Gilbert P, Alexander S, Thornley P et al (2014) Assessing economically viable carbon reductions for the production of ammonia from biomass gasification. J Clean Prod 64:581–589. doi:10.1016/j.jclepro.2013.09.011

Gírio FM, Fonseca C, Carvalheiro F et al (2010) Hemicelluloses for fuel ethanol: a review. Bioresour Technol 101(13):4775–4800. doi:10.1016/j.biortech.2010.01.088

Gouda MK, Swellam AE, Omar SH (2001) Production of PHB by a Bacillus megaterium strain using sugarcane molasses and corn steep liquor as sole carbon and nitrogen sources. Microbiol Res 156(3):201–207. doi:10.1078/0944-5013-00104

Gupta A, Verma JP (2015) Sustainable bio-ethanol production from agro-residues: a review. Renew Sust Energ Rev 41:550–567. doi:10.1016/j.rser.2014.08.032

Hassuani SJ, Leal MRLV, Macedo IC (eds) (2005) Biomass power generation—sugarcane bagasse and trash. CTC and PNUD, Piracicaba

He J, Zhang W (2008) Research on ethanol synthesis from syngas. J Zhejiang Univ-Sci A 9(5):714–719. doi:10.1631/jzus.A071417

Hofbauer H, Rauch R, Ripfel-Nitsche K (2007) Report on gas cleaning for synthesis applications work package 2E: Gas treatment. Institute of Chemical Engineering—University of Technology—Vienna. http://ec.europa.eu/energy/intelligent/projects/sites/iee-projects/files/projects/documents/thermalnet_report_on_syngas_cleaning.pdf. Accessed May 2015

Hugot E (1986) Handbook of cane sugar engineering. Elsevier, Amsterdam

Humbird D, Davis R, Tao L (2011) Process design and economics for biochemical conversion of lignocellulosic biomass to ethanol. National Renewable Energy Laboratory—NREL. http://www.nrel.gov/biomass/pdfs/47764.pdf. Accessed Apr 2015

IEA (2008) IEA bioenergy task 42 biorefinery. http://www.biorefinery.nl/fileadmin/biorefinery/docs/Brochure_Totaal_definitief_HR_opt.pdf. Accessed May 2015

IEA (2011) Technology roadmaps: biofuels for transport. http://www.iea.org/publications/freepublications/publication/Biofuels_Roadmap_WEB.pdf. Accessed May 2015

Informa Economics FNP (2013) Usina pioneira de etanol de milho consolida a produção em escala comercial em Mato Grosso.IEA http://www.informaeconfnp.com/noticia/9530. Accessed Jul 2014

Jenekar GM, Karne TM, Thangamuthu P (2005) Separation of tramp iron and ferric materials from sugarcane to achieve efficient milling. In: Proceedings of the 79th annual congress of South African Sugar Technologists' Association, 19–22 July 2005. Mount Edgecombe, South Africa, pp 267–272

Junqueira TL (2010) Simulação de colunas de destilação convencional, extrativa e azeotropica no processo de produção de bioetanol atraves da modelagem de não equilibrio e da modelagem de estagios de equilibrio com eficiencia. Dissertation, Universidade Estadual de Campinas—UNICAMP

Junqueira TL, Dias MOS, Jesus CDF et al (2011) Simulation and evaluation of autonomous and annexed sugarcane distilleries. Chem Eng Trans 25:941–946. doi:10.3303/CET1125157

Kamm B, Kamm M (2004a) Principles of biorefineries. Appl Microbiol Biotechnol 64(2):137–145. doi:10.1007/s00253-003-1537-7

Kamm B, Kamm M (2004b) Biorefinery-systems. Chem Biochem Eng Q 18(1):1–6

Kamm B, Gruber PR, Kamm M (eds) (2006) Biorefineries—industrial processes and products, vol 1. Wiley-VCH, Weinheim, p 441

Karagöz P, Ozkan M (2014) Ethanol production from wheat straw by Saccharomyces cerevisiae and Scheffersomyces stipitis co-culture in batch and continuous system. Bioresour Technol 158:286–293. doi:10.1016/j.biortech.2014.02.022

Karatzos S, McMillan J, Saddler J (2014) The potential and challenges of drop-in biofuels. A report by IEA bioenergy task 39. http://task39.org/files/2014/01/Task-39-drop-in-biofuels-report-summary-FINAL-14-July-2014-ecopy.pdf. Accessed 15 May 2015

Kersten S, Garcia-Perez M (2013) Recent developments in fast pyrolysis of ligno-cellulosic materials. Curr Opin Biotechnol 24:414–420. doi:10.1016/j.copbio.2013.04.003

Koppejan J, Sokhansanj S, Melin S et al (2012) Status overview of torrefaction technologies. In: IEA bioenergy task 32 report. http://www.ieabcc.nl/publications/IEA_Bioenergy_T32_Torrefaction_review.pdf. Accessed 15 May 2015

Kesler MG, Kessler MM (1958) Flexible Flow. World Pet. 29, p 60

Laird DA, Rogovska NP, Garcia-Perez M et al (2011) Pyrolysis and biochar—opportunities for distributed production and soil quality enhancement. In: Braun R, Karlen D, Johnson D (eds) Sustainable alternative fuel feedstock opportunities, challenges and roadmaps for six U.S. Regions. Soil and Water Conservation Society, Ankeny

Leal MRLV (2010) Tecnological evolution of sugarcane processing for ethanol and electric power generation. In: Cortez LAB (ed) Sugarcane bioethanol—R&D for productivity and sustainability. Edgard Blücher, São Paulo, pp 561–576

Leibbrandt NH, Knoetze JH, Görgens JF (2011) Comparing biological and thermochemical processing of sugarcane bagasse: an energy balance perspective. Biomass Bioenerg 35 (5):2117–2126. doi:10.1016/j.biombioe.2011.02.017

Limayem A, Ricke SC (2012) Lignocellulosic biomass for bioethanol production: current perspectives, potential issues and future prospects. Prog Energ Combust 38(4):449–467. doi:10.1016/j.pecs.2012.03.002

Lora EES, Venturini OJ (eds) (2012) Biocombustíveis. Interciência, Rio de Janeiro

Luque R, Lin CSK, Du C et al (2009) Chemical transformations of succinic acid recovered from fermentation broths by a novel direct vacuum distillation-crystallisation method. Green Chem 11(2):193–200. doi:10.1039/b813409j

Maas RHW, Bakker RR, Boersma AR et al (2008) Pilot-scale conversion of lime-treated wheat straw into bioethanol: quality assessment of bioethanol and valorization of side streams by anaerobic digestion and combustion. Biotechnol Biofuels 1(14):1–13. doi:10.1186/1754-6834-1-14

Macrelli S, Mogensen J, Zacchi G (2012) Techno-economic evaluation of 2nd generation bioethanol production from sugar cane bagasse and leaves integrated with the sugar-based ethanol process. Biotechnol Biofuels 5(22). doi:10.1186/1754-6834-5-22

Mafe OAT, Davies SM, Hancock J et al (2015) Development of an estimation model for the evaluation of the energy requirement of dilute acid pretreatments of biomass. Biomass Bioenerg 72:28–38. doi:10.1016/j.biombioe.2014.11.024

Mahishi LH, Tripathi G, Rawal SK (2003) Poly(3-hydroxybutyrate) (PHB) synthesis by recombinant *Escherichia coli* harbouring *Streptomyces* aureofaciens PHB biosynthesis genes: effect of various carbon and nitrogen sources. Microbiol Res 158(1):19–27. doi:10.1078/0944-5013-00161

Mantelatto PE (2005) Estudo do processo de cristalização de soluções impuras de sacarose de cana-de-açúcar por resfriamento. Dissertation, Universidade Federal de São Carlos—UFSCAR

Mantelatto PE (2009) Process for juice treatment in industrial large scale plant. In: BIOEN workshop on process for ethanol production, São Paulo

Mantelatto PE (2010) Information about the sugarcane industry. Private communication

Mariano AP, Maciel Filho R, Ezeji TC (2012) Energy requirements during butanol production and in situ recovery by cyclic vacuum. Renew Energ 47:183–187. doi:10.1016/j.renene.2012.04.041

Mariano AP, Dias MOS, Junqueira TL et al (2013) Utilization of pentoses from sugarcane biomass: techno-economics of biogas vs. butanol production. Bioresour Technol 142:390–399. doi:10.1016/j.biortech.2013.05.052

Martín MM (2014) Introduction to software for chemical engineers. CRC Press/Taylor & Francis, Boca Raton, Florida

Matsuoka S, Kennedy AJ, Santos EGD et al (2014) Energy cane: its concept, development, characteristics, and prospects. Adv Bot 2014:1–13. doi:10.1155/2014/597275

Meirelles AJA (2006) Expansão da produção de bioetanol e melhoria tecnológica da destilação alcoólica. In: Workshop Tecnológico sobre Produção de Etanol, Lorena. http://www.apta.sp.gov.br/cana/anexos/PPaper_sessao_4_Antonio_Meirelles.pdf

Menon V, Rao M (2012) Trends in bioconversion of lignocellulose: biofuels, platform chemicals and biorefinery concept. Prog Energ Combust 38(4):522–550. doi:10.1016/j.pecs.2012.02.002

Milanez AY, Nyko D, Valente MS et al (2014) A produção de etanol pela integração do milho-safrinha às usinas de cana-de-açúcar: avaliação ambiental, econômica e sugestões de política. Revista do BNDES 41:147–208

Milanez AY, Nyko D, Valente MS et al (2015) De promessa a realidade: como o etanol celulósico pode revolucionar a indústria da cana-de-açúcar—uma avaliação do potencialcompetitivo e sugestões de política pública. BNDES Setorial 41:237–29

Modelica (2014) Modelica—a unified object-oriented language for systems modeling—language specification version 3.3 Revision 1. Modelica Association, 11 July 2014

Modesto M, Zemp RJ, Nebra SA (2009) Ethanol production from sugar cane: assessing the possibilities of improving energy efficiency through exergetic cost analysis. Heat Transf Eng 30(4):272–281. doi:10.1080/01457630802380386

Moraes BS, Zaiat M, Bonomi A (2015) Anaerobic digestion of vinasse from sugarcane ethanol production in Brazil: challenges and perspectives. Renew Sust Energ Rev 44:888–903. doi:10.1016/j.rser.2015.01.023

Novacana (2013) Sorgo sacarino é alternativa na entressafra de cana. http://www.novacana.com/n/etanol/alternativas/sorgo-sacarino-alternativa-entressafra-cana-220413. Accessed Jul 2013

Obando DS, Guialdrón MA, Reno MLG (2010) Techno-economic indicators for the thermo-chemical and biochemical routes for biofuels production using sugarcane bagasse and feedstock. In: XXVII Congress of the International Society of Sugar Cane Technologists. Cidade de Mexico

Oliveira LRM, Nascimento VM, Gonçalves AR et al (2014) Combined process system for the production of bioethanol from sugarcane straw. Ind Crops Prod 58:1–7. doi:10.1016/j.indcrop.2014.03.037

Olivério JL, Ferreira FM (2010) Cogeneration—a new source of income for sugar and ethanol mills or bioelectricity—a new business. Proc Int Soc Sugar Cane Technol 27:1–12

OpenModelica (2015) OpenModelica—introduction. In: OpenModelica website. https://openmodelica.org/index.php. Accessed Jun 2015

OSI (2015) OSI—Open source initiative. http://opensource.org. Accessed May 2015

Padula AD, Santos MS, Ferreira L (2012) The emergence of the biodiesel industry in Brazil: current figures and future prospects. Energ Policy 44:395–405. doi:10.1016/j.enpol.2012.02.003

Pantelides CC, Urban ZE (2004) Process modelling technology: a critical review of recent developments. In: 6th International foundations of computer aided process design (FOCAPD'04), Princeton, New Jersey, 11–16 July 2004

Payne JH (1989) Operações unitárias na produção de açúcar de cana. Nobel S.A, São Paulo

Pedra Mill (2006) Data on the ethanol production process from sugarcane. Serrana

Pelaez-Samaniego MR, Yadama V, Garcia-Perez M et al (2014) Effect of temperature during wood torrefaction on the formation of lignin liquid intermediates. J Anal Appl Pyrol 109:222–233. doi:10.1016/j.jaap.2014.06.008

Pfaltzgraff LA, Clark JH (2014) Green chemistry, biorefineries and second generation strategies for re-use of waste: an overview, In: Waldron K (ed) Advances in biorefineries: biomass and waste supply chain exploitation. Elsevier B.V., Amsterdam, pp 3–33

Pourbafrani M, McKechnie J, Shen T et al (2014) Impacts of pre-treatment technologies and co-products on greenhouse gas emissions and energy use of lignocellulosic ethanol production. J Clean Prod 78:104–111. doi:10.1016/j.jclepro.2014.04.050

Puig-Arnavat M, Bruno JC, Coronas A (2010) Review and analysis of biomass gasification models. Renew Sust Energ Rev 14:2841–2851. doi:10.1016/j.rser.2010.07.030

Rabelo SC, Amezquita Fonseca NA, Andrade RR et al (2011a) Ethanol production from enzymatic hydrolysis of sugarcane bagasse pretreated with lime and alkaline hydrogen peroxide. Biomass Bioenerg 35(7):2600–2607. doi:10.1016/j.biombioe.2011.02.042

Rabelo SC, Carrere H, Maciel Filho R et al (2011b) Production of bioethanol, methane and heat from sugarcane bagasse in a biorefinery concept. Bioresour Technol 102(17):7887–7895. doi:10.1016/j.biortech.2011.05.081

Raman R (1985) Chemical process computations. Elsevier Applied Science Publishers Ltd., New York

Ramos A, Marques LD, Santos JJ et al (2011) Atual estágio de desenvolvimento da tecnologia GTL e perspectivas para o Brasil. Quim Nova 34(10):1704–1716

Reddy MM, Vivekanandhan S, Misra M (2013) Biobased plastics and bionanocomposites: current status and future opportunities. Prog Polym Sci 38(10–11):1653–1689. doi:10.1016/j.progpolymsci.2013.05.006

Rein P (2007) Cane sugar engineering. Verlag Dr, Albert Bartens KG, Berlin

Ribeiro PR (2003) A Usina de Açúcar e sua Automação. SMAR, Araçatuba

Richardson Y, Blin J, Julbe A (2012) Review: a short overview on purification and conditioning of syngas produced by biomass gasification: catalytic strategies, process intensification and new concepts. Prog Energy Combust Sci 38:765–781. doi:10.1016/j.pecs.2011.12.001

Rocha GJM, Gonçalves AR, Oliveira BR et al (2010) Compositional variability of raw, steam exploded and delignificated sugarcane bagasse. In: AGRENER—8º Congresso Internacional sobre Geração Distribuída e Energia no Meio Rural, Campinas, Dez 2010

Rodríguez-Zúñiga UF, Cannella D, Giordano RDC et al (2015) Lignocellulose pretreatment technologies affect the level of enzymatic cellulose oxidation by LPMO. Green Chem 17:2896–2903. doi:10.1039/C4GC02179G

Rossell CEV (2011) Data for the ethanol production process. Private communication

Sathitsuksanoh N, George A, Zhang YHP (2013) New lignocellulose pretreatments using cellulose solvents: a review. J Chem Technol Biotechnol 88(2):169–180. doi:10.1002/jctb.3959

Scherrer R, Berlin E, Gerhardt P (1977) Density, porosity, and structure of dried cell walls isolated from bacillus megaterium and saccharomyces cerevisiae. J Bacteriol 129(2):1162–1164

Seabra JEA (2008) Avaliação técnico-econômica de opções para o aproveitamento integral da biomassa de cana no Brasil. PhD Thesis, Universidade Estadual de Campinas – UNICAMP

Silva AS, Inoue H, Endo T et al (2010) Milling pretreatment of sugarcane bagasse and straw for enzymatic hydrolysis and ethanol fermentation. Bioresour Technol 101(19):7402–7409. doi:10.1016/j.biortech.2010.05.008

Silva VFN, Arruda PV, Felipe MGA et al (2011) Fermentation of cellulosic hydrolysates obtained by enzymatic saccharification of sugarcane bagasse pretreated by hydrothermal processing. J Ind Microbio Biotechnol 38(7):809–17. doi:10.1007/s10295-010-0815-5

Singh J, Suhag M, Dhaka A (2015) Augmented digestion of lignocellulose by steam explosion, acid and alkaline pretreatment methods: a review. Carbohydr Polym 117:624–631. doi:10.1016/j.carbpol.2014.10.012

Spath P, Aden A, Eggeman T et al (2005) Biomass to hydrogen production detailed design and economics utilizing the battelle columbus laboratory indirectly heated gasifier. National Renewable Energy Laboratory—NREL. http://www.nrel.gov/docs/fy05osti/37408.pdf. Accessed Apr 2015

Spath PL, Dayton DC (2003) Preliminary Screening—technical and economic assessment of synthesis gas to fuels and chemicals with emphasis on the potential for biomass-derived syngas. National Renewable Energy Laboratory—NREL. http://www.nrel.gov/docs/fy04osti/34929.pdf. Accessed May 2015

Spivey JJ, Egbebi A (2007) Heterogeneous catalytic synthesis of ethanol from biomass-derived syngas. Chem Soc Rev 36:1514–1528. doi:10.1039/B414039G

Stanmore BR (2010) Generation of energy from sugarcane bagasse by thermal treatment. Waste Biomass Valorization 1(1):77–89. doi:10.1007/s12649-009-9000-3

Stephanopoulos G, Gintaras VR (2011) Process systems engineering: from Solvay to modern bio and nanotechnology: a history of development, successes and prospects for the future. Chem Eng Sci 66(19):4272–4306. doi:10.1016/j.ces.2011.05.049

Stevens DJ (2001) Hot gas conditioning: recent progress with larger-scale biomass gasification systems: update and summary of recent progress. National Renewable Energy Laboratory—NREL. http://www.nrel.gov/docs/fy01osti/29952.pdf. Accessed May 2015

Subramani V, Gangwal SK (2008) A review of recent literature to search for an efficient catalytic process for the conversion of syngas to ethanol. Energ Fuel 22:814–839. doi:10.1021/ef700411x

Swayze S (2009) The sweet sorghum opportunity: a complementary source of low-cost fermentable sugars for biofuel. Int Sugar J 111:691–695

Tao L, Aden A, Elander RT et al (2011) Process and technoeconomic analysis of leading pretreatment technologies for lignocellulosic ethanol production using switchgrass. Bioresour Technol 102(24):11105–11114. doi:10.1016/j.biortech.2011.07.051

Tripathi AD, Srivastava SK, Singh RP (2013) Statistical optimization of physical process variables for bio-plastic (PHB) production by *Alcaligenes* sp. Biomass Bioenerg 55:243–250. doi:10.1016/j.biombioe.2013.02.017

Vaz Júnior S (ed) (2013) Biomassa para química verde. Embrapa Agroenergia, Brasília

Vijayakumar J, Aravindan R, Viruthagiri T (2008) Recent trends in the production, purification and application of lactic acid. Chem Biochem Eng Q 2(2):245–264

Walter A (2010) Potential of electricity production from sugarcane residues. In: Cortez LAB (ed) Sugarcane bioethanol—R&D for productivity and sustainability. Edgard Blücher, São Paulo, pp 577–582

Walter A, Dolzan P, Quilodrán O et al (2008) A sustainability analysis of the Brazilian ethanol. UK Embassy and Department for Environment, Food & Rural Affairs—DEFRA. http://www.globalbioenergy.org/uploads/media/0811_Unicamp_-_A_sustainability_analysis_of_the_Brazilian_ethanol.pdf

Wang H, Pan J, Wang J (2014) Succinic acid production from xylose mother liquor by recombinant *Escherichia coli* strain. Biotechnol Biotechnol Equip 28(6):1042–1049. doi:10.1080/13102818.2014.952501

Werpy T, Petersen G (2004) Top value added chemicals from biomass: volume 1—results of screening for potential candidates from sugars and synthesis gas. National Renewable Energy Laboratory—NREL. http://www.nrel.gov/docs/fy04osti/35523.pdf. Accessed May 2015

Westerberg AW, Hutchinson HP, Motard RL et al (1979) Process flowsheeting. Cambridge University Press, Cambridge

Westerberg AW (1998) Process engineering: part II—a system view. Institute for Complex Engineered Systems. ICES, Carnegie Mellon University Pittsburgh, PA 15213

Westereng B, Agger JW, Horn SJ et al (2013) Efficient separation of oxidized cello-oligosaccharides generated by cellulose degrading lytic polysaccharide monooxygenases. J Chromatogr A 1271(1):144–152. doi:10.1016/j.chroma.2012.11.048

Weusthuis RA, Aarts JMMJG, Sanders JPM (2011) From biofuel to bioproduct: is bioethanol a suitable fermentation feedstock for synthesis of bulk chemicals? Biofuel Bioprod Biorefin 5:486–494. doi:10.1002/bbb

Williams RH, Larson ED (1991) Advanced biomass power generation: the biomass-integrated gasifier/gas turbine and beyond. In: Kuliasha MA, Zucker A, Ballew KJ (eds) Technologies for a greenhouse constrained society: proceedings of the technologies for a greenhouse constrained society conference, Oak Ridge, Tennessee, pp 105–158

Wooley RJ, Putsche V (1996) Development of an aspenplus physical property database for biofuels components. National Renewable Energy Laboratory—NREL. http://www.nrel.gov/docs/legosti/old/20685.pdf. Accessed May 2015

Wyman CE, Dale BE, Elander RT et al (2005) Coordinated development of leading biomass pretreatment technologies. Bioresour Technol 96(18):1959–1966. doi:10.1016/j.biortech.2005.01.010

Yoneda N, Kusano S, Yasui M (2001) Recent advances in processes and catalysts for the production of acetic acid. Appl Catal A-Gen 221(1–2):253–265. doi:10.1016/S0926-860X(01)00800-6

Yu Q, Zhuang X, Lv S et al (2013) Liquid hot water pretreatment of sugarcane bagasse and its comparison with chemical pretreatment methods for the sugar recovery and structural changes. Bioresour Technol 129:592–598. doi:10.1016/j.biortech.2012.11.099

Zhou Z, Wang C, Chen Y (2015) Increasing available NADH supply during succinic acid production by *Corynebacterium glutamicum*. Biotechnol Prog 31(1):12–19. doi:10.1002/btpr.1998

Chapter 5
Biorefinery Products Logistics, Commercialization, and Use

Mateus F. Chagas, Otávio Cavalett, Lucas Gonçalves Pereira and Antonio Bonomi

5.1 Products Logistics

The main objective of distribution logistics and final use modeling in the Virtual Sugarcane Biorefinery is to provide information in order to complete the assessment of the biorefinery products life cycles. Different options for transport services have to be included in a model for logistic evaluation, taking into consideration several criteria such as geographical operation, transported goods, and vehicle size.

This chapter presents a model, called Log&UsoSoft, intended for simulation and measurement of the most important parameters of distribution logistics—transportation and commercialization—and use of biorefinery products, with focus on ethanol. In the context of the Virtual Sugarcane Biorefinery, different modes of transport are intended to be included in the construction of the Log&UsoSoft model to account for economic, environmental, and social impacts of products transportation between their production centers to final consumers. Although covering economic, environmental, and social aspects, similar to the modeling of agricultural and industrial stages presented on Chaps. 3 and 4, the end-use modeling is of fundamental importance particularly for environmental assessment. A complete LCA requires the assessment of the entire life cycle of a product, including its final use and emissions. In comparative LCA studies, the inclusion of final use is mandatory to compare different fuel alternatives, since only in this phase the efficiencies of fuels use in vehicles are included.

M.F. Chagas · O. Cavalett · L.G. Pereira · A. Bonomi (✉)
Brazilian Bioethanol Science and Technology Laboratory (CTBE), Brazilian Center of Research in Energy and Materials (CNPEM), Campinas, SP, Brazil
e-mail: antonio.bonomi@bioetanol.org.br

O. Cavalett
e-mail: otavio.cavalett@bioetanol.org.br

L.G. Pereira
Department of Civil Engineering, University of Toronto, Toronto, ON, Canada

A. Bonomi et al. (eds.), *Virtual Biorefinery*, Green Energy and Technology,
DOI 10.1007/978-3-319-26045-7_5

The focus of this model is on ethanol distribution system, mainly performed by road transportation. Modeling of ethanol transportation by trucks will be detailed described, considering the ethanol logistics in Brazilian context. Other transportation modes are briefly discussed in the end of this section. Modeling of commercialization agents—distributors and fuel stations—is not included in this first version of the model.

5.1.1 Ethanol Distribution Logistics

According to the Brazilian National Agency of Petroleum, Natural Gas, and Biofuels regulations (ANP 2014) for fuel commercialization in national market, production plants are not allowed to sell ethanol directly to the final consumers. Only distributors, the intermediary agents between producers and final consumers, are allowed to sell it to the fuel stations. Distributors can also trade fuels among themselves.

Fuels available in Brazilian fuel stations for light vehicles include hydrated ethanol (93 % w/w) and grade "C" gasoline (up to 27.5 % anhydrous ethanol blended with pure gasoline, according to governmental regulation). The mixture of anhydrous ethanol with pure gasoline is also an activity under the responsibility of the distributors (ANP 2014).

A single production plant generally does not have a production scale that enables the use of other modes of transport other than road transportation. The same limitation occurs for transport between distributors and fuel stations. As a result, practically the total volume of ethanol produced in the country is transported by using road system (Milanez et al. 2010). In few cases, there are short road sections to collect ethanol from a set of industrial plants to a transference terminal, using other modes to transport ethanol to a secondary distributor, and finally using road transportation to deliver ethanol to fuel stations. Figure 5.1 illustrates the typical Brazilian ethanol distribution system.

5.1.2 Modeling Ethanol Road Transportation

The Log&UsoSoft model is structured to calculate the operational costs, life cycle inventories, and labor use for ethanol road transportation by vehicles performing specific transport activities. Road transportation modeling in Log&UsoSoft model is comprised of a scenario definition module, truck and trailer databanks, and an integrated calculation module, where the operational parameters are defined and the calculations of the transport system operation are performed. Model outputs include the economic, environmental, and social inventories for ethanol distribution by road transportation.

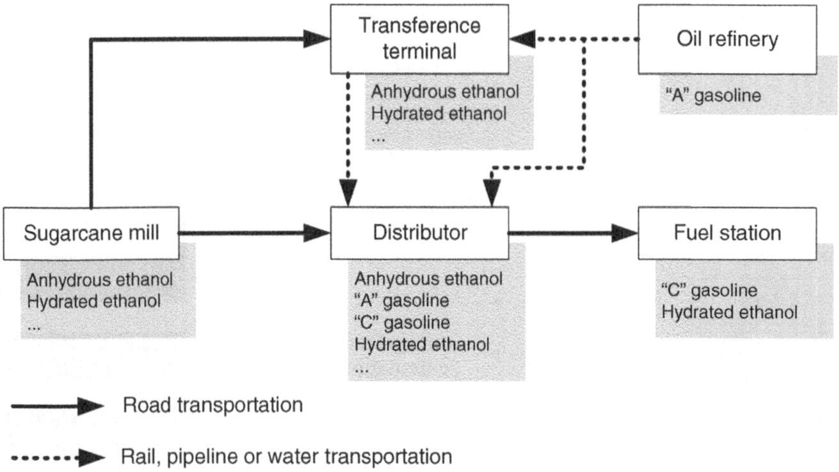

Fig. 5.1 Typical Brazilian ethanol distribution system

In the scenario definition module, the user defines the total amount of fuel to be transported and the distances between loading and unloading sites, as well as the type of truck and trailer used. Simulations can be performed considering each section of the logistic system (Fig. 5.1), with their own characteristics, and results are always related to fuel at the end of each section. Truck and trailer databanks contain basic information about equipment used in road transportation. In the system operation module, the efficiencies, materials, and labor uses and costs of transportation are calculated based on information from scenario description and equipment databanks. Infrastructure and operation of distributors and fuel stations are not included in this first version of the model. Finally, model outputs are organized in economic, environmental, and social inventories.

The general framework used for development of the ethanol transport by road system in Log&UsoSoft model is illustrated in Fig. 5.2.

5.1.2.1 Scenario Definition

In scenario definition module, the total volume and transport distances are defined, as well as the type of truck and trailer used, for each logistic system section.

Total volume is used to determine the number of freight operations required to transfer the total fuel amount from its production place to its final destination, according to Eq. 5.1. This information will also be useful in future versions of the model to define investments and operational costs of distribution bases and fuel

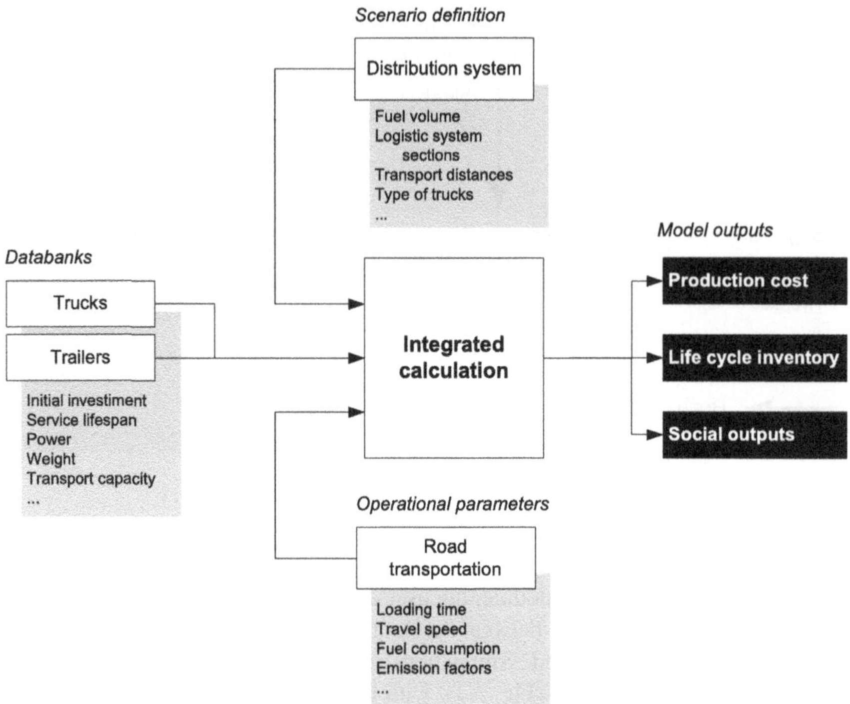

Fig. 5.2 Ethanol distribution modeling scheme showing different calculation modules

stations. Transport distances and type of trucks are basic information to system operation module, which will be further detailed in Sect. 5.1.2.3.

$$N_{\text{freight}} = \frac{\text{TV}_{\text{fuel}}}{V_{\text{truck}}} \quad (5.1)$$

N_{freight} total number of freight operations,
TV_{fuel} total volume of fuel to be transported (m^3), and
V_{truck} volume transported per truck per freight operation (m^3).

Basis for calculations in the system operation module is one freight operation. Final economic and environmental scores are obtained by multiplying outputs from system operation module by total number of freight operations.

5.1.2.2 Truck and Trailer Databanks

The truck and trailer databanks contain basic information for modeling the transport operations in Log&UsoSoft model, such as truck and trailer weights, load

Table 5.1 Part of road transport databank for trucks

Trucks (hp)	Weight (kg)	Number of tires	Initial investment (US$)
Truck 310	9214	10	117,400.00
Truck 360	9342	10	143,500.00
Truck 400	9549	10	169,600.00
Truck 440	9986	10	173,900.00
Truck 480	10112	10	187,000.00

Table 5.2 Part of road transport databank for trailers

Trailers	Volume (m^3)	Number of tires	Initial investment (US$)
Tank, semitrailer, 2 shafts	20	8	38,100.00
Tank, semitrailer, 3 shafts	30	12	44,800.00
Tank, 2 semitrailers, 4 shafts	45	16	76,100.00
Tank, 2 trailer, 6 shafts	60	24	89,600.00

capacities, number of wheels per vehicle, initial investments, and lifespan. Most of this information is obtained from truck and trailer manufacturers' specification. Initial investments, estimated by market price and mean values from literature, are considered for lifespan. Some examples of trucks and trailers information from Log&UsoSoft databanks are shown, respectively, in Tables 5.1 and 5.2.

Service lifespan for trucks, trailers, and tires are assumed to be independent on equipment specification. For trucks and trailers, it was assumed a lifespan of 550,000 km (Spielmann et al. 2007). Tires lifespan of 150,000 km for a tire after 2 retreads is assumed (Banchi 2006a, b), and the mean lifespan (50,000 km) is used in Log&UsoSoft model.

5.1.2.3 Road Transport System Operation

Parameters used to calculate economic, environmental, and social outputs in road transport system operation are detailed in this section. Basis for calculations is one freight operation, i.e., transportation per one truck and trailer unit of its volumetric capacity between loading and unloading sites.

Transportation Distances

Transportation distance is defined as the road length between loading and unloading sites, following a specific route. Considering this definition, it comprises only the delivery route, but in a freight operation, the return route must be accounted for as well. Normally, in dedicated transport systems, return distances are equal to that for delivery. Nevertheless, return freight—when the same truck and trailer unit is used for other loads in return trip—is a common practice on road transportation systems, and therefore, return distance needs to be proportionally shared. Total traveled distance of the freight is thus given by Eq. 5.2.

$$D_{\text{freight}} = D_{\text{delivery}} + f_{\text{return}} D_{\text{return}} \tag{5.2}$$

D_{freight} total road distance for a freight operation (km/freight),
D_{delivery} road distance for delivery travel (km),
D_{return} road distance for return travel (km), and
f_{return} factor for return travel partition.

Transportation Time

Time used in transport operations is calculated by summing up all the time required to perform a travel. Times spent on effective trip, on maintenance, on waiting in lines and performing operations of loading and unloading are accounted for. Based on the transportation distance, travel speeds, and total loading and unloading times, the total freight operation time can be calculated by Eq. 5.3.

$$t_{\text{freight}} = \frac{D_{\text{delivery}}}{v_{\text{load}}} + \frac{D_{\text{return}}}{v_{\text{unload}}} + t_{\text{waiting for load}} + t_{\text{loading}} + t_{\text{waiting for unload}} + t_{\text{unloading}} + t_{\text{maintenance}} \tag{5.3}$$

t_{freight} total time required for a freight operation (h),
D_{delivery} road distance for delivery travel (km),
D_{return} road distance for return travel (km),
v_{load} travel speed for delivery travel (km h^{-1}),
v_{unload} travel speed for return travel (km h^{-1}),
$t_{\text{waiting for load}}$ total time spent on waiting lines for load (h),
t_{loading} total time required for loading operation (h),
$t_{\text{waiting for unload}}$ total time spent on waiting lines for unload (h),
$t_{\text{unloading}}$ total time required for unloading operation (h), and
$t_{\text{maintenance}}$ total time spent on maintenance (h).

Average travel speed and time required for loading and discharging are estimated based on average conditions for Brazilian fuel transportation sector. Default values in Log&UsoSoft model are shown in Table 5.3. Distances for delivery and return travels are strictly dependent on loading and unloading sites and transportation route between them.

Table 5.3 Parameters for time requirement calculation

Parameter	Unit	Default value
v_{load}	km h^{-1}	45
v_{unload}	km h^{-1}	55
$t_{\text{waiting for load}} + t_{\text{loading}}$	h	3
$t_{\text{waiting for unload}} + t_{\text{unloading}}$	h	7
$t_{\text{maintenance}}$	h	10 % of effective travel time

Truck and Trailer Consumption

According to the basis for calculation of road transport operation, one transportation unit—consisting of one truck and one or more trailers—is required to perform a freight operation, transporting its nominal volumetric capacity. Consumption of trucks and trailers in a freight operation accounts for the physical depreciation of the transport equipment. Portion of vehicles consumed in a freight operation, considering the total distance and the expected lifetime for the vehicles, is given by Eqs. 5.4 and 5.5.

$$\mathrm{TC}_{\text{truck}} = \frac{D_{\text{freight}}}{\mathrm{LS}_{\text{truck}}} \tag{5.4}$$

$$\mathrm{TC}_{\text{trailer}} = \frac{D_{\text{freight}}}{\mathrm{LS}_{\text{trailer}}} \tag{5.5}$$

$\mathrm{TC}_{\text{truck}}$ consumption of truck unities in a freight operation,
$\mathrm{TC}_{\text{trailer}}$ consumption of trailer unities in a freight operation,
D_{freight} total road distance for a freight operation (km),
$\mathrm{LS}_{\text{truck}}$ truck lifespan (km), and
$\mathrm{LS}_{\text{trailer}}$ trailer lifespan (km).

Labor Consumption

Labor use in transport operations is considered to be equal to total time required to perform a trip, according to Eq. 5.3.

Diesel Consumption

Diesel consumption in road transportation is calculated using the specific diesel consumption, according to the capacity of the truck, and total transport distance (Eq. 5.6). Mean values adopted in Log&UsoSoft model for specific diesel

Table 5.4 Diesel consumption parameters in road transport systems

Truck class	Total weight (t)	Fuel economy (km L^{-1})	Specific diesel consumption (L km^{-1})
Medium duty	W < 25	2.5	0.40
Semi-heavy duty	25 < W < 40	1.9	0.53
Heavy duty	W > 40	1.5	0.67

Based on data from Santos (2012)

consumption are shown in Table 5.4. Values may change according to truck's engine technology, transport load, road conditions, among several other factors, and default values can be changed to represent each specific situation.

$$C_{\text{diesel}} = \text{sc}_{\text{diesel}} D_{\text{freight}} \tag{5.6}$$

C_{diesel} diesel consumption per freight operation (L/freight),
$\text{sc}_{\text{diesel}}$ specific diesel consumption (L km^{-1}), and
D_{freight} total road distance for a freight operation (km/freight).

Tires Consumption

Like the consumption of transport equipment, tires consumption in Log&UsoSoft model accounts for the physical depreciation of the tires. Number of wheels per truck, average tire lifespan, and freight transportation distances are considered to determine tires consumption in one road transportation freight. The number of tires consumed per freight operation is calculated according to Eq. 5.7.

$$C_{\text{tires}} = N_{\text{wheels}} \frac{D_{\text{freight}}}{\text{LS}_{\text{tire}}} \tag{5.7}$$

C_{tires} tires consumption per freight (number of tires),
D_{freight} road distance between loading and unloading places (km),
N_{wheels} number of wheels per truck (number of tires), and
LS_{tire} tire lifespan (km).

5.1.2.4 Transportation Costs

All fixed and variable costs related to the operation of transport vehicles are considered in the model. The main components of transport costs are presented as follows.

Trucks and Trailers

The costs related to trucks and trailers consider the uniform recovery of initial investment, accounting for physical depreciation of equipments in their lifetime, plus interest of 12 % per year on invested capital. Salvage value of 20 % over initial investment is considered in the end of working lifespan for trucks.

Diesel and Lubricants

Diesel costs are calculated multiplying the amount of fuel used and the fuel market price. Costs related with lubricating oil are defined as a fraction of expenditures on diesel. Default value in Log&UsoSoft model for lubricants costs is 3 % over diesel costs.

Tires

Expenditures on tires are calculated multiplying tires' use by its cost, estimated considering the market price of a new tire with 2 retreading costs and a total lifespan of 150,000.0 km.

Labor

Labor costs are calculated by multiplying the labor use in transport operation by the manpower cost per hour.

Maintenance and Taxes

Annual maintenance costs include the expenditures on repairs and maintenance of trucks and trailers. It accounts for expenditures on replacement of pieces and garage labor, and also on payments for maintenance performed by third parties. It is normally presented as a fraction of initial costs of equipment, and default value in Log&UsoSoft is 25 % for transport trucks and trailers. This value can be changed if required in a specific assessment.

Other costs related to transport operations include expenditures with insurance and taxes—considered 8 % over costs of equipment are considered in Log&UsoSoft.

Highway Tolls

The last component of road transportation costs are the expenditures with highway tolls. These costs are highly dependent on transportation routes and need to be priced for each route included in the model.

5.1.2.5 Transportation Emissions

Road transportation emissions include airborne exhaust emissions, dependent on fuel combustion, and non-exhaust emissions from tire and brakes abrasion. Emissions to soil and water due to abrasion are also included.

Default values for diesel based emissions are shown in Table 5.5 and the abrasion based emissions in Table 5.6. Figures are based on ecoinvent database (Spielmann et al. 2007), considering trucks with Euro 3 engine technology and load capacity greater than 35 t. Changes to represent other engines technologies and load capacities are included as option in the model.

Table 5.5 Airborne emissions from diesel use in road transportation

Substance	Unit	Value
Carbon dioxide	$g\ kg^{-1}{}_{diesel}$	3171.7
Sulfur dioxide	$g\ kg^{-1}{}_{diesel}$	1.0
Cadmium	$g\ kg^{-1}{}_{diesel}$	1.0E−05
Copper	$g\ kg^{-1}{}_{diesel}$	1.7E−03
Chromium	$g\ kg^{-1}{}_{diesel}$	5.0E−05
Nickel	$g\ kg^{-1}{}_{diesel}$	7.0E−05
Zinc	$g\ kg^{-1}{}_{diesel}$	1.0E−03
Lead	$g\ kg^{-1}{}_{diesel}$	1.1E−10
Selenium	$g\ kg^{-1}{}_{diesel}$	1.0E−05
Mercury	$g\ kg^{-1}{}_{diesel}$	2.0E−08
Chromium IV	$g\ kg^{-1}{}_{diesel}$	1.0E−07
Carbon monoxide	$g\ km^{-1}$	1.6
Nitrogen oxides	$g\ km^{-1}$	7.9
Particulate matter	$g\ km^{-1}$	1.2E−01
Non-methane hydrocarbons	$g\ km^{-1}$	1.7E−01
Methane	$g\ km^{-1}$	7.6E−02
Toluene	$g\ km^{-1}$	1.7E−08
Benzene	$g\ km^{-1}$	1.2E−07
m, p, o Xylene	$g\ km^{-1}$	1.5E−06
Formaldehyde	$g{\cdot}km^{-1}$	1.5E−05
Acetaldehyde	$g\ km^{-1}$	8.0E−06
Dinitrogen monoxide	$g\ km^{-1}$	3.0E−02
Ammonia	$g\ km^{-1}$	5.0E−03
Polyhydroxyalkanoates	$g\ km^{-1}$	1.0E−06

Values based on ecoinvent database (Spielmann et al. 2007). Diesel sulfur content was assumed 500 ppm, according to ANP (2013a) regulation

Table 5.6 Emissions from tires abrasion in road transportation

Substance	Unit	Value
Airborne emissions		
TSP-PM10	g km^{-1}	5.6E−02
PM10-PM2.5	g km^{-1}	6.2E−02
PM2.5	g km^{-1}	3.5E−02
Waterborne emissions		
Zinc	g km^{-1}	7.2E−04
Copper	g km^{-1}	1.7E−05
Cadmium	g km^{-1}	2.5E−07
Chrome	g km^{-1}	1.2E−06
Nickel	g km^{-1}	3.3E−06
Lead	g km^{-1}	1.0E−05
Soil emissions		
Zinc	g km^{-1}	7.2E−04
Copper	g km^{-1}	1.7E−05
Cadmium	g km^{-1}	2.5E−07
Chrome	g km^{-1}	1.2E−06
Nickel	g km^{-1}	3.3E−06
Lead	g km^{-1}	1.0E−05

Values based on ecoinvent database (Spielmann et al. 2007). Tire mass was assumed 47.5 kg, with mass losses of 20 % in tire lifespan

5.1.2.6 Example of Log&UsoSoft Application for Ethanol Distribution System

An example of scenario definition information is shown in Table 5.7. In this example, 100,000 m^3 of hydrous ethanol is defined as a basis for calculations. Ethanol is transported between the cities of Ribeirão Preto—the largest ethanol producer region—and São Paulo—the largest consumer market. It is considered also an intermediary distributor in Paulínia city and transport trailers with volumetric capacity of 60.0 m^3 in both logistical sections.

Table 5.7 Example of scenario definition information

Section	1	2
Origin	Production plant (Ribeirão Preto)	Distributor (Paulínia)
Destination	Distributor (Paulínia)	Final consumer (São Paulo)
Transported volume (m^3 year^{-1})	100,000	100,000
Transport distance (km)	220	125
Type of truck	Truck 480 hp + Tank, 45 m^3	Truck 480 hp + Tank, 45 m^3

Table 5.8 Ethanol transportation costs by road system in assessed scenario

Cost component	Section			
	1		2	
	US$ m^{-3} km^{-1}	%	US$ m^{-3} km^{-1}	%
Truck and trailers	0.0319	30.4	0.0387	33.6
Labor	0.0138	13.2	0.0191	16.6
Diesel	0.0342	32.6	0.0342	29.7
Tires	0.0094	9.0	0.0094	8.2
Maintenance	0.0080	7.6	0.0097	8.4
Lubricating	0.0010	1.0	0.0010	0.9
Taxes and insurance	0.0026	2.4	0.0031	2.7
Highway tolls	0.0039	3.7	0.0039	3.4
Total	0.1048	100.0	0.1152	100.0

Considering default values for system operation variables, and 0.09 US$ km^{-1} for expenditures with highway tolls, economic results calculated with Log&UsoSoft model are shown in Table 5.8.

According to EsalqLog (2013), freight prices in 2012 for Ribeirão Preto–Paulínia route was US$ 0.0933 per m^3 km, and for Paulínia–São José dos Campos (route similar to the Paulínia – São Paulo), was US$ 0.1071 per m^3 km. Adjusted values for July 2014 are US$0.1003 and US$ 0.1151 per m^3 km, respectively, using IGP-M price index. Transportation costs calculated by Log&UsoSoft model are slightly higher than real freight prices. It can be justified by the expressive share of self-employed workers in Brazilian transportation service sector. Self-employed workers generally do not include important cost components in freight prices, such as capital costs and insurances, underestimating real transport costs (Araújo et al. 2014). In Log&UsoSoft model, all fixed and variable costs related to the operation of transport vehicles are considered.

5.1.3 Other Transportation Modes

5.1.3.1 Pipeline Transportation

Although it represents an alternative for only about 5 % of the transported goods in Brazil (CNT 2015), pipeline transportation is a promising alternative for ethanol. This mode of transport is primarily used in the country to transport crude oil and derivatives and natural gas.

Despite of the high investments in pipeline system implementation, reduced operational costs and energy requirements are attractive advantages of pipelines over road system. Transport of ethanol through pipelines indeed is the basis of a new Brazilian ethanol multimodal logistics system that started to operate its first section in 2014. Designed by Logum Logistica SA, this system is based on pipelines use associated with an integrated logistics system of collection and distribution terminals, covering also the interconnection of transport modes (Logum 2014). Planned to transport 20 billion liters of the biofuel per year when achieved its full capacity, the ethanol pipeline network have the potential to become one of the most important systems for ethanol transportation in Brazil (Logum 2014).

5.1.3.2 Railway Transport

For intermediate and long transport distances, the railway transport modal has potential to be more efficient than road transportation. However, the lack of investment in recent decades imposes several limitations to this mode of transport in Brazil (Logum 2014). Today, only about 20 % of goods are transported using railways; the country has 30,500 km of railways and more than 200,000 km of paved roads (CNT 2015), evidencing the importance of roads in detrition of rails.

The option of transporting ethanol production by railways implies the need for investment in acquisition of new wagons, the construction of new railways, the creation of transshipment structures, and implementation of modern operating systems, since the vast majority of Brazilian railways was established more than 50 years ago, and has serious limitations to be fully utilized as an option for ethanol transportation (Logum 2014).

5.1.3.3 Inland Waterway Transport

Although there is a great potential for the fluvial transport in Brazil, inland waterway transport is still rather limited, accounting for about 15 % of the total freight transportation in the country (CNT 2015). With more than 41,500 km of navigable rivers, only 21,000 km is economically viable for transport use (CNT 2015).

The waterway network is restricted to the occurrence of rivers with navigability potential, and expanding this mode of transport requires investments in barges and also in infrastructure to increase the extension of the waterways and its transport capacity. Due to geographical restrictions, the waterway transport is also extremely dependent on interconnection with other transport models.

Transportation of ethanol by the waterway system is almost negligible. Increased volumes are expected to be transported with the implementation of the ethanol multimodal logistics system designed by Logum Logistica SA, which foresees the use of the Tiete-Paraná waterway (Logum 2014).

5.1.3.4 Example of Log&UsoSoft Application for Transport System Comparison

In order to illustrate Log&UsoSoft model application to compare environmental performance of different transport systems, greenhouse gas emissions of ethanol transportation were calculated for two scenarios based on road and pipeline transportation. For both scenarios, the same transportation route presented in Sect. 5.1.2.6, between Ribeirão Preto and São Paulo cities, is considered.

Scenario 1 is based only in road transportation, using the same scenario definition shown in Table 5.7. Scenario 2 is mainly based on pipeline transportation option. In this scenario, three sections were defined to describe the transportation route, according to Table 5.9.

Life cycle datasets for road transportation sections include trucks, trailers, tires, and diesel consumption, as well as emissions from diesel combustion and tires and breaks abrasion, according to information presented in Sects. 5.1.2.3 and 5.1.2.5, respectively. For pipeline transportation, the electricity used to pump ethanol and the pipeline infrastructure were considered, according to the information shown in Table 5.10.

Table 5.9 Scenario definition information for pipeline transportation alternative (Scenario 2)

Section	1	2	3
Origin	Production plant (Ribeirão Preto region)	Distributor (Ribeirão Preto)	Distributor (Paulínia)
Destination	Distributor (Ribeirão Preto)	Distributor (Paulínia)	Final consumer (São Paulo)
Transported volume (m^3 $year^{-1}$)	100,000	100,000	100,000
Transport distance (km)	15	205	125
Type of truck	Truck 480 hp + Tank, 45 m^3	–	Truck 480 hp + Tank, 45 m^3
Pipeline diameter (in)	–	28	–

Table 5.10 Inventory for ethanol pipeline construction and operation

Parameter	Unit	Value
Electricity[a]	kWh tkm^{-1}	0.02
Carbon steel[a,b]	kg tkm^{-1}	4.4E−4
Polyethylene[a,b]	kg tkm^{-1}	8.9E−5

[a]The pipeline is projected to transport 16.9 billion liters of ethanol per year, with an annual operation period of 330 days

[b]Infrastructure material consumption consider a pipeline with inside diameter of 28 in, 12.7 mm wall thickness, lined with polypropylene (10 mm), with a lifespan of 25 years

Table 5.11 Greenhouse gas emissions for different ethanol transport alternatives, in $kg_{CO2\ eq}m^{-3}$

	Scenario 1 (road transportation)	Scenario 2 (road and pipeline transportation)
Diesel production	6.8	2.7
Electricity production	–	0.8
Truck construction	3.5	1.4
Pipeline construction	–	0.3
Truck operation	49.9	20.3
Pipeline operation	–	0.0
Total	60.2	25.5

Greenhouse gas emissions of ethanol transportation scenarios are shown in Table 5.11. Values include total emissions in transport system life cycles and show an expressive reduction when ethanol is transported by pipelines in part of transportation route (Scenario 2), mainly due to smaller amount of diesel used for road transportation.

5.2 Biorefinery Products Use

In order to complete the assessment of the entire sugarcane production chain, the last sector to be simulated is the one that includes use of the different products produced in the biorefinery.

According to Weidema et al. (2004), all products are produced to perform one or more functions, providing one or more services, fulfilling one or more customer requirements. In view of that, to define a functional unit for the biorefineries' products may be quite subjective, since it requires a quantified description of the performance of the product under the end-user perspective. Sugar and electricity, for example, may have a very wide range of applications, satisfying the end user's needs in different ways. Besides the use as a fuel, ethanol can be used as a feedstock for diverse industrial processes, from plastic production to the synthesis of intermediates for the chemical industry.

In view of the wide variety of uses for the products of a biorefinery, the main focus of Log&UsoSoft model is on the environmental assessment of ethanol use as an automotive fuel. This choice does not limit the assessment possibilities, since a traveler may use its car with different purposes, following different routes, under different traffic conditions. All these factors affect the use performance of the fuel, and each specific situation could be assessed, but it is not the purpose of the first version of the model. In this assessment tool, a simplified functional unit—the use of vehicles to transport passengers by a certain distance—was defined, with no specific concerns about the motivations, which lead the user to need the vehicle.

The assessment of the use of fuels in motor vehicles using Log&UsoSoft model allows identifying the different sources for airborne emissions, ranking the main pollutants. It is also possible to assess potential measures for increasing fuel economy and reducing pollutant emissions, by evaluating the effects of implementation of new technologies and measures for emission control.

Combined with historical data, fleet estimates and projection of technological developments, the Log&UsoSoft model can be used to determine trends in fuel consumption and emissions for a specific geographic region, serving as an important tool for both the production sector and for the formulation of public policies.

5.2.1 Ethanol Use in Passenger Vehicles

Besides some experiments on heavy-duty vehicles, ethanol is primary envisioned to be used as fuel for passenger cars. In the Brazilian context, ethanol is used either as a fuel itself (hydrated ethanol), or in mixture with gasoline (gasohol), according to Brazilian National Agency of Petroleum, Natural Gas and Biofuels (ANP) regulations (ANP 2013b, 2015). Gasohol alternative, named "*C*" *gasoline* in Brazil, consists of a mixture of pure gasoline from oil refining process ("*A*" *gasoline*) and anhydrous ethanol, in the proportions defined by legislation.

Both commercial fuels are included as default options in Log&UsoSoft model. The model contains databanks for fuel consumption and emissions for different vehicles and for the fleet average in Brazil, which will be further presented in this section. Final impacts are presented as fuel consumption and emissions per kilometer.

The Log&UsoSoft model user is able to choose whether to evaluate a specific vehicle or the average fleet of vehicles registered in the most recent year. Another option is the analysis of the current national fleet including also older vehicles with deterioration factors for emissions. The user choice defines the scenario to be evaluated. Results are generated coupling scenario definition with the information from databank, as shown in Fig. 5.3.

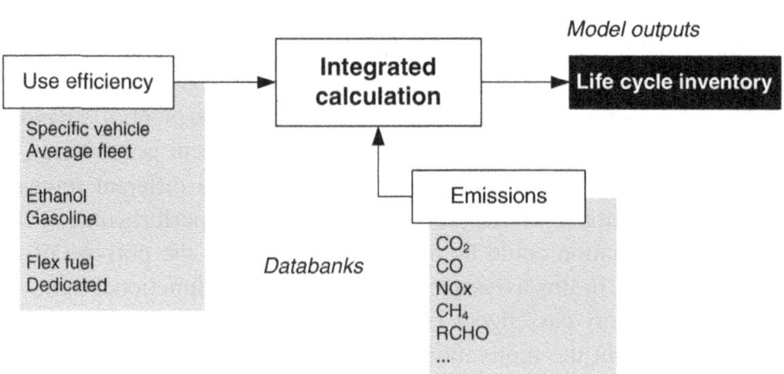

Fig. 5.3 Fuel use modeling scheme

Fuel use is modeled considering all processes that are directly connected with the operation of the vehicles, with focus on fuel consumption and airborne emissions. The reference unit for analysis is one kilometer traveled. Alternatively, fuel energy content (MJ), volume (L), or mass (kg) may be used as reference flows.

It is important to mention that default values for fuel use and emissions factors included in Log&UsoSoft model databanks are average data for hydrous ethanol and standard Brazilian "C" gasoline used in regular Brazilian vehicles. As long as the necessary adjustments are introduced, other fuels, mixtures, or improvements in vehicles technologies can be assessed using the tool.

5.2.1.1 Fuel Consumption Databank

Log&UsoSoft model contains a fuel consumption databank, with information from official literature (INMETRO 2015; CETESB 2011; MMA 2011, 2014) about the Brazilian vehicles and fleets efficiencies. Data are provided by fuel—hydrated ethanol or "C" gasoline—and engine—dedicated or flex fuel.

Fuel consumption per vehicle model were taken from the database of the Brazil's Vehicle Labeling Program (INMETRO 2015), and some examples for the vehicles homologated in 2015 are shown in Table 5.12.

Annual fleet efficiency describes the average efficiency of the totality of vehicles commercialized in a year. Data from the Brazilian Ministry of the Environment (MMA 2014) are used; values for some years are shown in Table 5.13. It is important to mention that the reduction on fuel economy observed in some years for fleet average does not necessarily mean a regression on vehicles efficiencies. For

Table 5.12 Some examples of fuel economy databank for Brazilian vehicles

Vehicle	Fuel economy (km L^{-1})			
	Ethanol—flex		Gasoline—flex	
	Urban	Road	Urban	Road
Citroen C3 Origine 1.5–8 V	7.5	9.3	11.9	14.7
Fiat Palio Fire 1.0–8 V	8.8	10.3	12.3	15.0
Ford Ecosport SE 1.6–16 V	7.0	8.2	10.2	11.8
Ford New Fiesta Hatch S 1.5–16 V	7.6	9.4	10.7	13.4
Honda Civic LXS 1.8–16 V	6.8	9.6	10.5	14.4
Hyundai HB20 Comfort 1.6–16 V	7.7	9.3	11.5	13.0
Hyundai HB20 Plus 1.6–16 V	7.7	9.3	11.5	13.0
Nissan New March SV 1.0–16 V	8.7	10.4	12.5	14.8
Renault Duster Dynamique 1.6–16 V	6.7	7.4	10.0	10.7
Renault Novo Sandero Expression 1.6–8 V	7.3	8.7	10.6	12.5
Toyota Corolla Gli 1.8–16 V	7.3	9.1	10.7	13.2
Volkswagen Gol City 1.0–8 V	7.7	9.6	11.6	13.9

Data from INMETRO (2015)

Table 5.13 Fuel economy of fleet average, per manufacturing year, fuel, and engine

Manufacturing year	Fuel economy (km L^{-1})		
	Ethanol—flex	Gasoline—flex	Gasoline—dedicated
2003	8.5	10.3	11.2
2004	8.6	10.8	11.4
2005	8.3	11.5	11.3
2006	7.8	11.7	11.3
2007	7.4	11.7	11.3
2008	7.8	11.7	9.74
2009	7.8	11.5	9.5
2010	7.7	12.2	9.5
2011	7.3	12.2	9.5
2012	6.9	12.2	9.5

Data from MMA (2014)

some years, fleet average comprises a higher percentage of vehicles with larger engine displacement, which require more fuel. The same occurs for dedicated fleet when compared to flex fuel fleet running with gasoline: Normally, vehicles with lower engine displacement are available to consumers only with flex fuel engines, while dedicated gasoline engines are the only alternative for larger vehicles. Technological advances in vehicles may be offset in the fleet average by a change in buyer habits with a propensity for heavier vehicles, which are less efficient.

5.2.1.2 Emissions Databank

Vehicle use emissions account for airborne emissions due to fuel burning in engines (exhaust emissions), losses from fuel evaporation (evaporative emissions), and particulate emissions due to abrasion of tires, brakes, and road. It includes carbon dioxide (CO_2), carbon monoxide (CO), nitrogen oxides (NO_X), methane (CH_4), non-methane hydrocarbons (NMHC), aldehydes (RCHO), particulate matter (PM), and nitrous oxide (N_2O).

Average emission factors are considered in Log&UsoSoft databank for each particular vehicle model, according to the database of the Brazil's Vehicle Labeling Program (INMETRO 2015). In Table 5.14, some examples for 2015 commercialized vehicles are presented.

Alternatively to specific vehicles, fleet average can be assessed in Log&UsoSoft model. Databank for annual fleet average emissions was calculated considering information from the Brazilian Ministry of the Environment (MMA 2014) and from the Environmental Company of the State of São Paulo (CETESB 2011). Emission factors for annual fleet average in Log&UsoSoft model also accounts for incremental emissions due to vehicles wear and tear, considering average annual use

Table 5.14 Emission factors for some vehicles commercialized in 2015

Vehicle	Pollutants (g km^{-1})		
	NMHC	CO	NOx
Citroen C3 Origine 1.5–8 V	0.009	0.607	0.041
Fiat Palio Fire 1.0–8 V	0.020	0.388	0.019
Ford Ecosport SE 1.6–16 V	0.019	0.475	0.033
Ford New Fiesta Hatch S 1.5–16 V	0.014	0.370	0.020
Honda Civic LXS 1.8–16 V	0.021	0.145	0.030
Hyundai HB20 Comfort 1.6–16 V	0.021	0.325	0.011
Hyundai HB20 Plus 1.6–16 V	0.021	0.325	0.011
Nissan New March SV 1.0–16 V	0.033	0.269	0.047
Renault Duster Dynamique 1.6–16 V	0.032	0.450	0.054
Renault Novo Sandero Expression 1.6–8 V	0.027	0.583	0.035
Toyota Corolla Gli 1.8–16 V	0.021	0.730	0.040
Volkswagen Gol City 1.0–8 V	0.029	0.370	0.020

Data from INMETRO (2015)

according to MMA (2014). Some examples of annual fleet average emission factors are shown in Tables 5.15, 5.16, and 5.17, by fuel and engine type.

The current fleet average emissions, which consider new and older vehicles being used in the country, are calculated from annual fleet data and vehicle sales history, with scrapping rates for older vehicles. Vehicle sales history is obtained from statistical data of the Brazilian Automotive Industry Association (Anfavea 2015) and scrapping criteria from MMA (2014).

For gasoline, all the emissions referred to the standard Brazilian "C" gasoline (for emissions and fuel consumption tests, a standard mixture with 22 % of anhydrous ethanol is adopted) and, for ethanol, to hydrous ethanol (93 %). In

Table 5.15 Fleet average emission factors for hydrated ethanol use in flex fuel vehicles, in g L^{-1}

Manufacturing year	CO	NOx	RCHO	NMHC	CH_4	MP	N_2O
2003	3.931	0.570	0.113	0.905	0.238	–	1.228
2004	7.421	0.544	0.117	0.756	0.361	–	1.257
2005	6.961	0.576	0.111	0.624	0.415	–	1.171
2006	6.340	0.431	0.909	0.580	0.203	–	1.034
2007	6.976	0.524	0.134	0.481	0.074	–	0.926
2008	5.116	0.675	0.127	0.779	0.234	–	1.034
2009	4.719	0.640	0.122	0.737	0.234	–	1.034
2010	3.637	0.827	0.116	0.838	0.308	–	1.008
2011	3.562	1.040	0.105	0.752	0.292	–	0.906
2012	3.519	0.966	0.138	0.759	0.276	–	0.809

Emissions factors were calculated based on data from MMA (2014)
Incremental emissions due to vehicles wear and tear are included for vehicles older than 2012

Table 5.16 Fleet average emission factors for "C" gasoline use in flex fuel vehicles, in g L^{-1}

Manufacturing year	CO	NOx	RCHO	NMHC	CH_4	MP	N_2O
2003	7.806	0.882	0.027	0.707	0.144	0.011	2.440
2004	7.753	0.863	0.022	0.759	0.086	0.012	2.566
2005	7.658	0.851	0.028	0.745	0.104	0.013	2.777
2006	7.629	0.794	0.032	0.738	0.070	0.013	2.738
2007	9.198	0.837	0.031	0.868	0.234	0.013	2.738
2008	8.170	0.876	0.041	1.159	0.234	0.013	2.738
2009	7.342	0.783	0.039	1.079	0.230	0.013	2.645
2010	6.669	0.745	0.040	1.079	0.366	0.013	2.828
2011	5.159	0.656	0.038	0.767	0.244	0.013	2.828
2012	6.100	0.488	0.049	0.488	0.122	0.013	2.828

Emissions factors were calculated based on data from MMA (2014)
Incremental emissions due to vehicles wear and tear are included for vehicles older than 2012

Table 5.17 Fleet average emission factors for "C" gasoline use in gasoline dedicated vehicles, in g L^{-1}

Manufacturing year	CO	NOx	RCHO	NMHC	CH_4	MP	N_2O
2003	9.944	1.967	0.058	1.407	0.302	0.012	0.235
2004	8.982	1.595	0.058	1.349	0.342	0.013	0.239
2005	8.203	1.514	0.056	1.285	0.226	0.012	0.237
2006	7.481	1.332	0.032	1.006	0.226	0.012	0.237
2007	6.849	1.260	0.030	0.951	0.226	0.012	0.237
2008	5.730	0.632	0.019	0.478	0.097	0.011	0.234
2009	3.785	0.362	0.021	0.350	0.057	0.010	0.238
2010	3.008	0.390	0.017	0.299	0.086	0.010	0.219
2011	2.782	0.321	0.020	0.284	0.124	0.010	0.200
2012	2.375	0.285	0.016	0.133	0.247	0.010	0.200

Emissions factors were calculated based on data from MMA (2014)
Incremental emissions due to vehicles wear and tear are included for vehicles older than 2012

addition to the emission factors shown in Tables 5.14, 5.15, 5.16, and 5.17, the CO_2 emissions from combustion are still taken into account: 1178 g L^{-1} for hydrated ethanol and 2041 g L^{-1} for "C" gasoline. These factors were calculated considering fuels properties and carbon content. It is worth noting that other emissions could be included, such as emissions from sulfur and metals from fuel composition, especially for gasoline, but only regulated emissions were accounted for.

Emissions of particulate matter due to tire and brakes abrasion (0.0138 g km^{-1}) and road wear (0.0075 g km^{-1}), independents on the fuel type and vehicle efficiency, are also accounted for (MMA 2014).

It is important to mention that carbon-based emissions for ethanol are classified as biogenic emissions in life cycle inventories modeling. For "C" gasoline, the carbon content partially comes from anhydrous ethanol in the mixture. Since it is not possible to identify the origin of each elementary flow for emissions, these are divided into biogenic and fossil origins based on the proportion of carbon in the standard “C” gasoline mix: 16 % of carbon from ethanol and 84 % from gasoline, considering an average molecular formula C_8H_{18} for “A” gasoline.

5.2.2 Model Adaptation to Other Fuels

It needs to be emphasized that besides Log&UsoSoft model was constructed for ethanol use in average Brazilian vehicles, it also can be used for evaluation of other fuels and vehicles technologies by changing some parameters.

Specific fuel consumption and emissions factors for other fuels, either pure or in mixtures, and also for new automotive technologies, can be included in Log&UsoSoft model. In order to be consistent with the basic data in the model, results from standardized road tests should be preferred to include the new emission factors and fuel use efficiencies. Alternatively, they can be estimated based on preliminary tests, the literature review or targets proposed by innovation programs.

In the study from Pereira et al. (2014), Log&UsoSoft model was used to assess butanol use as a biofuel in an average flex engine vehicle. Adapted information from average flex fuel fleet was used, considering that energy efficiency of engines does not change as a function of the fuel used. The amount of butanol used per kilometer was estimated by means of its energy content, and exhaust emissions are considered to be similar to that from gasoline.

5.3 Log&UsoSoft Application

The assessment tool presented in this chapter, Log&UsoSoft model, is a structured tool to account for economic, environmental, and social impacts of biofuel transportation, commercialization, and use. Results from this model can be used to support decision of producers, commercialization agents, or even end users, providing information, for example, about the best option to locate a mill, the cheapest transportation mode or the fuel with better environmental performance.

Applied at regional or national scales, this model has the potential to provide scientific-based information for public decisions related to the expansion of biofuels distribution and commercialization structures, previously to their effective implementation. It can also be used to calculate sustainability effects of gradual implementation of new energy and greenhouse gases policies for biofuels, or even to measure the benefits of new technologies to reduce fuel consumption or emissions in vehicle operation.

References

Anfavea (2015) Brazilian Automotive Industry Yearbook—2015. Associação Nacional dos Fabricantes de Veículos Automotores

Araújo MP, Bandeira RAM, Campos VBG (2014) Custos e fretes praticados no transporte rodoviário de cargas: uma análise comparativa entre autônomos e empresas. J Transp Lit 8 (4):187–226

ANP (2013a) Resolution ANP N° 50, from December 23, 2013, about diesel composition. Brazilian National Agency of Petroleum, Natural Gas and Biofuels

ANP (2013b) Resolution ANP N° 40, from October 25, 2013, about the regulation and specifications of automotive gasolines. Brazilian National Agency of Petroleum, Natural Gas and Biofuels

ANP (2014) Resolution ANP N° 58, from October 17, 2014, about regulation of the market of fuels. Brazilian National Agency of Petroleum, Natural Gas and Biofuels

ANP (2015) Resolution ANP N° 19, from April 15, 2013, about the regulation and specifications of anhydrous and hydrated ethanol fuel. Brazilian National Agency of Petroleum, Natural Gas and Biofuels

Banchi AD, Lopes JR, França LS (2006a) Gerenciamento técnico de pneus em frotas Parte I. Revista Agrimotor, junho 2006, pp 8–9

Banchi AD, Lopes JR, França LS (2006b) Gerenciamento técnico de pneus em frotas Parte II. Revista Agrimotor, julho 2006, pp 8–10

CETESB (2011) 1° Inventário de emissões antrópicas de gases de efeito estufa diretos e indiretos do Estado de São Paulo. Secretaria do Meio Ambiente. Companhia Ambiental do Estado de São Paulo, São Paulo

CNT (2015) Boletim Estatístico. Confederação Nacional do Transporte Transporte Atual 21 (236):71–74

EsalqLog (2013) Anuário 2013. Grupo de Pesquisa e Extensão em Logística Agroindustrial. Departamento de Economia, Administração e Sociologia. Escola Superior de Agricultura "Luiz de Quiroz", Universidade de São Paulo

INMETRO (2015) Programa Brasileiro de Etiquetagem Veicular. Instituto Nacional de Metrologia, Qualidade e Tecnologia, 2015

Logum (2014) Estudo de Impacto Ambiental - Projeto Logum, Trecho Paulínia - RMSP – Santos, vol 1, Fevereiro de 2014

Milanez AY, Nyko D, Garcia JLF, Xavier CEO (2010) Logística para o etanol: situação atual e desafios futuros. BNDES Setorial 31:49–98

MMA (2011) 1° Inventário Nacional de Emissões Atmosféricas por Veículos Automotores Rodoviários – Relatório Final. Ministério do Meio Ambiente, Brasil

MMA (2014) Inventário Nacional de Emissões Atmosféricas por Veículos Automotores Rodoviários, 2013 – Ano Base 2012 – Relatório Final. Ministério do Meio Ambiente, Brasil

Pereira LG, Dias MOS, Junqueira TL, Pavanello LG, Chagas MF, Cavalett O, Maciel Filho R, Bonomi A (2014) Butanol production in a sugarcane biorefineryusing ethanol as feedstock. Part II: Integration to asecond generation sugarcane distillery. Chem Eng Res Des 92:1452–1462

Santos CE (2012) Competitividade da intermodalidade no transporte de açúcar na região de Araçatuba. Grupo de Pesquisa e Extensão em Logística Agroindustrial—ESALQLOG. Universidade de São Paulo Escola Superior de Agricultura "Luiz de Queiroz", Piracicaba, p 22. http://goo.gl/P3uBg3

Spielmann M, Bauer C, Dones R, Tuchschmid M (2007) Transport services. ecoinvent report No. 14. Swiss Centre for Life Cycle Inventories, Dübendorf

Weidema, B, Wenzel, H, Petersen, C, Hansen, K (2004) The product, functional unit and reference flows in LCA. Environmental News No. 70, Danish Ministry of the Environment, Environmental Protection Agency, 2004

Chapter 6
Sustainability Assessment Methodologies

Marcos D.B. Watanabe, Lucas G. Pereira, Mateus F. Chagas, Marcelo Pereira da Cunha, Charles D.F. Jesus, Alexandre Souza, Elmer C. Rivera, R. Maciel Filho, Otávio Cavalett and Antonio Bonomi

One of the challenges in respect of evaluating the sustainability of new—and especially alternative—technologies is to provide an assessment (and a comparison) of their impacts, mainly due to two aspects: (i) the inter-connections among the three pillars of sustainability (social, environmental, and economic) and (ii) the interactions over the production chain involved in any product of the economy.

It is worth to mention the continued development and improvement of methodologies and approaches in this field. Part of the advances comes from the combination and integration of methodologies, including the linkages among economic and environmental models.

This chapter presents some of the most useful techniques applied in the Virtual Sugarcane Biorefinery (VSB) to support both the sustainability assessment and the decision-making regarding many products and process based on biomass, especially those obtained from sugarcane.

The first Sect. 6.1 describes the techno-economic analysis used to evaluate and compare the economic competitiveness of the different technologies. The second Sect. 6.2 presents methodologies to assess direct and indirect socioeconomic

M.B.D. Watanabe · L.G. Pereira · M.F. Chagas · C.D.F. Jesus · A.M. Souza · E.C. Rivera · O. Cavalett · A. Bonomi (✉)
Brazilian Bioethanol Science and Technology Laboratory (CTBE), Brazilian Center of Research in Energy and Materials (CNPEM), Campinas, São Paulo, Brazil
e-mail: antonio.bonomi@bioetanol.org.br

O. Cavalett
e-mail: otavio.cavalett@bioetanol.org.br

L.G. Pereira
Department of Civil Engineering, University of Toronto, Toronto, ON, Canada

M.P. Cunha
Institute of Economics (IE), University of Campinas (UNICAMP), Campinas, São Paulo, Brazil
e-mail: marcelocunha@eco.unicamp.br

R.M. Filho
School of Chemical Engineering, University of Campinas (UNICAMP), Campinas, São Paulo, Brazil

A. Bonomi et al. (eds.), *Virtual Biorefinery*, Green Energy and Technology,
DOI 10.1007/978-3-319-26045-7_6

effects, mostly relied on Input-Output Analysis (IOA). In Sect. 6.3, Social Assessment is discussed under the Social Life Cycle Assessment (S-LCA) concept, based on IOA. Regarding the environmental assessment, Sect. 6.4.1 shows one of the most used methodology for this purpose, the Life Cycle Assessment (LCA), which identifies and quantifies the energy, materials used, and waste released during the entire life cycle of the product analyzed. Section 6.4.2 deals with the Hybrid LCA based on Input–Output Analysis, a powerful strategy taking the advantage of both methodologies. Issues involving uncertainties and risk analysis are covered in Sect. 6.5. Finally, Sect. 6.6 shows an Optimization Strategy tool looking for assessing the sustainability attractiveness of different technologies, taking into consideration the three pillars of sustainability.

6.1 Techno-Economic Analysis

Although environmental concerns are always in the core of sustainability debate, economic issues related to the decision-making process are key elements that are considered for investors and policy makers for adopting and/or promoting alternative products and technologies.

A traditional and well-established approach for this subject is based on engineering economics, a technique which combines elements of engineering and microeconomics to evaluate the time value of money based on the cash flow analysis. This section details most of the applied parameters and variables in this area and which have been largely explored by VSB studies.

The concepts of engineering economics introduced in this chapter are applied in the techno-economic analysis within Chaps. 7 and 8. The same concepts have also been applied in previous studies. The assessment of first- (Cavalett et al. 2012) and second-generation biorefineries (Dias et al. 2011, 2012) and further studies related to straw recovery systems (Cardoso et al. 2013, 2015), n-butanol production (Dias et al. 2014; Mariano et al. 2013a, b; Pereira et al. 2014a, b, 2015a), and use of vinasse for anaerobic digestion (Moraes et al. 2014, 2015) as well as other assessments associated with a wide variety of technology routes were important to consolidate the approach used to assess the economic sustainability of diverse biorefinery technologies in the VSB.

In order to provide a comparison among biorefinery projects, the main parameters used in engineering economics such as payback period, net present value (NPV), internal rate of return (IRR), and the production costs are employed in the VSB. The inputs for such evaluations are based on the engineering economics principles (Grant et al. 1990; Newnan et al. 2004). The assessment is based on a cash flow analysis for each project alternative, taking into account the investment and all expenses and revenues for an expected project lifetime.

The assessment based on a cash flow analysis depends on data associated with the investment required to build a biorefinery as well as on its main cash inflows (revenues) and outflows (expenses). These are based on technical parameters from

Table 6.1 Main parameters considered in the techno-economic analysis

Item	Value	Reference
Expected plant lifetime	25 years	Assumption
Discount rate	12 % per year	Assumption
Reference date	July 2014	Assumption
Exchange rate	2.30 BRL per US$	Market data
Depreciation	10 years, linear	Assumption
Anhydrous ethanol price	1.34 BRL per L	CEPEA (2014)
Hydrous ethanol price	1.16 BRL per L	CEPEA (2014)
Sugar price	1.00 BRL per kg	CEPEA (2014)
Electricity price	132 BRL per MWh	MME (2014)

biorefinery alternatives modeling and historical data observed over the last decade for ethanol, electricity, and sugar production costs and market prices (CEPEA 2014; MME 2014; Pecege 2014). A minimum acceptable rate of return (MARR) must be assumed—12 % per year normally would be reasonable for projects of sugarcane mills in Brazil (Pinto 2010; Rangel et al. 2010; Bini and Madail 2011). Table 6.1 presents other examples of important parameters that must be assumed in the cash flow analysis. The values presented in Table 6.1 were used in the economic assessments of the Chaps. 7 and 8.

Concerning the biomass costs, most of the studies deal with the concept of vertically integrated model. In the VSB, this definition means that the biomass cost, which is an output of CanaSoft model, is considered as an input to the industrial cash flow analysis. It is worth to mention that, in this case, the agricultural system is fully integrated to the industrial scenarios, i.e., biomass production costs are directly affected by the industrial scenarios since they consider parameters such as the average distance between farm and industry, for example, or even the influence of industrial residues such as vinasse—which is recycled to the sugarcane field, thus decreasing the expenses with fertilizer, for example—on the biomass production cost.

6.1.1 Investment Estimations

Depending on the biorefinery project information, several approaches and levels of accuracy can be observed when estimating the capital costs during a project evaluation. As a result, there is a trade-off between the knowledge about the conceptual project of the new plant and how accurate the cost estimate needs to be in order to achieve its purpose (Peters et al. 2002; Turton et al. 2009). Thus, performing a good capital cost estimation of a sugarcane biorefinery is a very important step toward its actual implementation.

Regarding the quality and accuracy of such estimate, Perry and Green (2008) present the following classification: 'Order of Magnitude,' 'Study Estimate,' 'Preliminary Estimate,' 'Definitive Estimate,' and 'Detailed Estimate.' Each class has its own assumptions and appropriate methods. The range of the expected accuracy may differ from source to source, but this classification is pretty usual.

In case of lack of information or even new, immature and yet unavailable technologies, the Order of Magnitude Estimate is expected. That is the case, for example, of most of cellulosic ethanol technologies. On the other hand, whenever more data are available, the Study Estimate is more likely to be adequate. First-generation sugarcane biorefineries are well-known plants and its conceptual design is, in some ways, standardized. Despite this fact, only Order of Magnitude estimates of the investment are available in the literature and there is a lack of a detailed cost database for assessing the impact of integrating a new technology to an existing plant (brownfield project).

For this reason, three distinct levels of detail are used in VSB. The first one is less detailed because it breaks the 1G plant in its main processing areas and estimates their capital cost as a function of some quantitative value (e.g., processed sugarcane, amount of steam, production of ethanol per hour). The second level is more detailed than the first because cost estimations are based on the cost of the main equipment of each processing area. In the second approach, more flexibility is obtained to compare alternative scenarios and assess technological changes related to specific operations, for example, the use of molecular sieves or extractive distillation to produce anhydrous ethanol.

Finally, there is a more detailed level in which the capital cost data are associated with individual equipment. This approach is employed when support to the previous levels is needed in a given analysis. In this case, classes of equipment, e.g., pumps, heat exchangers, distillation columns, are assessed aiming the development of cost models that take into account the equipment's size, capacity, and construction material to provide the cost estimate.

The VSB's capital cost database was developed by gathering data from literature and from industrial equipment suppliers, engineering offices, experts, and consultant specialists. These data were compiled and processed in order to provide cost models for specific classes of equipment and total capital cost estimates for each processing area of the process (for example, juice extraction and treatment, sugar production, fermentation, and distillation).

For 2G technologies, the investment estimates were performed after defining size and number of units of the main (or more expensive) equipment, based on the conceptual process design and information from process diagrams, mass, and energy balances, as well as operational parameters of the studied scenarios.

The rules used to calibrate and validate this methodology were presented and discussed with specialists and engineering companies to ensure that the capital cost estimates were accurate, credible and in accordance with industry estimates.

6.1.2 Payback Period

Payback period is the period of time required for the profit or other benefits from an investment to equal the cost of the investment. The optimum in all situations is to minimize the payback period. This metric can either be based on simple or discounted cash flows.

6.1.3 Net Present Value

The NPV or net present worth is the difference between the cash inflows and the present value of cash outflows. NPV is an indicator of how much value an investment or project adds to the firm or business. It depends on the number of time periods (n), the net cash flow during the period (C_n, which can be either positive or negative), and the discount rate (r), following Eq. 6.1:

$$\mathrm{NPV} = \sum_{n=0}^{n=N} \frac{C_n}{(1+r)^n} \tag{6.1}$$

The project is rejected when NPV is negative. When NPV ≥ 0, the alternative should not be rejected because value would be added to the firm. When considering mutually exclusive projects and NPV alone, the decision rule is to invest in the project with the greatest NPV.

6.1.4 Internal Rate of Return

IRR is the average interest rate paid per year by the evaluated project. The IRR of an investment is discount rate at which the NPV of costs (negative cash flows) of the investment equals the NPV of the benefits (positive cash flows) of the investment. In other words, IRR can be found when NPV equals zero. In general, the Eq. 6.2 cannot be solved analytically but only iteratively.

$$0 = \sum_{n=0}^{n=N} \frac{C_n}{(1+\mathrm{IRR})^n} \tag{6.2}$$

For a project to be accepted, the IRR must be greater than or equal to the MARR.

6.1.5 Production Costs

Evaluating a set of feasible alternatives requires that many costs be analyzed (Newnan et al. 2004). In the VSB, operating and capital expenses are taken into consideration in order to compute the total production cost of biorefinery products. The operating expenses are associated with variable (e.g., sugarcane, chemical inputs, utilities, and enzymes) and fixed costs (mainly maintenance and labor costs) of a biorefinery, in yearly basis.

The total production cost, however, depends also on the capital expenses associated with buildings, equipment, and infrastructure. These expenses will depend on the project lifetime and company's financial leverage (which indicates the proportion of equity and debt the firm is using to finance its assets). Most of the studies using the VSB assume a 25-year project lifetime and no financial leverage, i.e., the firm is financed by equity. Therefore, the yearly capital cost of a biorefinery is estimated by taking into account the annual payment that would be necessary to remunerate the total investment as it was a loan (12 %-per-year interest rate over a 25-year period).

The total production costs are obtained by summing operating and capital expenses. In the case of ethanol production, the cost per liter would be the yearly total cost divided by the number of liters of ethanol produced over the year. In the example illustrated in Table 6.2, the biorefinery is producing a single product: ethanol.

An alternative methodology is to calculate the average production cost based on the cash flow. Once its structure is modeled in electronic spreadsheets (MS Excel, for instance), it is possible to calculate the levelized production cost based on the average ethanol price by which the IRR would equal the MARR. Usually, a

Table 6.2 Illustration of ethanol production cost breakdown

Cost elements	Total production cost
Variable costs	x
Sugarcane	…
Straw	…
Chemical inputs	…
Utilities	…
Enzymes	…
…	…
Fixed costs	y
Labor	…
Maintenance	…
…	…
Capital expenses	z
Total production cost (US\$ year^{-1})	$x + y + z$
Overall production (L year^{-1})	p
Average cost (US\$ L^{-1})	$(x + y + z)/p$

biorefinery has other coproducts (sugar and electricity, for instance). In this case, the production cost would be calculated using a price factor (*k*) which proportionally adjusts ethanol and other coproduct's prices in order to calculate the IRR.

6.1.6 Cost Allocation

In the VSB, the total cost and its elements are usually allocated among the biorefinery products according to their share on the total revenues. This approach is particularly important to determine the cost breakdown of ethanol, sugar, electricity, and other biorefinery products. This allocation approach has been applied to estimate the second-generation ethanol costs in the integrated first- and second-generation ethanol production (Milanez et al. 2015; Macrelli et al. 2012). A good illustration of the procedure is presented in three steps (see Fig. 6.1). First, a biorefinery whose main products are 1G ethanol and bioelectricity is represented. All operating and capital expenses are allocated according to the ethanol and electricity's share on the total revenues. For instance, if 80 % of revenues come from ethanol and 20 % from electricity production, this proportion will be maintained to allocate the expenses associated with sugarcane, chemical inputs, maintenance, human labor, capital costs, and so on.

In the second step, a scenario whose production integrates first- and second-generation ethanol (1G2G) and electricity is evaluated. In this case, the 1G ethanol production system is identical to the 1G biorefinery in the first step. Therefore, it is possible to determine the 1G2G ethanol cost and its main components. In this case, other inputs such as straw and enzyme costs can also be included as elements of the total costs. These costs are allocated between electricity and ethanol according to their share on the revenues.

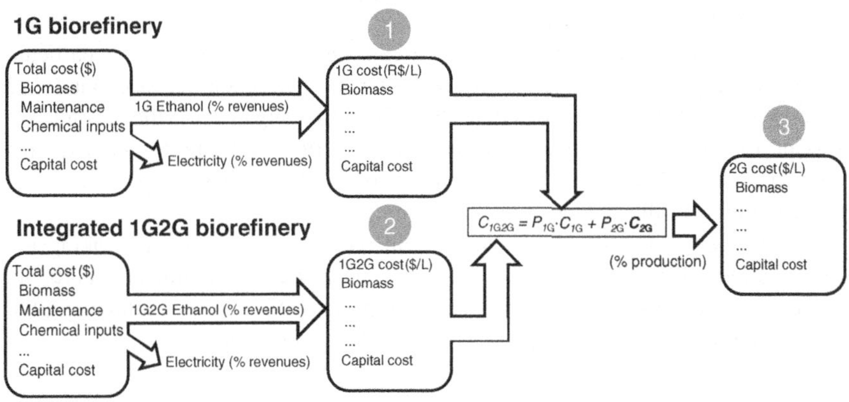

Fig. 6.1 Illustration of cost allocation for first- and second-generation ethanol

The third and final step is to calculate the second-generation ethanol production cost (2G cost). It is assumed that 1G2G ethanol cost is the weighted average of first- and second-generation ethanol costs according to their share on the total ethanol production. This premise is represented by Eq. 6.3:

$$C_{1G2G} = P_{1G}C_{1G} + P_{2G}C_{2G} \tag{6.3}$$

where

P_{1G} percentage share of first generation on the total 1G2G ethanol production (%),

P_{2G} percentage share of second generation on the total 1G2G ethanol production (%),

C_{1G} first-generation ethanol production cost (US\$ L^{-1}),

C_{2G} second-generation ethanol production cost (US\$ L^{-1}), and

C_{1G2G} integrated first- and second-generation ethanol production cost (US\$ L^{-1}).

6.2 Input–Output Analysis

The evaluation of socioeconomic impacts of any economic activity[1] needs to take into account the interdependence of industries in an economy. This section describes the principles and applications of IOA, one of the most used frameworks in applied economics which has been developed by 1937 Nobel Prize in Economic Science Wassily Leontief.

The assessment of socioeconomic impacts on a particular economic activity typically includes quantifying the impacts on (i) the level of sectoral production (i.e., the total output), (ii) jobs created, and (iii) the income generated throughout the whole economy. Evaluating across the whole economy means taking into account the need for production of all sectors involved in the whole production chain, i.e., evaluate the inter-industry flow of goods and services considering all sectors, starting from the first one.

As an example, consider evaluating socioeconomic indicators of a particular sugarcane ethanol technology. In the first step, it is necessary to account for all distillery's direct inputs (for example, the amount of sugarcane); the second is to account for all inputs required by the suppliers of direct inputs (for example, fertilizers and fuel used at sugarcane production); the third is to account for all inputs required in the second step, and so on.

The IOA quantifies the total output in all sectors of the economy, taking into consideration all direct and indirect inputs to meet the production of a particular sector. The quantification of each sectoral production allows estimating other socioeconomic variables of interest, such as jobs created and income.

[1]In this chapter, economic activity (or just activity), sector, and industry are synonyms.

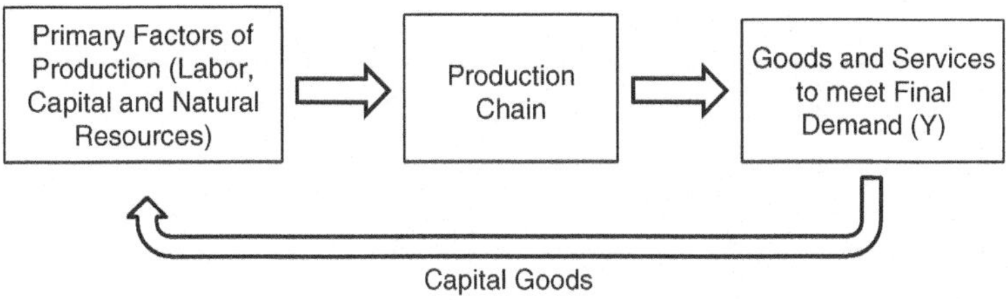

Fig. 6.2 A simplified schematic view of an economy

In the VSB, IOA is a powerful and applied framework to compare socioeconomic indicators and impacts related to different biorefineries routes. The description of IO methodology is made as follows.

The development of IOA is attributed to Professor Wassily Leontief in 1930s (Miller and Blair 2009). This section presents the main principles of this framework.[2]

Consider Fig. 6.2, which shows a simplified description of an economy. This figure illustrates that all products and services destined to meet Final Demand (composed by household consumption, exports, government purchases, and domestic investment—capital goods) come from the production chain, which, in turn, depends on the use of primary factors of production (labor, capital, and natural resources). It is important to notice that capital goods (part of the sectoral production destined to final demand) return as primary factors (capital) to replace capital depreciation and to enlarge the production capacity of the economy.

The simplified scheme in Fig. 6.2 suggests that all goods and services to meet the final demand, in certain time period (typically one year), have embodied primary factors, such as labor and capital. In this sense, it is very comprehensive to think about 'how many jobs are necessary to produce one million liters of biofuels' or 'how many jobs are embodied in one million liters of biofuels.' Of course, this answer will depend on the technology route and on the production chain involved in that biofuel.

The 'Production Chain' box in Fig. 6.2 does not explicit the inter-industry flows of goods and services which are used as inputs to the economic activities. Looking for accounting the total output of all sectors to meet final demand, Fig. 6.3 shows a typical transactions table showing an economy aggregated in three sectors.

Usually, the transactions table presents monetary values with respect to one year. In Fig. 6.3, the sectors (Sector 1, Sector 2, and Sector 3) are the same both in rows and in columns. When observing the columns, sectors can be interpreted as consumers. For example, $Z_{2,3}$ corresponds to the inputs consumed by Sector 3 (S_3) and supplied by Sector 2 (S_2). Therefore, the elements $Z_{i,j}$ (or matrix Z in gray color) correspond to the intermediate consumption (or the inter-industry flows of good and services) of the

[2]Miller and Blair (2009) describe in more detail the principles, extensions, and applications of IOA.

Transactions Table	Sector 1 – S_1	Sector 2 – S_2	Sector 3 – S_3	Final Demand (Y)	Output (X)
Sector 1 – S_1	$Z_{1,1}$	$Z_{1,2}$	$Z_{1,3}$	Y_1	X_1
Sector 2 – S_2	$Z_{2,1}$	$Z_{2,2}$	$Z_{2,3}$	Y_2	X_2
Sector 3 – S_3	$Z_{3,1}$	$Z_{3,2}$	$Z_{3,3}$	Y_3	X_3
Value Added (VA)	VA_1	VA_2	VA_3		
Output (X)	X_1	X_2	X_3		

Fig. 6.3 A typical transactions table showing an economy aggregated in three sectors

economy in a given year. The 'Value-Added' row identifies all taxes paid by each sector, employee compensation, and capital remuneration. In this sense, both $Z_{i,j}$ and VA_j elements of a Sector j (S_j) (read on the columns) are the expenses of this sector, whose sum is equal to the output of the activity (X_j); mathematically:

$$X_j = \sum_{i=1}^{n} z_{i,j} + VA_j \tag{6.4}$$

for all $1 \leq j \leq n$where n denotes the number of the sectors on the economy (three in Fig. 6.3).

Returning to the Transactions Table in Fig. 6.3, the elements identified on the rows (Sector 1 to Sector 3) are interpreted as the producers of goods and services; for example, $Z_{1,3}$ represents that Sector 1 providing goods and services to Sector 3; Y_1 identifies the portion of Sector 1 production to meet final demand. Therefore, the elements identified on the rows are understood, as well, as the 'revenues' of those sectors; mathematically:

$$X_i = \sum_{j=1}^{n} z_{i,j} + Y_i \tag{6.5}$$

for all $1 \leq i \leq n$.

The main assumption on IOA is that the use of intermediate consumption and primary factors is in fixed proportions to the output in each sector; mathematically:

$$a_{i,j} = \frac{z_{i,j}}{X_j} \tag{6.6}$$

Matrix $A = [a_{i,j}]$ defined in Eq. 6.6 is called matrix of direct technical coefficients. From Eqs. 6.5 and 6.6,

$$X_i = \sum_{j=1}^{n} a_{i,j} \cdot X_i + Y_i \tag{6.7}$$

for all $1 \leq i \leq n$.

Transactions Table	Agriculture - S_1	Industry - S_2	Services - S_3	Final Demand (*Y*)	Output (*X*)
Agriculture - S_1	17	165	14	137	333
Industry - S_2	75	1,043	325	1,730	3,173
Services - S_3	27	533	870	2,504	3,934
Value Added (VA)	214	214	214		
Output (*X*)	333	3,173	3,934		

Fig. 6.4 A simplified transactions table for Brazil in 2011 (billion Reais)

Equation 6.7 makes explicit that the output of sector i needs to meet the intermediate consumption by the all sectors and final demand. Equation 6.7 can be written in a matrix form as follows:

$$X = AX + Y \tag{6.8}$$

Solving Eq. 6.8 for *X*, the solution is given as:

$$X = (I - A)^{-1}Y \tag{6.9}$$

Equation 6.9 corresponds to the well-known (linear) Open Leontief Model. This model captures all direct and indirect effects on total sectoral production (for each sector on the economy) to meet a given final demand. Matrix $(I - A)^{-1}$ is named 'Leontief Inverse Matrix'; their elements depend, of course, on the *A* Matrix, which reflects the economic structure of a given economy.

For example, Fig. 6.4 shows an estimation of a simplified Transactions Table for the Brazilian Economy in 2011 based on three aggregated sectors; all values in this table are accounted in billion Reais (Real—BRL—is the Brazilian currency[3]).

Using the concepts described on Eqs. 6.6 and 6.9:

$$(I - A)^{-1} = \begin{bmatrix} 1.076 & 0.087 & 0.014 \\ 0.385 & 1.561 & 0.167 \\ 0.195 & 0.346 & 1.322 \end{bmatrix} \tag{6.10}$$

In matrix (6.10), it is possible to conclude that using Eq. 6.9 to meet final demand in US$ 1.000 billion for Brazilian industrial sector in 2011, taking into consideration all direct and indirect effects on the Brazilian production chain, the total output would be US$ 0.087 billion for agriculture, US$ 1.561 billion for industry, and US$ 0.346 billion for services. From the description of technical coefficients related to jobs for each sector, it is possible, of course, to quantify the total impacts on jobs creation.

[3]In 2011, US$ 1.00 = R$ 2.00, on average.

The main source of information to provide an IOA for any country is the Transactions Table; in this sense, it is important to mention that the construction of such table is based on the information of the System of National Accounts (SNA), which is standardized by United Nations.

For the purposes of VSB, Eq. 6.9 is often used to evaluate and compare the main socioeconomic impacts of a biorefinery route including the direct technical coefficients of this technology on an original Matrix A; using this approach, another Matrix A (A^*) is obtained and, from this Matrix A^*, another $(I - A^*)^{-1}$ is calculated and the comparisons are estimated. In conclusion, this approach allows evaluating changes in socioeconomic indicators and/or impacts due to the inclusion of this technology on the economy, but taking into account the whole intersectoral structure of that economy and computing all direct and indirect effects on the production chain—with the inclusion of a new biorefinery technology.

6.3 Social Assessment

A social impact can be either positive or negative depending on the changes that people experience (UNEP/SETAC 2009). The adoption of a new technology by an organization, for example, may cause changes in the number of job positions, working conditions, community health quality, and so on. Although these social disturbances directly and indirectly affect human well-being, the social impact assessments are still much less explored when compared to environmental and techno-economic viability assessments (Ribeiro 2013).

According to Burdge (2002), the assessment of social impacts has not been widely adopted due to the lack of proper tools. Indeed, the complexity of the social behavior makes the quantification of social impacts a hard task. In order to deal with the complexity involved in social assessments, the main concepts of S-LCA are applied along with a variety of tools in the VSB, as shown below.

6.3.1 Social LCA Concept

The S-LCA is a product-oriented method that aims at assessing social and socio-economic aspects of products, including their potential positive and negative impacts along their life cycle (UNEP/SETAC 2009, 2011). Once the S-LCA is in its infancy, there is not defined common framework and there is a long way to go through (Jørgensen et al. 2008; Jørgensen 2013; Macombe and Loillet 2013).

Macombe and Loillet (2013) highlight the S-LCA as a tool to forecast social effects considering base and future scenarios (Macombe 2013). Therefore, S-LCA would be able to anticipate the social consequences of changes such as the adoption of a new technology. This particular feature is extremely important to the Virtual Sugarcane Biorefinery.

In the VSB, the S-LCA is being implemented to predict social effects associated with technological changes in both agricultural and industrial scenarios. The S-LCA relies both on outputs from the VSB models (CanaSoft, for instance) and their correlation with social effects. For instance, the total number of workers involved in sugarcane production can be estimated in simulations performed in CanaSoft. Then, a given social effect—total number of accidents, for instance—of a certain technology can be quantified once a quantitative correlation between social effect and output from the inventory is stablished.

It is important to point out that the S-LCA should not be restricted to the organization level, i.e., the social effects observed in the agricultural and industrial phases should be extended to a broader scope, such as the entire sugarcane production chain. This kind of approach would add more information to the decision-making processes which would consider the effects on the sugarcane and its supporting sectors, such as the transportation, fertilizer, and chemical industry.

In order to account for the broader social effects of a given technology, the IOA is being integrated to other methods—such as the LCA inventories—in order to make the quantification of social effects possible in the VSB, considering the entire production chain.

6.3.2 Social Assessment Based on Input–Output Analysis

In the VSB framework, the IOA has been initially applied to assess traditional impacts associated with the introduction of a new biorefinery technology into the Brazilian economy—such as changes in industrial outputs and value-added. However, the IOA can also be explored in a broader perspective and give support to other assessments dealing with environmental and social aspects.

Leontief (1970) has originally explored the importance of considering undesirable—or desirable—external effects of productive or consumptive activities of the economic system in the context of IOA. Therefore, pollution and other externalities would be incorporated into the conventional IO picture of a national economy. The technical interdependence between the levels of desirable and undesirable outputs would be described in terms of structural coefficients similar to those used to trace the structural interdependence between all the regular branches of production and consumption.

According to Miller and Blair (2009), the generalization of IOA techniques to a wider conceptual level, such as accounting for social indicators in so-called social accounting matrices, began with simpler efforts to link IO models and other national income accounting techniques with many measurable quantities associated with inter-industry activity such as energy use, environmental pollution, and employment.

The Economic Input–Output Life Cycle Assessment (EIO-LCA) is another example of extending the IOA to a broader conceptual view. The EIO-LCA technique has been initially conceived in the Carnegie Mellon University's Green Design Institute (Hendrickson et al. 2006) to determine the main environmental burdens associated with sectorial production in the US economy. As mentioned in Sect. 6.2, the Leontief inverse $(I - A)^{-1}$ allows calculating the vector including all suppliers' outputs (X) based on an incremental change in final demand (Y) given the direct requirements table (A), according to the Eq. 6.9:

$$X = (I - A)^{-1} Y \tag{6.9}$$

A vector of the total external impacts (E) can be obtained according to the expression (6.11), where R is a diagonal matrix used to denote the external impact in sectors:

$$E = r_1 X_1 + r_2 X_2 + \cdots + r_n X_n = RX = R[I - A]^{-1} Y \tag{6.11}$$

A wide variety of impacts can be included in the matrix R; resource inputs such as energy and environmental burdens can be accounted in case enough information is available. In other words, monetary results can be translated into a variety of categories of environmental and socioeconomic effects. According to Hendrickson et al. (2006), even occupational safety risks can be determined using the IO framework. For instance, it is possible to calculate the occupational injuries embodied in the car-manufacturing sector. Such effects are not restricted to assembling the vehicle. Rather, the injuries accumulate from the mining and transport of coal and iron ore, making, rolling, and casting steel, and making the components that are assembled into the automobile. Then, assembling the car represents a small fraction of the total occupational injuries from the entire supply chain (Hendrickson et al. 2006).

The EIO-LCA thinking can provide the S-LCA with a quantitative approach which is ready to capture social effects embodied in a product or a service. In the VSB, a hybrid model integrating LCA and IOA was developed based on the EIO-LCA approach. Such hybrid model is used to assess the impacts of a new biorefinery on the entire supply chain considering metrics such as the total job creation, occupational accidents, salary profile, and education profile.

In order to quantify such impacts at sector level, official databases were consulted. Data related to job creation, for instance, are available in previous studies (Guilhoto and Sesso Filho 2005, 2010) based on the Brazilian SNA (IBGE 2015). Data associated with wages, gender, and education profiles for industries are available on the Ministry of Labor and Employment Web site database (MTE 2015). Scenarios were modeled using results from previous studies with the VSB. Working hours and the production costs were used to estimate the number of jobs and the wage profile in sugarcane sector. Educational degree profile was estimated considering the average degree profile for sugarcane sector according to MTE (2015). The number of occupational accidents in the sugarcane sector was estimated

$X = [I - A]^{-1}Y$

Sector	Sector Outputs ($)
Biorefinery	X_1
Sugarcane	X_2
Diesel	X_3
Chemicals	X_4
Transportation	X_5
...	...
Total	$X_1+X_2+\cdots+X_n$

$E = R{\cdot}X = R[I - A]^{-1}Y$

Sector	Number of workers (*E*)
Biorefinery	E_1
Sugarcane	E_2
Diesel	E_3
Chemicals	E_4
Transportation	E_5
...	...
Total	$E_1+E_2+\cdots+E_n$

$E' = RXZ = R[I - A]^{-1}YZ$

Sector	Number of accidents (*E'*)
Biorefinery	E'_1
Sugarcane	E'_2
Diesel	E'_3
Chemicals	E'_4
Transportation	E'_5
...	...
Total	$E'_1+E'_2+\cdots+E'_n$

Fig. 6.5 Quantification of occupational accidents in the biorefinery supply chain based on the EIO-LCA framework (*X* is the vector of industry outputs; $(I - A)^{-1}$ is the Leontief Inverse; *Y* is the vector of final demands; *R* is the diagonal matrix of workers per dollar output, and *Z* is the vector of accidents per worker)

in two steps. First, a linear correlation between the incidence of accidents (number of accidents per worker) in the sugarcane production sector (MPS 2015) and its level of mechanization (IDEA 2012) was established. With this correlation, it was possible to estimate the number of accidents in each biorefinery scenario. A similar procedure was carried out to estimate the parameters in the industrial plants. Additional data related to ethanol production and other sectors of the ethanol supply chain were available in Brazilian official databases (MTE 2015; MPS 2015).

One example of application of EIO-LCA approach is to calculate the number of occupational accidents in a biorefinery (see Fig. 6.5). The first step is to assume an increase in final demand associated with a given biorefinery's product: ethanol, for instance. Its production requires a variety of direct and indirect inputs such as raw materials and services to be produced. Using the Leontief inverse, it is possible to calculate all the industry outputs associated with an increase in ethanol final demand. Furthermore, it is also possible to estimate the number of workers in each sector of the ethanol supply chain. In order to do so, a diagonal matrix '*R*' describing the number of jobs created per dollar output of each sector is required. The following step is to build the vector of occupational accidents per worker (*Z*). Those sectoral data are available on the Web site of the Ministry of Social Security (MPS 2015). Therefore, it is possible to calculate the overall occupational accidents (*E'*) according to the expression (Eq. 6.12):

$$E' = R \cdot X \cdot Z = R[I - A]^{-1}Y \cdot Z \tag{6.12}$$

The example in Fig. 6.5 describes the general methodology to quantify social effects using the EIO-LCA approach, which makes use of industry-by-industry square matrices. In the VSB, however, non-square commodity-by-industry tables are used because more disaggregated information is available on the official databases. Moreover, the assessment of biorefinery technologies implies on adding a set of new industries and commodities into the original IO table in order to represent the technology scenarios. This additional level of complexity implies on the use of a hybrid approach integrating LCA and IO rather than the EIO-LCA. Details of the hybrid approach using commodity-by-industry tables will be described in the next Sect. 6.5.1.

6.4 Environmental Analysis

6.4.1 Environmental Analysis Using Life Cycle Assessment

The evaluation of environmental impacts associated with the production of biofuels and materials from biomass within the VSB framework is performed using the LCA. LCA is a recognized methodology for estimating the environmental burden associated with a product, process, or activity, by the identification and quantification of energy and materials used and waste released, during the entire life cycle, from extraction of raw materials through manufacturing, logistics, use and final disposal, or recycling of the product. The general framework for conducting a LCA is found in the 14040 and 14044 ISO standards (ISO 2006a, b). In Brazil, ABNT has published two standards for LCA: ABNT NBR ISO 14040 (ABNT 2009a) and ABNT NBR ISO 14044 (ABNT 2009b). The method consists of four main steps: goal and scope definition, inventory analysis, impact assessment, and interpretation.

6.4.1.1 Goal and Scope Definition

Goal and scope definition provides the context for the assessment establishing to whom and how results obtained are to be communicated. This step includes the detailing of technical information—such as defining the functional unit, the system boundaries, the assumptions and limitations of the study, the impact categories and the methods that will be used to allocate environmental burdens in cases where there is more than one product or function.

There can be a series of objectives within a LCA study. In the VSB, those objectives are normally focused on the production and use of biofuels and bioproducts derived from biomass feedstock (mostly sugarcane). First objective of the method applied in this context is to compare the environmental performance of technologies and the set of process configurations that characterize the investigated scenarios. Second objective may be defined as the comparison of biobased products generated by the processes studied with fossil-based equivalent products.

Boundaries and functional units for the study are therefore established according to the objectives proposed. In the case of the comparison of technological biorefinery scenarios, boundaries are usually set from field to factory gate and the

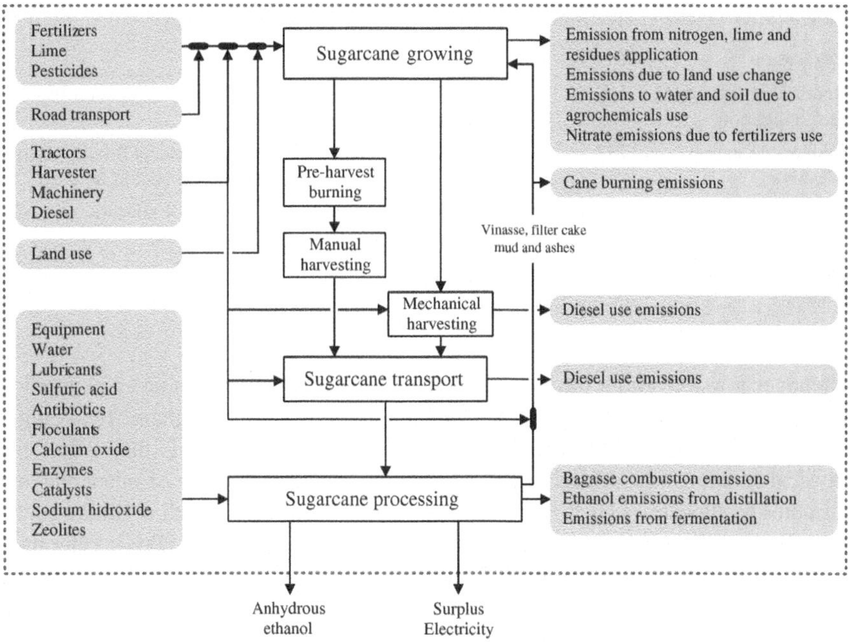

Fig. 6.6 System boundaries considering a field-to-gate approach for the production of anhydrous ethanol and surplus electricity from sugarcane

functional unit is defined as one unit of mass or energy of the product of interest (e.g., anhydrous ethanol, *n*-butanol, and gasoline).

The life cycle of biofuels considers all stages from production, harvesting and transportation of biomass until its final use, including industrial conversion processes and transportation stages when required. Figure 6.6 depicts an example of the system boundary considering a cradle-to-gate approach for the production of anhydrous ethanol and electricity from sugarcane. Direct local emissions are presented on the right side and the main inputs considered in the assessment are shown on the left side. The 'sugarcane processing' box represents the industrial conversion stage.

In the case of comparison of production and use of biofuels versus gasoline as transportation fuel, product system boundaries should be expanded to a well to wheel approach, and the functional unit defined as 1 km travelled by a car (usually a flex fuel vehicle). Detailed explanation on the logistics and use of biofuels as well as gas end-pipe emissions from combustion of such fuels according to the different car engines can be found in Chap. 5.

Besides anhydrous ethanol and electricity, the biorefinery may also produce a series of other materials, chemicals, and biofuels as coproducts. In this case, it bears noting that the allocation procedure in a multiproduct process is a critical issue in LCA. The ISO documents (ISO 2006a, b) recommend avoiding allocation

Table 6.3 Overview of allocation rules used in selected bioenergy directives

Legislation	Region	Allocation method
Renewable transport fuel obligation	UK	Substitution whenever possible, if not allocation based on economic value
Renewable energy directive	EU	Allocation based on energy content except for electricity co-production for which substitution should be used
Low carbon fuel standard	California	Substitution whenever possible, if not allocation based on energy content
Renewable fuel standard	USA	Substitution

whenever possible either through subdivision of certain processes or by expanding the system limits to include the additional functions related to them. If avoiding allocation is not possible, recommendation is the use of methods that reflect the physical relationship such as mass and energy content or the use of other relevant variables to allocate, such as the economic value of products, which is similar to the cost allocation methods in managerial accounting (Guinée et al. 2004). Cherubini et al. (2011), Luo et al. (2009) and Svanes et al. (2011) have discussed the effect of different allocation on LCA studies with impact values varying up to 80 % for the same product in face of the method chosen. Additionally, Waardenar et al. (2012) point out that the choice of allocation method in policy directives has large influence on the outcomes: The uncertainty in the results hampers the optimal use of LCA in the policy context (especially for global warming). Table 6.3 shows the allocation methods chosen by the most important bioenergy directives.

Two allocation criteria are normally adopted in the calculations using the VSB:

- Economic: The splitting is based on economic properties of the multifunctional process products, such as the gross revenues or the expected economic gain. Additionally, as proposed by Pereira et al. (2014a, b), environmental impacts associated with a productive activity may be presented in economic terms. This would be useful for stakeholders concerned with both aspects of sustainability and would also help to identify the importance of policies such as mandated GHG reductions and carbon taxes. The approach also allows the comparative evaluation of scenarios, which generate products providing different services (or functions). The comparison makes sense having in mind that all productive activities have the common purpose of generating economic benefit (profit). In this case, functional unit is defined as one unit of monetary currency generated as revenue in the biorefinery.
- Energy (with coproducts displacement/substitution): Energy partitioning is applied to the main energy products generated considering their lower heating value. Coproducts considered as solvents and feedstock for the chemical industry are treated as avoided products using the product displacement method as discussed by Wang et al. (2011). Coproduct displacement (also termed substitution) is based on the concept of displacing the existing product (usually fossil-based) with the new product (biobased). For example, acetone produced

in the biorefinery process is considered as a renewable chemical: One kg of acetone produced was assumed to displace one kg of acetone from petroleum-based production. Applying the same method, the excess electricity displaces an equivalent amount of grid electricity, thus avoiding GHG emissions and other impacts involved on grid electricity production. This method is more aligned with the US fuel standard (EPA 2010) that establishes GHG emissions reduction targets for biofuels; therefore, it is usually adopted for the comparison of the use of biofuels with gasoline as transportation fuel.

6.4.1.2 Life Cycle Inventory (LCI) Modeling

Life Cycle Inventory (LCI) is the methodological step where an overview is given of the environmental interventions (energy use, resource extraction, or emission to an environmental compartment) caused by or required for the processes within the boundaries of the studied system.

Using the VSB framework, the LCI modeling is performed using different tools. Agricultural information is obtained from the CanaSoft model that generates a comprehensive inventory for sugarcane (and other biomass) production systems (Chap. 3). The sugarcane industrial biorefinery alternatives' inventories are based on mass and energy balances calculated using computer simulation platforms (Chap. 4). Inventories for logistics and use of ethanol, coproducts and derivatives are obtained in the Log&UsoSoft model (Chap. 5).

The life cycle inventories of the most impactful inputs used in the sugarcane sector were identified using life cycle inventories from the ecoinvent database as reference. Background processes such as fossil fuels, fertilizers, and electricity were, to some extent, updated to the Brazilian context, including the technologies employed, energy sources, and other important parameters related to these processes as explained by Chagas et al. (2012). The study shows that some environmental impact categories can vary more than 20 % when using updated life cycle inventories in comparison with the reference cases, demonstrating the relevance of using specific background life cycle data.

6.4.1.3 Life Cycle Impact Assessment (LCIA)

With the translation of the product system's environmental flows from the LCI phase into scores that represent their (potential) impacts on environment, Life Cycle Impact Assessment (LCIA) is essential for the results interpretation in relation to the questions posed in the goal definition (Finnveden et al. 2009). The challenge of LCIA is to evaluate the potential impact of the emitted substances by using a procedure that is ideally simple, applicable consistently to all substances, that uses a common unit of measure and gives results that are comparable between impact categories.

The methodologies CML 2 baseline 2000 v2.05 described by Guinée et al. (2002) and ReCiPe Midpoint (H) v1.05 are normally used for the environmental impact assessment (Goedkoop et al. 2008). This choice is explained by the fact that they have been widely used and consensually applied for the evaluation/comparison of impacts associated with the production of biofuels. Within the methods, the use of resources and the emissions are categorized in terms of equivalent values according to the potential environmental impact; a great range of categories are included in the methods such as: climate change (global warming) measured in kg of CO_2-eq; ozone layer depletion measured in kg of CFC-11-eq (trichlorofluoromethane); acidification measured in kg of SO_2-eq; photochemical oxidation measured in kg of C_2H_4-eq; particulate formation measured in kg of PM 10-eq; and fossil depletion measured in kg oil-eq. The software package SimaPro (PRé Consultants B.V.) has been used as a tool for the environmental impact assessment in the VSB framework.

6.4.1.4 Interpretation

A life cycle interpretation is necessary for identifying, quantifying, checking, and evaluating information from the results of the LCI and/or the LCIA. This interpretation should also raise significant environmental issues, including an evaluation of the study considering completeness, sensitivity and consistency checks, and limitations.

Regarding the limitations of the approach, several sources of uncertainties can be identified in the LCA method, which may eventually generate great variability in results obtained, as more deeply discussed in the section related to 'Uncertainties and Risk Analysis' of this chapter.

6.4.2 *Hybrid LCA Based on IOA*

Hybrid techniques for compiling life cycle inventories are also used in the VSB. These have emerged to overcome many of the main shortcomings related to process and IOA (Lenzen 2002) and also to maximize the advantages of these two methods (Suh et al. 2004). Although process-based LCA creates a very detailed and specific inventory, it generates a systematic truncation error by which either inclusion or exclusion of processes is decided based on exclusion rules or subjective choices. Since it tends to omit part of upstream inputs, process analysis can result in significant underestimation in the LCI (Suh et al. 2004).

Three main approaches for compiling hybrid life cycle inventories have been used so far: tiered hybrid analysis, integrated hybrid analysis model, and IO-based hybrid analysis. Suh et al. (2004) describe tiered hybrid analysis as an approach performed when the direct and downstream requirements and some low-order upstream requirements are acquired from process analysis while remaining high-order upstream requirements are covered by IOA. One of the strengths of such method is its easiness to use and also well-documented literatures, databases, and case studies. According to Suh et al. (2004), double counting can be considered a weakness compared to the other approaches.

In integrated hybrid analysis model, the process-based system is represented in a technological matrix by physical units per unit operation time of each process, whereas the IO system is represented by monetary units. This model is derived from a make and use framework for both the process-based and the IO-based systems by linking them through flows crossing the border between the two systems. This approach has a consistent mathematical framework to avoid double counting, although it is complex to use and depends on high data and time requirements (Suh et al. 2004).

On the other hand, in IO-based hybrid analysis, a new product or technology is assessed by disaggregating an existing sector when more detailed monetary data are available. Moreover, it deals with a national IO table representing only pre-consumer stages of a product life cycle and use and end-of-life stages should be added to the results from disaggregated IO table (Suh et al. 2004). In the VSB, this approach is very close to the method used to estimate socioeconomic and environmental impacts associated with biorefineries at national level.

In the VSB, a hybrid model integrating LCA and IOA has been developed to assess the environmental impacts of sugarcane biorefineries in Brazil whose assessment relied on data from the previous studies of first- (Cavalett et al. 2012) and second-generation biorefineries (Dias et al. 2011, 2012). The LCI of a biorefinery scenario includes all processes and emissions from sugarcane cultivation to its industrial processing. The LCI related to each biorefinery scenario is translated into a list of purchased commodities (goods and services) according to the Brazilian commodity-by-sector direct requirements table for 2009 (Guilhoto and Sesso Filho 2010). The resulting model is considered hybrid because process-based and IO-based approaches are used to simulate the effects of biorefinery scenarios.

The hybrid method is described and deeply explored in a previous study of ethanol production technologies using the hybrid LCA approach (Watanabe et al. 2015). So far, the IO model comprises 65 industries and 120 commodities. In order to solve a system with rectangular tables, the mixed technology approach (Miller and Blair 2009) is employed. This approach relies on a group of matrices, which quantitatively describe the relationships among industries and commodities of a national economy (see Fig. 6.7). Matrix B represents the typical direct requirements table, i.e., its coefficients describe the technical coefficients. Based on the industry-based approach, matrix D represents the group of coefficients related to proportions of each industry's contribution to the total commodity output. Based on the commodity-based approach, the matrix C provides a group of coefficients related to different biorefinery configurations making a constant proportion of ethanol, electricity, and other products.

There are 189 equations in the system presented in Fig. 6.7: 120 to describe commodity outputs ($Q = B \cdot X + E$), 61 equations to describe industry-based technology for industry outputs ($X = D \cdot Q$), and 8 equations related to

		Variables								
Equations		X_1	...	X_{65}	E_1	...	E_{120}	Q_1	...	Q_{120}
BX+E−Q=0	P_1	matrix B			1	0	0	−1	0	0
...	...				0	...	0	0	...	
BX+E−Q=0	P_{120}				0	0	1	0	0	−1
−X+DQ=0	S_5	0	0	0	Zeros			matrix D		
...	...	0	...	0						
−X+DQ=0	S_{65}	0	0	−1						
CX−Q=0	P_5	matrix C'		0	Zeros			−1	0	0
...	...			0				0	...	
CX−Q=0	P_{12}			0				0	...	0

P_1 to P_{120} represent a list of equations for all products or commodities, according to the expression BX+E = Q.
E_1 to E_{120} represent the commodity final demands. S_5 to S_{65} represent a list of equations for sectors or industries related to the industry-based technology approach (X=DQ). P_5 to P_{12} represent product or commodity outputs related to the commodity-based approach (Q=C.X). Q_1 to Q_{120} represent the commodity outputs.

Fig. 6.7 Structure of the system of equations (mixed technology approach) *Source* Watanabe et al. (2015)

commodity-based technology for commodity outputs ($Q = C \cdot X$). Also, 305 variables are observed in the system by summing up those related to 65 sector outputs (X_1 to X_{65}), 120 commodity outputs (Q_1 to Q_{120}), and 120 commodity final demands (E_1 to E_{120}). The number of rows and columns represents the number of equations and variables in the system, respectively.

Once a change in the ethanol final demand is simulated, for instance, it is possible to estimate the direct and indirect outputs from each sector of the economy required to produce it. The details on the resolution of such system of equations are discussed in the previous studies (Watanabe et al. 2015; Cunha 2012).

6.5 Uncertainties and Risk Analysis

Nilsen and Aven (2003) state that the main objective of performing risk analyses was to support decision-making processes. The procedure enables the analyst to take both certain and uncertain quantities into account and calculate to what extent specific events or scenarios can be expected to occur in the future, providing basis for comparing alternative concepts, actions, or system configurations under uncertainty. The stochastic approach is the probabilistic counterpart to a deterministic model. Instead of describing a process which can only evolve in one way, in a stochastic or random process there is indeterminacy or uncertainty.

Risk is often viewed as the probability of an undesired or harmful event. The connection to probability is implied by the uncertainty in the occurrence of the event. The definition of uncertainty—as the lack of surety or certainty—is readily defined in a statistical or probabilistic context as the implication that uncertainty exists when the probability of an event occurring is not zero or one (Good 1994).

6.5.1 Risk Analysis Framework

Risk analysis is the methodology by which the uncertainty encompassing the main variables projected in a predictive model is processed in order to estimate the impact of risk on the results. It is a technique by which a mathematical model is subjected to a number of simulation runs. During the simulation process, successive scenarios are built up using input values for the key uncertain variables selected from multi-value probability distributions. The simulation may be controlled so that the random selection of values from the specified probability distributions does not violate the existence of known correlation relationships among the variables. The results are collected and analyzed statistically in order to arrive at a probability distribution of the potential outcomes and to estimate measures of risk. The methodology for the assessment of uncertainties can be found in a well-known publication by Morgan et al. (1990).

Risk variables

A risk variable is defined as critical to the project viability in the sense that a small deviation from its projected value is both probable and potentially damaging to the project worth. In order to select risk variables, sensitivity and uncertainty analysis can be applied.

Sensitivity analysis is used to identify the most important variables in a model. It measures the responsiveness of the result vis-à-vis a change (usually a fixed percentage deviation) in the value of a given project variable. The problem with sensitivity analysis as it is applied in practice is that there are no rules as to the extent to which a change in the value of a variable is tested for its impact on the projected result. For example, a 10 % increase in labor costs may be very likely to occur while a 10 % increase in revenues may be very unlikely to be obtained. The sensitivity test applied uniformly on a number of variables does not take into account how realistic or unrealistic the projected change is in the value of a tested variable. In order for sensitivity analysis to yield meaningful results, the impact of uncertainty should be incorporated into the test. Uncertainty analysis is the attainment of some understanding of the type and magnitude of uncertainty encompassing the variables to be tested, and using it to select risk variables.

The reason for including only the most crucial variables in a risk analysis application is twofold. First, the greater the number of probability distributions employed in a random simulation, the higher the likelihood of generating inconsistent scenarios because of the difficulty in setting and monitoring relationships for correlated variables. Second, the cost (in terms of expert time and money) needed to define accurate probability distributions and correlation conditions for many variables with a small possible impact on the result is likely to outweigh any benefit to be derived. Hence, rather than extending the breadth of analysis to cover a larger number of project variables, it is more productive to focus attention on adding more depth to the assumptions regarding the few most sensitive and uncertain variables.

Probability distributions

The ability to narrow the margins of uncertainty of a forecast depends on the understanding of the nature and level of uncertainty regarding the variable in question and the quality and quantity of information available. Often such information is embedded in the experience of the analyst or the person making the prediction.

In defining the uncertainty encompassing a given variable, one should widen the uncertainty margins to account for the lack of sufficient data or the inherent errors contained in the database used in making the prediction. While it is almost impossible to forecast accurately the actual value that a variable may assume, it should be possible to include the true value within the limits of a sufficiently wide probability distribution. The analyst should make use of the available data and expert opinion to define a range of values and probabilities that are capable of capturing the outcome of the event in question.

Probability distributions are used to regulate the likelihood of selection of values within the defined ranges. Some of the probability distributions used in the application of risk analysis are illustrated in Fig. 6.8.

Simulation runs

During the simulation, values of the 'risk variables' are selected randomly within specified ranges and in accordance with the set probability distributions. The results of the model are thus computed and stored following each run. Each run generates a different result, because the input values for the risk variables are selected randomly from their assigned probability distributions. The result of each run is calculated and stored away for statistical analysis (the final stage of risk analysis).

Analysis of results

The final stage in the risk analysis process is the interpretation of the results collected during the simulation runs stage. As depicted in Fig. 6.9, different graph analyses can be generated. Results obtained can be useful for: (a) the comparison of scenarios in terms of a specific output; (b) ranking of variables correlation with the output of interest; or (c) analysis of the cumulative probability.

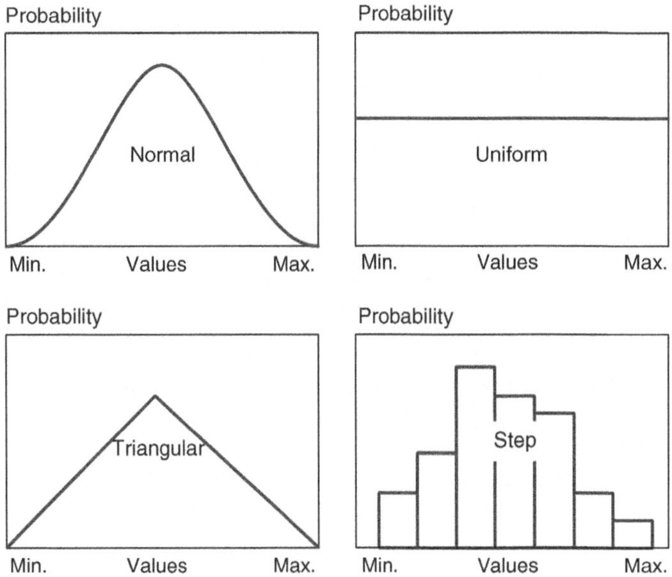

Fig. 6.8 Examples of probability distributions

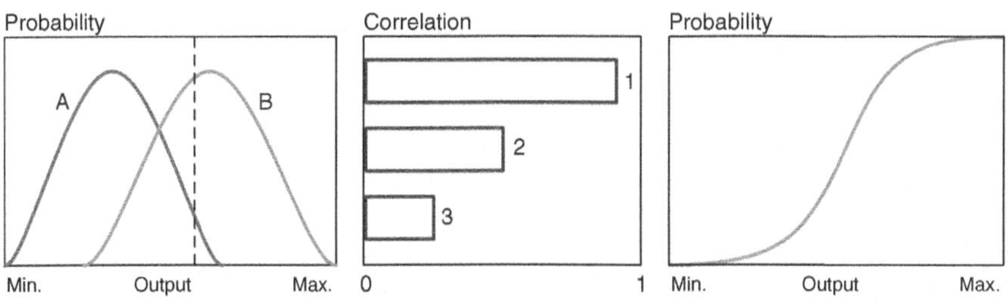

Fig. 6.9 Examples of results obtained in the risk analysis

6.5.2 Implementing Risk Analysis Within the VSB

The construction and development of the VSB are directly focused on the prediction, calculation, and comparison of parameters of various technology alternatives. The main advantage proposed by the tool is the ability to assess potential impacts (economic, environmental, and social) in order to determine the viability or level of success of such technologies before their practical implementation.

The assessment in the VSB deals with complex and integrated models consisting of thousands of parameters. Information used in the models may be source of uncertainties which could affect the accuracy of the impact results. Uncertainties can be associated with insufficient knowledge and/or over-simplification of systems, mechanisms, or parameters. The sole use of a deterministic approach within the framework is troublesome for two reasons: first, using only single values disregards the ranges and uncertainty of data used to generate the point estimate (such as a mean value) and second, a great share of new technologies investigated (such as second-generation ethanol), can only be predicted or estimated, not measured.

A risk analysis within the VSB may include the assignment of uncertainties to several parameters in order to assess the influence of such indeterminacies in the model outputs. However, it is important to highlight that the platform integrates the various models and evaluation methods; therefore, any risk analysis performed in one of the modules will be influenced or will end up influencing other modules of the framework (Pereira et al. 2015b) as follows:

- CanaSoft model: The agricultural simulation module provides outputs such as feedstock cost and number of workers, which are used as important inputs to financial, environmental, and socioeconomic analysis within the VSB. Uncertainties can be assigned to several parameters of the CanaSoft model depending on the objective of the evaluation to be performed.
- Industrial process simulation: developed in Aspen Plus software; can be translated into Excel spreadsheet, for instance, in order to be used in the evaluation of the uncertainties of processes technical parameters. The models should include the variables impacting the production (ethanol, electricity, and chemicals) along with their assigned distribution of probabilities. It is important to mention that not all technical parameters should be considered as variables in the stochastic approach; especially those related to iterative calculation or that could invalidate the models.
- Log&UsoSoft model: This model can be implemented in analogous way as the CanaSoft model. The Log&UsoSoft model is intended for simulation and measurement of the most important parameters of distribution logistics—transportation and commercialization—and use of biorefinery products.
- Financial analysis: Biorefineries normally require large upfront capital investment and are characterized by large product demand and price uncertainties. Consequently, this type of industry is fraught with considerable risks associated with the financial viability. Uncertainties for the economic return output can be assigned to most critical parameters such as capital investment (CAPEX), operational expenses (OPEX), and market prices of products. Moreover, for being dependent on the process simulation and the agricultural operations, the risk analysis of the financial impacts will be highly influenced by the assumptions assigned in those previous modules.
- LCA: The sources of uncertainties that can be identified in the LCA are, for example, the construction of the LCI, allocation methods, and the assumptions and characterization factors within the impacts assessment methods. The uncertainties associated with those parameters may have great effects in the environmental impacts calculated, eventually generating misleading results.

6.6 Optimization Strategy

Sustainability impacts of products and technology options are evaluated through VSB, as shown throughout this book. Within this context, appropriate analyzing tools and design methods, as well as optimization procedures coupled with the VSB can be used to properly assess and optimize sustainability's triple bottom line, which assembles economic, social, and environmental performance metrics. An optimization strategy would allow having a better guideline for assessing the sustainability attractiveness of different perspectives of technology configurations.

VSB is a reliable tool to assess sustainability's triple bottom line as showed in numerous study cases along the past few years (Pereira et al. 2015b; Dias et al. 2012; among others). However, depending on the system complexity under study, VSB requires detailed inputs and large computational efforts to perform an optimization procedure. These drawbacks make the optimization procedure more complex since the strategies require many simulations to cover a number of alternative scenarios.

An optimization methodology was implemented in VSB applying process system engineering tools such as surrogate modeling and optimization algorithms. These tools, along with VSB simulations, are used to assist synthesis of optimized sugarcane production chain.

6.6.1 Surrogate Modeling

VSB-based simulations offer high level of accuracy to predict sustainability impacts, requiring specialized simulation software. Its configuration provides accurate prediction although may impose challenges when used in an optimization procedure. Thus, it is plausible to use surrogate modeling to optimize a certain technology route. A surrogate model or reduced-order model is a regression to a set of input-output data obtained from the VSB based on Design of Experiments. It can facilitate analysis of complex systems such as sugarcane production chain by significantly reducing computational efforts.

As shown in Fig. 6.10, the surrogate modeling can be applied for: (i) the whole VSB (Fig. 6.10a); in this case, the entire production chain is simplified into a surrogate model and also (ii) for its subsystems separately, resulting in the so-called hybrid VSB model (Fig. 6.10b). The industrial subsystem (subsystem 2) simulated using Aspen Plus is surrogated by a reduced-order model because its sequential modular approach is not adequate (difficult to integrate) for an iterative optimization procedure.

The construction of a surrogate model consists of the following steps:

- *Determining the design range of the most sensitive variables.* The aim was to develop a surrogate model, which accurately captures the behavior of the overall system or subsystem in VSB in a more synthesized way. Under a range of

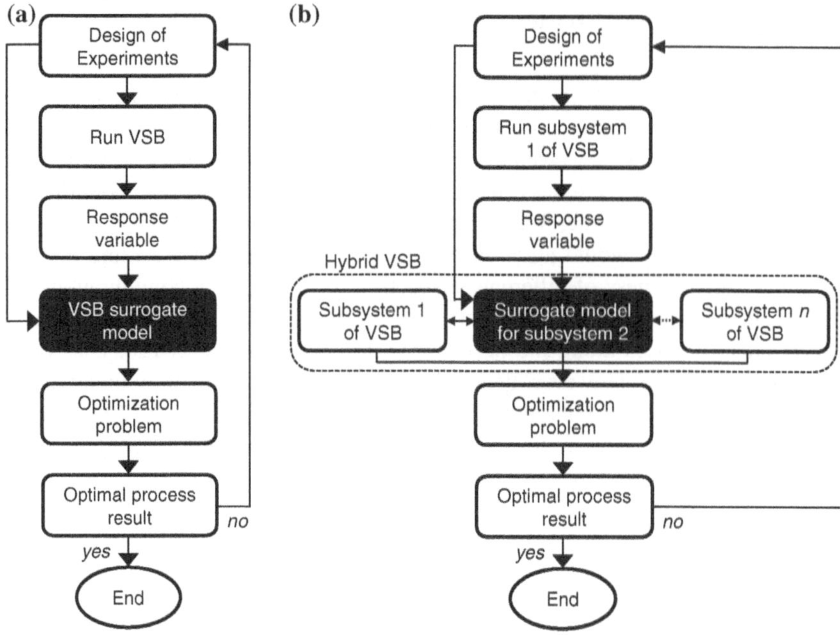

Fig. 6.10 Scheme for the surrogate model construction for **a** the entire sugarcane production chain emulated by the VSB and **b** a subsystem of the VSB

operating and design conditions, the surrogate model can be used for optimization purposes. In the surrogate model construction, the selected input variables and their design ranges are determined from expert knowledge, which could be supported by the risk analysis results presented above.

- *Data sampling*. It refers to the Design of Experiments, which is used to optimize the choice of design points for the construction of a surrogate model. In this procedure, the Central Composite Design (CCD) was used. For a case study applied to the VSB involving n independent variables (inputs), the design matrix for a CCD includes the factorial computational experiments (two-level design coded as +1 and −1) (Matrix F), one center point (coded as a vector of zeros of size n) (Matrix C), and axial point, with $2n$ computational experiments (Matrix A). Each variable is sequentially positioned at $\pm\alpha$ level and all other variables at zero. The α value is determined based on the number of variables and corresponds to $(2^n)^{1/4}$. Then D is the vertical concatenation to set up the design matrix for a CCD:

$$D = \begin{bmatrix} F \\ A \\ C \end{bmatrix} \tag{6.13}$$

The possibility of generating all the data using the VSB becomes highly time consuming. Therefore, it is suitable to construct and use surrogate model based on CCD either to substitute part (industrial simulation) or to the whole VSB.

- *Construction of the surrogate model.* For the modeling problem, a set of n computational experiments has to be defined following the CCD procedure; each computational experiment has a set of input data and a set of responses. It is desired to generate a model for each response that persistently tries to achieve high accuracy for modeling different systems. From previous studies applied only to the industrial sector (Rivera et al. 2014) and also to a vertically integrated production system coupling the agricultural and industrial sectors (Cardoso et al. 2015), a good accuracy was obtained when assuming that the underlying functional form of the response surfaces is quadratic (i.e., a second-order polynomial). Thus, a quadratic surrogate model showed to have sufficient complexity to represent the VSB or a subsystem of it, while maintaining the degree of effectiveness to be integrated into a model-based optimization algorithm.

A second-order polynomial can be written as follows:

$$z = \beta_0 + \sum_{i=1}^{n} \beta_i x_i + \sum_{i=1}^{n} \beta_{ii} x_i^2 + \sum_{i=1}^{n} \sum_{j=i+1}^{n} \beta_{ij} x_i x_j \tag{6.14}$$

where β_O, β_i, β_{ii} and β_{ij} are the model parameters (regression coefficients), x_i and x_j are the design variables, and z is the estimation of the response variable.

In general, the model described by Eq. 6.14 may be written in matrix form as Eq. 6.15.

$$Z = \beta X \tag{6.15}$$

where β is the vector of the regression coefficients, X is the design matrix, and Z is the vector of the responses.

The least-squares method (Eq. 6.16) is used to calculate the regression coefficients, where X^T indicates the transpose of X and $(X^T X)^{-1}$ is the covariance matrix (for details on the method, see Montgomery (2001)).

$$\beta = (X^T X)^{-1} X^T Y \tag{6.16}$$

In order to avoid the overfitting problem, a standardized Pareto Chart is used to examine the significance of the coefficients of the variables and their interactions. Figure 6.11 illustrates an example of a standardized Pareto Chart for three variables. This chart indicates each of the estimated effects in decreasing order of magnitude. The size of each bar is relative to the standardized effect, calculated as the estimated effect (effect is twice the value of the corresponding regression coefficient) divided by its standard error, which is equivalent of computing a t-statistic for each effect. There is also a vertical line (critical t-value) that represents the effect's limit of

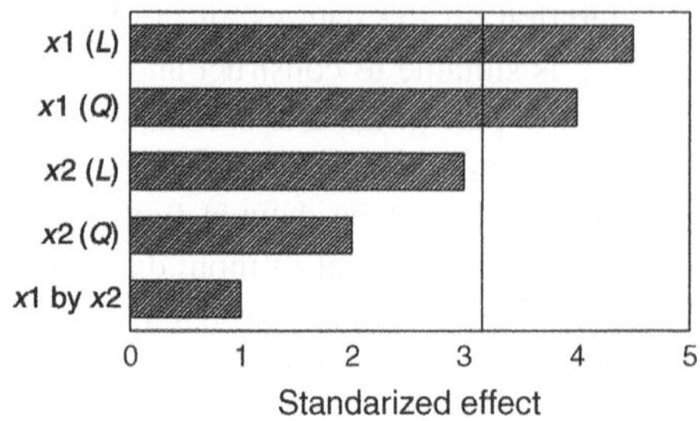

Fig. 6.11 Example of a typical standardized Pareto Chart of three variables. The linear and quadratic effects are, respectively, denoted by (*L*) and (*Q*)

significance. Bars extending beyond that line relate to effects that are statistically significant at 95 % significance level, which has been widely used in engineering design.

Finally, to guarantee that generated models have statistical significance, the Analysis of Variance (ANOVA) is performed.

6.6.2 *Optimization*

The developed surrogate model or even a hybrid VSB model obtained from the modeling step is optimized to calculate the optimal value of the sustainability indicators within the range of the economic, social, and environmental variables assessed. The objectives are the maximization of economic performance of a technological route, as well as maximizing its social benefits and minimizing its environmental impacts.

Therefore, the optimization problem would be:

$$\text{Min(or Max)}\, f(x) \tag{6.17}$$

$$\text{Subject to: } x_{L,k} \leq x_k \leq x_{U,k}, \quad k = 1, \ldots, N \tag{6.18}$$

where Eq. 6.18 refers to the box constraints, which include lower and upper bounds of the variables considered. The objective function *f* could be either the surrogate model of the whole VSB or a hybrid VSB model.

The generic form of the above optimization problem is purely continuous, which allows the use of traditional meta-heuristic (e.g., genetic algorithm) and derivative-based (e.g., quasi-newton) methods. Computational complexity to get the optimal solution is not a major issue for this application, as the resulting model is a second-order polynomial or polynomials coupled with Excel spreadsheets (which are the case of the hybrid VSB model). These model structures are far simpler than the VSB.

At this point, it is important to mention that the level of detail of the model considered for each phase of VSB, emulating the sugarcane production chain, is constantly validated. This validation is performed due to the need to develop reliable models, compatibility with optimization techniques to be used, as well as standards of quality of information required for a proper decision-making.

References

ABNT (2009a) NBR ISO 14040: Gestão ambiental - avaliação do ciclo de vida - princípios e estrutura. Associação Brasileira de Normas Técnicas (ABNT), Rio de Janeiro, Brasil

ABNT (2009b) NBR ISO 14044: Gestão ambiental - avaliação do ciclo de vida - requisitos e orientações. Associação Brasileira de Normas Técnicas (ABNT), Rio de Janeiro, Brasil

Bini DA, Madail JCM (2011) Viabilidade Econômica da Cultura de Cana-de-Açúcar no Rio Grande do Sul (Economic viability of sugarcane plantation in Rio Grande do Sul) In: Anais do XII Encontro de Pós-Graduação, Universidade Federal de Pelotas, Pelotas, 9-12 Nov 2010

Burdge R (2002) Why is social impact assessment orphan of the assessment process? Impact Assess Proj Appraisal 20(1):3–9

Cardoso T, Cavalett O, Chagas MF et al (2013) Technical and economic assessment of trash recovery in the sugarcane bioenergy production system. Sci Agric 70:353–360

Cardoso TF, Chagas MF, Rivera EC, Cavalett O, Morais ER, Geraldo VC, Braunbeck O, Cunha MP, Cortez LAB, Bonomi A (2015) A vertically integrated simplified model for straw recovery and use in sugarcane biorefineries. Biom Bioen. Accepted for publication (forthcoming)

Cavalett O, Junqueira TL, Dias MOS et al (2012) Environmental and economic assessment of sugarcane first generation biorefineries in Brazil. Clean Technol Environ 14(3):399–410. doi:10.1007/s10098-011-0424-7

CEPEA (2014) The CEPEA registers. Center for advanced studies on applied economics (CEPEA). http://www.cepea.esalq.usp.br. Accessed 10 Jul 2014

Chagas MF, Cavalett O, Silva CRU et al (2012) Adaptação de inventários de ciclo de vida da cadeia produtiva do etanol (Adaptation of the life cycle inventory of ethanol production chain). In: Ravagnani MASS, Morita AM (eds) Proceedings of the III Congresso Brasileiro em Gestão do Ciclo de Vida de Produtos e Serviços, Maringá, vol 1. DentalPress Publis-inhg, pp 1–6, 3–6 Sep 2012

Cherubini F, Strømman AH, Ulgiati S (2011) Influence of allocation methods on the environ-mental performance of biorefinery products—a case study. Resour Conserv Recycl 55:1070–1077

Cunha MP (2012) Avaliação socioeconômica e ambiental de rotas de produção de biodiesel no Brasil, baseada em análise de insumo-produto (Socioeconomic and environmental assessment of biodiesel production in Brazil, based on input-output analysis). Thesis, State University of Campinas

Dias MOS, Cunha MP, Jesus CDF et al (2011) Second generation ethanol in Brazil: can it compete with electricity production? Bioresour Technol 102(19):8964–8971

Dias MOS, Junqueira TL, Cavalett O et al (2012) Integrated versus stand-alone second genera-tion ethanol production from sugarcane bagasse and trash. Bioresour Technol 103:152–161

Dias MOS, Pereira LG, Junqueira TL et al (2014) Butanol production in a sugarcane biorefinery using ethanol as feedstock—part I: integration to a first generation sugarcane distillery. Chem Eng Res Des 92(8):1441–1451. doi:10.1016/j.cherd.2014.04.030

EPA (2010) Renewable fuel standard program (RFS) for 2010 and beyond. Environmental protection agency (EPA). http://www.epa.gov/otaq/renewablefuels/420f10007.pdf. Accessed 3 Feb 2015

Finnveden G, Hauschild MZ, Ekvall T et al (2009) Recent developments in life cycle assessment. J Environ Manage 91:1–21
Goedkoop MJ, Heijungs R, Huijbregts M et al (2008) A life cycle impact assessment method which comprises harmonised category indicator at the midpoint and the endpoint level—first edition Report 1: characterisation. http://www.lcia-recipe.net. Accessed 2 Nov 2014
Good IJ (1994) Reliability always depends on probability of course. J Stat Comput Simul 52: 192–193
Grant EL, Ireson WG, Leavenworth RS (1990) Principles of engineering economy. Wiley, New York
Guilhoto JJM, Sesso Filho UA (2005) Estimação da matriz insumo-produto a partir de dados preliminares das contas nacionais (Input-output table estimation from preliminary data on the system of national accounts). Economia Aplicada 9(2):277–299
Guilhoto JJM, Sesso Filho UA (2010) Estimação da matriz insumo-produto utilizando dados preliminares das contas nacionais: aplicação e análise de indicadores econômicos para o Brasil em 2005 (Input-output table estimation from preliminary data on the system of national accounts: application and assessment of economic indicators for Brazil in 2005). Economia e Tecnologia 6(23):53–62
Guinée JB, Gorrée M, Heijungs R et al (2002) Handbook on life cycle assessment—operational guide to the ISO standards. Kluwer Academic Publishers, Dordrecht
Guinée JB, Heijungs R, Huppes G (2004) Economic allocation—examples and derived decision tree. Int J Life Cycle Assess 9(1):23–33
Hendrickson C, Lave L, Matthews HS (2006) Environmental life cycle of goods and services: an input-output approach. RFF Press, Washington D.C
IBGE (2015) System of national accounts. brazilian institute of geography and statistics (IBGE). http://www.ibge.gov.br/home/estatistica/economia/matrizinsumo_produto/default.shtm. Accessed 14 May 2015
IDEA (2012) Indicadores de desempenho da agroindústria canavieira (Performance indicators of sugarcane agroindustry: seasons 2010/2011 and 2011/2012). Grupo IDEA, Ribeirão Preto, Brasil
ISO (2006a) ISO 14040: environmental management—life cycle assessment—principles and framework. The International Organization for Standardization (ISO), Geneva, Switzerland
ISO (2006b) ISO 14044: environmental management—life cycle assessment—requirements and guidelines. The International Organization for Standardization (ISO), Geneva, Switzerland
Jørgensen A (2013) Social LCA—a way ahead? Int J Life Cycle Assess 18:296–299
Jørgensen A, Le Bocq A, Nazarkina L et al (2008) Methodologies for social life cycle assessment. Int J Life Cycle Assess 13(2):96–103
Lenzen M (2002) A guide for compiling inventories in hybrid life-cycle assessments. J Clean Prod 10(6):545–572
Leontief W (1970) Environmental repercussions and the economic structure: an input-output approach. Rev Econ Stat 52(3):262–271
Luo L, van der Voet E, Huppes G (2009) Life cycle assessment and life cycle costing of bioethanol from sugarcane in Brazil. Renew Sustain Energy Rev 13(6–7):1613–1619
Macombe C (ed) (2013) Social LCAs: socio-economic effects in value chains. Cirad, Montpellier
Macombe C, Loeillet D (2013) Social life cycle assessment, for who and why? In: Macombe C (ed) Social LCAs: socio-economic effects in value chains. Cirad, Montpellier, pp 35–52
Macrelli S, Mogensen J, Zacchi G (2012) Techno-economic of 2nd generation bioethanol production from sugar cane bagasse and leaves integrated with the sugar-based ethanol process. Biotechnol Biofuels 5(22). doi:10.1186/1754-6834-5-22
Mariano AP, Dias MOS, Junqueira TL et al (2013a) Butanol production in a first-generation Brazilian sugarcane biorefinery: technical aspects and economics of greenfield projects. Bioresour Technol 135:316–323
Mariano AP, Dias MOS, Junqueira TL et al (2013b) Utilization of pentoses from sugarcane biomass: techno-economics of biogas vs. butanol production. Bioresour Technol 142:390–399

Milanez AY, Nyko D, Valente MS et al (2015) De promessa à realidade: como o etanol celulósico pode revolucionar a indústria da cana-de-açúcar. Uma avaliação do potencial competitivo e sugestões de política pública (From promise to reality: how cellulosic ethanol can revolutionize the sugarcane industry. An assessment of potential competition and suggestions of public policy). BNDES Setorial 41:265–322

Miller RE, Blair PD (2009) Input-output analysis: foundations and extensions. Cambridge University Press, New York

MME (2014) Energy auctions. Ministry of mines and energy (MME). http://www.mme.gov.br/programas/leiloes_de_energia/. Accessed Jul 2014

Montgomery DC (2001) Design and analysis of experiments, 5th edn. Wiley, New York, p 684

MPS (2015) Anuário Estatístico da Previdência Social—AEAT. Ministry of Social Security. http://www.previdencia.gov.br/estatisticas/. Accessed 20 Jan 2015

MTE (2015) Base de dados on-line. Programa de Disseminação de Estatísticas do trabalho – PDET: ministry of labor and employment (MTE). http://bi.mte.gov.br/bgcaged/login.php. Accessed 20 Jan 2015

Moraes BS, Junqueira TL, Pavanello LG et al (2014) Anaerobic digestion of vinasse from sug-arcane biorefineries in Brazil from energy, environmental, and economic perspectives: profit or expense? Appl Energy 113:825–835. doi:10.1016/j.apenergy.2013.07.018

Moraes BS, Zaiat M, Bonomi A (2015) Anaerobic digestion of vinasse from sugarcane ethanol production in Brazil: challenges and perspectives. Renew Sustain Energy Rev 44:888–903

Morgan MG, Henrion M, Small M (1990) Uncertainty: a guide to dealing with uncertainty in quantitative risk and policy analysis. Cambridge University Press, UK

Newnan DG, Eschenbach TG, Lavelle JP (2004) Engineering economic analysis. Oxford University Press, New York

Nilsen T, Aven T (2003) Models and model uncertainty in the context of risk analysis. Reliab Eng Syst Saf 79:309–317

Pecege (2014) Custos de produção de cana-de-açúcar, açúcar e etanol no Brasil: fechamento da safra 2013/2014. Programa de Educação Continuada em Economia e Gestão de Empresas (Pecege), Piracicaba, Brasil

Pereira LG, Dias MOS, Junqueira TL et al (2014a) Butanol production in a sugarcane biorefinery using ethanol as feedstock—part II: integration to a second generation sugarcane distillery. Chem Eng Res Des 92(8):1452–1462. doi:10.1016/j.cherd.2014.04.032

Pereira LG, Dias MOS, Junqueira TL et al (2014b) Production of butanol and other high valued chemicals using ethanol as feedstock integrated to a first and second generation sugarcane distillery. Chem Eng Trans 37:805–810. doi:10.3303/CET1437135

Pereira LG, Chagas MF, Dias MOS et al (2015a) Life cycle assessment of butanol production in sugarcane biorefineries in Brazil. J Clean Prod 96:557–568. doi:10.1016/j.jclepro.2014.01.059

Pereira LG, Dias MOS, MacLean HL et al (2015b) Investigation of uncertainties associated with the production of n-butanol through ethanol catalysis in sugarcane biorefineries. Bioresour Technol 190:242–250. doi:10.1016/j.biortech.2015.04.095

Perry RH, Green DW (2008) Perry's chemical engineers' handbook, 8th edn. McGrawHill, New York

Peters M, Timmerhaus K, West R (2002) Plant design and economics for chemical engineers. McGraw-Hill Science, New York

Pinto FHPB (2010) Etanol Celulósico: Um Estudo de Viabilidade Econômico-financeira (Cellulosic ethanol: an economic-financial feasibility study). Dissertation, FGV-EESP, Embrapa and ESALQ/USP

Rangel LC, Lopes RS, Joana RV (2010) Cogeração e comercialização de excedentes de eletricidade em uma usina sucroalcooleira (Cogeneration and commercialization of surplus electricity in a sugarcane mil). Perspectivas Online 4(13):1–19

Ribeiro BE (2013) Beyond commonplace biofuels: social aspects of ethanol. Energy Policy 57:355–362

Rivera EC, Geraldo VC, Sanghikian N et al (2014) A screening design to analyze the influence of technological configurations on techno-economic parameters for autonomous distilleries in Brazil. Chem Eng Trans 37:307–312

Suh S, Lenzen M, Treolar GJ et al (2004) System boundary selection in life-cycle inventories using hybrid approaches. Environ Sci Technol 38(3):657–664

Svanes E, Vold M, Hanssen OJ (2011) Environmental assessment of cod (Gadus morhua) from autoline fisheries. Int J Life Cycle Assess 16:611–624

Turton R, Bailie RC, Whiting WB, Shaeiwitz JA (2009) Analysis, synthesis and design of chemical processes, 3rd edn. Prentice Hall, New Jersey

UNEP/SETAC (2009) Guidelines for social life cycle assessment of products. United Nations Environment Programme, Paris

UNEP/SETAC (2011) The methodological sheets for sub-categories in social life cycle assessment (S-LCA). United Nations Environment Programme, Paris

Waardenar T, Ruijven T, Beltran AM et al (2012) Differences between LCA for analysis and LCA for policy: a case study on the consequences of allocation choices in bio-energy policies. Int J Life Cycle Assess 17:1059–1067

Wang M, Huo H, Arora S (2011) Methods of dealing with co-products of biofuels in life cycle analysis and consequent results within the US context. Energy Policy 39(10):5726–5736

Watanabe MDB, Chagas MF, Cavalett O et al (2015) Hybrid input-output life cycle assessment of first and second generation ethanol production technologies in Brazil. J Ind Ecol Accepted for publication (doi: 10.1111/jiec.12325)

Chapter 7
Use of the VSB to Assess Biorefinery Strategies

Marina O.S. Dias, Tassia L. Junqueira, Isabelle L.M. Sampaio, Mateus F. Chagas, Marcos D.B. Watanabe, Edvaldo R. Morais, Vera L.R. Gouveia, Bruno C. Klein, Mylene C.A.F. Rezende, Terezinha F. Cardoso, Alexandre Souza, Charles D.F. Jesus, Lucas G. Pereira, Elmer C. Rivera, R. Maciel Filho and Antonio Bonomi

The VSB has been used to assess several configurations of sugarcane biorefineries, including first- and second-generation facilities with alternative features, such as harvesting extension with other crops and integrated production of sugar, ethanol, electricity, and chemicals like *n*-butanol, as well as thermochemical routes. The analyses provide results for sustainability impact assessment of these different configurations and may indicate, for example, on which conditions one sugarcane biorefinery configuration will provide the highest economic return or lower environmental impact in a specific category, or, in other words, which route would perform better than the others taking into consideration certain criteria. Therefore, public institutions and private companies can make use of the data and direct their efforts in the appropriate direction, and policy makers and financing agencies have grounds to support their decisions.

In this chapter, in-depth examples of selected routes are provided. These examples were chosen in order to illustrate the different methods that can be applied as well as results obtained with the VSB allocation to several case studies.

M.O.S. Dias
Institute of Science and Technology, Federal University of São Paulo (ICT/UNIFESP), São José Dos Campos, SP, Brazil

T.L. Junqueira · I.L.M. Sampaio · M.F. Chagas · M.D.B. Watanabe · E.R. Morais
V.L.R. Gouveia · B.C. Klein · M.C.A.F. Rezende · T.F. Cardoso · A.M. Souza · C.D.F. Jesus
L.G. Pereira · E.C. Rivera · A. Bonomi (✉)
Brazilian Bioethanol Science and Technology Laboratory (CTBE), Brazilian Center of Research in Energy and Materials (CNPEM), Campinas, SP, Brazil
e-mail: antonio.bonomi@bioetanol.org.br

L.G. Pereira
Department of Civil Engineering, University of Toronto, Toronto, ON, Canada

R.M. Filho
School of Chemical Engineering, University of Campinas (UNICAMP), Campinas, SP, Brazil

A. Bonomi et al. (eds.), *Virtual Biorefinery*, Green Energy and Technology,
DOI 10.1007/978-3-319-26045-7_7

Case 1 in Sect. 7.1 describes first-generation sugarcane facilities, including both autonomous and annexed plants with both standard and optimized configurations. This example therefore represents a today commercial reality and can provide information regarding the impact of optimization features on the current technology of sugarcane facilities.

Second-generation ethanol production through biochemical route, comprised of pretreatment and enzymatic hydrolysis, is assessed in Case 2 in Sect. 7.2. Different technological levels (current and advanced technologies) and process configurations (stand-alone versus integrated with first-generation facilities) were evaluated. Although second-generation plants exist at industrial scale, this technology is currently at early stage of development, so analyses that allow appraisal of better operational conditions and process designs are of utmost importance.

Extended operation through inclusion of additional feedstock, other than sugarcane (which harvest lasts around 8 months per year in Brazil), is described in Case 3 in Sect. 7.3. Since sugarcane cannot be stored, increased operation and reduction of equipment idle time is possible if other sugary (e.g., sweet sorghum) or starchy (e.g., corn) feedstocks are processed.

A more complex biorefinery concept is demonstrated in Case 4 in Sect. 7.4 through the assessment of the integrated production of biofuels, chemicals and electricity. Integration of *n*-butanol production, through different routes, within the sugarcane biorefinery was assessed as an example in this case study.

Case 5 in Sect. 7.5 addresses the production of second-generation biofuels through the thermochemical route, providing a comparison with the biochemical route.

Examples of integration between agricultural and industrial sectors are provided in Case 6 in Sect. 7.6, where different straw harvesting methods are compared. In this section, scenarios with different fractions of straw recovery through integral or baling systems were assessed.

Sustainability impacts of these scenarios are evaluated using economic, social and environmental methods described in Chap. 6. Input–Output Analysis (combined to Life Cycle Assessment) is used to evaluate sugarcane and ethanol production impacts on the economic activity of Brazilian sectors in Sect. 7.7. Results of this integrated analysis are also employed in a social assessment in Sect. 7.8.

The agricultural sector is modeled using CanaSoft model described in Chap. 3, while the industrial conversion phase is based on process simulation using Aspen Plus detailed in Chap. 4. Differences from the configuration described in Chap. 4 from the scenarios presented here are highlighted in all the next sections.

7.1 Case 1: First-Generation Sugarcane Facilities

As described in Chap. 4, sugarcane can be processed either in autonomous distilleries (in which all sugarcane juice is diverted to ethanol production) or in annexed plants (part of the sugarcane juice is used for sugar production, while the remaining fraction, along with molasses, is sent to ethanol fermentation).

The possibility of the sugarcane distillery to divert a fraction of its feedstock either for sugar or for ethanol production is one of the reasons for the success of this industry in Brazil: When sugar market prices are favorable, the sugarcane facility can produce more sugar by decreasing the amount of juice sent to fermentation, once the factory has some flexibility, therefore maximizing its revenues.

Although sugarcane mills are energy self-sufficient, most facilities have been using the same technology for decades. These equipments date back to when sale of surplus electricity was not an objective; therefore, most mills have high energy demand and cogeneration systems that result in almost complete combustion of bagasse without generating significant electricity surpluses. These characteristics are present in the standard sugarcane facility in Brazil.

In this case, results for a comparison between standard and optimized annexed and autonomous distilleries are presented.

7.1.1 Scenario Description

Process flow sheet and parameters are those described in Chap. 4. The main differences on the operational characteristics between standard and optimized and autonomous distilleries are shown in Table 7.1.

Table 7.1 Main parameters for standard and optimized autonomous distilleries

Parameter	Standard distillery	Optimized distillery
Processing capacity (10^6 TC $year^{-1}$)[a]	2	4
Sugarcane cleaning	None	Dry cleaning
Mill engines	Direct drives	Electric drives
Surplus bagasse use	Sold	Burnt for production of electricity
Straw recovery (%)	None	50 (considering vegetal impurities and straw bales)
Fermentation efficiency (%)	90	90
Wine ethanol content (g L^{-1})	70	80
Dehydration	Azeotropic distillation	Molecular sieves
Boiler pressure (bar)/efficiency (%, LHV basis)	22/75	65/87.7
Type of Turbines/isentropic efficiency (%)	Back-pressure/72	Extraction-condensing/85
Reduction in 2.5 bar steam consumption due to heat integration (%)	None	20
Energy demand—mechanical/electric (kWh TC^{-1})	16 /12	0 /30

[a]An effective period of operation of 200 days was considered

In optimized autonomous distilleries, all sugarcane bagasse and straw available are burnt for production of steam and electricity. Optimized distilleries have modern, more efficient cogeneration systems, thus allowing the production of large amounts of steam and electricity. For standard distilleries, depending on how energy efficient is the plant, surplus bagasse may be available for selling as a by-product.

Mill engines for standard distilleries are driven by steam turbines, which are less efficient than electrified drives used in the optimized plants. The standard distillery uses a dehydration process for anhydrous ethanol production based on azeotropic distillation with cyclohexane. In this process, cyclohexane is used as entrainer and forms a heterogeneous azeotrope with ethanol and water, which allows the separation of ethanol from the mixture. This process is used in most of the sugarcane mills in Brazil, in spite of being energy intensive (around 2.5 $kg_{steam}\ L^{-1}{}_{anhydrous\ ethanol}$) when compared with more recent alternatives such as molecular sieves (0.6 $kg_{steam}\ L^{-1}{}_{anhydrous\ ethanol}$), considered in the optimized distillery.

A reduction of 20 % in process steam consumption (2.5 bar) was assumed for optimized distilleries, which would be obtained with thermal integration. In practice, the 2.5-bar steam demand calculated in the simulation is multiplied by 0.8 to account for this reduction.

For optimized annexed plants, the cogeneration and ethanol producing sectors of the plant present the same previously cited improvements and parameters of optimized autonomous distilleries, with the exception of fermentation efficiency. For annexed plants, the conversion of sugars to ethanol was considered lower (89.5 %), due to the presence of inhibitory compounds in the fermentation medium derived from molasses. No alterations for the sugar-producing sector of the plant are considered. The reduction of 20 % of the total steam consumption of the plant due to thermal integration is also assumed for the optimized annexed plant (in this case considering both sugar and ethanol producing sectors).

7.1.2 *Techno-economic Analysis*

7.1.2.1 Technical Results

Results for ethanol, sugar and electricity production for autonomous and annexed first-generation sugarcane distilleries are shown in Table 7.2.

From the results presented in Table 7.2, it is possible to notice that sugar and ethanol production for the standard and optimized annexed distilleries are very similar. This also applies to the production of ethanol for both standard and optimized autonomous distilleries. This is because the main improvements made in the optimized scenarios constructed are aimed at energetic performance, rather than ethanol or sugar production.

As previously shown in Table 7.1, the destination of the surplus bagasse is different for standard and optimized plants. In standard distilleries, the surplus of

Table 7.2 Outputs for first-generation sugarcane facilities

Scenarios[a]	1G–Anx–St	1G–Anx–Op	1G–Aut–St	1G–Aut–Op
Anhydrous ethanol production (L TC^{-1})	53.5	53.7	85.0	85.4
Sugar production (kg TC^{-1})	51.3	51.4	–	–
Surplus electricity (kWh TC^{-1})	10.9[b]	185.8	12.3[b]	185.8
Surplus bagasse, dry basis (kg TC^{-1})	27.8	–	24.1	–

[a]Scenarios: *Anx* annexed plant; *Aut* autonomous distillery; *St* standard configuration; *Op* optimized configuration

[b]Surplus electricity is not commercialized in the standard configuration scenarios

bagasse is sold, and in optimized distilleries, this material is burnt, generating steam and subsequently electricity. The energetic improvements made in the optimized plants and the fact that all bagasse, in addition to straw, is burnt in these scenarios lead to a great difference in the amount of electricity produced when compared to the standard scenarios.

7.1.2.2 Economic Assessment: Annexed Plants

The economic analysis was performed to assess annexed distilleries firstly considering the estimation on capital expenditures, as summarized in Table 7.3. It is clear that the industrial areas related to higher investments are the combined heat and power (CHP), sugarcane reception and juice extraction. As long as the 1G optimized annexed plant (1G–Anx–Op) processes twice as much the amount of sugarcane of 1G standard annexed plant (1G–Anx–St), a comparison between scenarios has limitations. The relative importance of the CHP area, however, is more preponderant in the optimized plant due to additional process of straw and adoption of equipment to maximize electricity production, such as high-pressure boilers.

The results of 1G–Anx-St and 1G–Anx–Op scenarios are presented in Table 7.4. It is important to highlight that the cash flow analysis considers a vertically integrated production model where biomass production costs associated with sugarcane and straw, calculated according to CanaSoft model (Chap. 3), were considered as

Table 7.3 Summary of capital expenditures associated with the main industrial areas of annexed plants

CAPEX (in US$ million)	1G–Anx–St	1G–Anx–Op
Administrative infrastructure, engineering and utilities	35.3	79.0
Sugarcane reception, juice extraction and evaporation	40.5	70.2
Sugar production (crystallization and drying)	15.3	30.7
Ethanol production (fermentation, distillation/dehydration)	25.4	54.4
Combined heat and power (steam and electricity generation)	39.3	190.5
Total	155.8	424.8

Table 7.4 Techno-economic analysis of annexed distilleries

Parameter	1G–Anx–St	1G–Anx–Op
IRR (% per year)	15.8	16.0
NPV (US$ mi)	47	134
Sugar cost (US$ t^{-1})[a]	354	338
Ethanol cost (US$ m^{-3})[a]	471	450

[a]Total cost = operating costs + cost of investment

operating costs in its respective industrial stage. Regarding 1G–Anx–St scenario, sugarcane cost was US$27.97 per tonne. In the 1G–Anx–Op scenario, sugarcane and dry straw production costs were US$27.90 and US$36.38 per tonne, respectively. These small changes in sugarcane production cost refer to specific conditions of industrial scenarios which affect agricultural stage such as the amount of vinasse to be returned to the sugarcane field for fertirrigation.

The results from the techno-economic analysis for annexed plants (see Table 7.4) highlight that 1G–Anx–Op scenario is related to the highest internal rate of return, roughly 16 % per year. Considering a minimum acceptable rate of return of 12 % per year, this scenario is associated with a positive net present value (NPV) of approximately US$134 million. These results indicate that this scenario is an interesting project alternative. This outcome was expected because of economies of scale and technological improvements that make higher electricity surplus possible in the 1G–Anx–Op scenario when compared to 1G–Anx–St. Because of that, the 1G–Anx–Op scenario is also related to lower sugar and ethanol production costs: US$0.338 per kg and US$0.450 per L, respectively. The 1G–Anx–St scenario was related to an IRR of 15.8 % which is very close to the optimized scenario; however, a significantly lower NPV—approximately 47 million dollars—is observed. Although scenarios lead to similar IRRs, the yearly net cash flows (net revenues subtracted from cash outflows such as operating costs and income taxes) of 1G–Anx–Op are significantly higher than the 1G–Anx–St scenario, because of the plant scale-up and exports of surplus electricity. This effect can be observed when comparing NPV of both projects. Also, higher sugar and ethanol production costs are related to the standard scenario: US$0.354 per kg and 0.471 per L, respectively.

In order to understand the impact of different scenarios on the ethanol production costs, Table 7.5 shows the ethanol cost breakdown represented by the expenses on biomass, capital cost, maintenance, labor and other costs (such as chemicals and facilities). It is possible to observe that main components of ethanol costs are related to biomass—from 60 to 65 % of the total cost—and the cost of investment (capital cost) which ranges from 25 to 30 % of the total ethanol cost.

Maintenance, labor and other costs are less important for 1G ethanol production and contribute from 11 to 12 % to the total cost. When compared to the 1G–Anx–St, the optimized distillery is also associated with lower biomass and capital costs because of cost allocation to surplus electricity. Consequently, 1G ethanol production costs become lower due to such technological improvements in 1G–Anx–Op.

Table 7.5 Comparison of 1G ethanol cost in annexed distilleries

Cost elements (US$ however, a sm^{-3})	1G–Anx–St	1G–Anx–Op
Biomass	297	267
Capital cost	116	135
Maintenance	25	29
Labor	23	10
Other costs	10	9
Total	471	450

7.1.2.3 Economic Assessment: Autonomous Distilleries

Similar to the previous section, the techno-economic analysis was performed to assess autonomous distilleries considering evaluation of capital expenditures as summarized in Table 7.6. For these plants, the assumptions about the estimated CAPEX are the same of the annexed plant scenarios. The investments in the optimized autonomous distillery (1G–Aut–Op) are roughly twice as much as those observed in the standard autonomous distillery (1G–Aut–St) scenario due to the processing capacity, except the combined heat and power (CHP) area that is multiplied by a factor of approximately five due to the use of straw as a complementary fuel.

Concerning biomass costs, sugarcane production cost related to the 1G–Aut–St was US$28.52 per tonne. For the optimized scenario, sugarcane and dry straw production costs were US$28.26 and 36.38 per tonne, respectively.

A comparison between 1G–Aut–St and 1G–Aut–Op scenarios was performed in order to quantify the impact of scaling-up and technological improvements associated with the autonomous distillery. As Table 7.7 shows, the 1G–Aut–Op is related to the highest internal rate of return, roughly 14.1 % per year and a positive NPV of approximately US$68 million. Once again, these results indicate a feasible project alternative that would add value to a company and rates of return above the minimum acceptable over the project lifetime. Also, because of the economies of scale, 1G–Aut–Op presents the lowest ethanol production costs: US$0.480 per L. It is important to highlight that the share of operating costs allocated to electricity also

Table 7.6 Summary of capital expenditures associated with the main industrial areas of autonomous distilleries

CAPEX (in US$ million)	1G–Aut–St	1G–Aut–Op
Administrative infrastructure, engineering and utilities	34.5	78.1
Sugarcane reception, juice extraction and evaporation	37.9	66.0
Ethanol production (fermentation, distillation/dehydration)	37.9	82.9
Combined heat and power (steam and electricity generation)	39.7	190.8
Total	150.0	417.8

Table 7.7 Techno-economic results of first-generation autonomous distilleries

Parameter	1G–Aut–St	1G–Aut–Op
IRR (% per year)	13.1	14.1
NPV (US$ mi)	13	68
Ethanol cost (US$ m^{-3})	511	480

Table 7.8 Comparison between autonomous distilleries considering the ethanol cost

1G cost (US$ m^{-3})	1G–Aut–St	1G–Aut–Op
Biomass	328	288
Capital cost	121	141
Maintenance	26	30
Labor	25	11
Other costs	10	10
Total	511	480

contributes to decrease the ethanol cost besides the scaling-up effect. The 1G–Aut–St, on the other hand, was related to lower IRR (13.1 % per year) and lower NPV (US$13 million). Although these results are less favorable, the standard distillery is still economically feasible.

Similar to the annexed plants, the main cost components of 1G autonomous distilleries are biomass, capital cost and maintenance, as described in Table 7.8. Once again, the autonomous 1G–Aut–Op is associated with lower biomass and capital costs mainly because of economies of scale and cost allocation between ethanol and electricity; therefore, 1G ethanol production costs become lower and more competitive.

7.1.2.4 Economic Assessment: Autonomous Versus Annexed Plants

A comparison among different autonomous and annexed distilleries is summarized in Fig. 7.1. The highest IRRs are related to the 1G–Anx–Op and 1G–Anx–St scenarios whose values are around 16.0 and 15.8 % per year, as previously discussed. These scenarios are favored by historical sugar prices whose values remained relatively high over the years when compared to the current prices. Autonomous distilleries, on the other hand, are heavily reliant on ethanol prices whose values are historically low. Consequently, better economic results are observed in annexed plants when compared with autonomous scenarios.

The NPV associated with the 1G–Aut–Op scenario, however, is still the second largest among the scenarios, mainly because of the scaling-up effect. Although its investment is high, net revenues are also high. On the other hand, the 1G–Aut–St scenario is associated with the lowest IRR and NPV values among all scenarios.

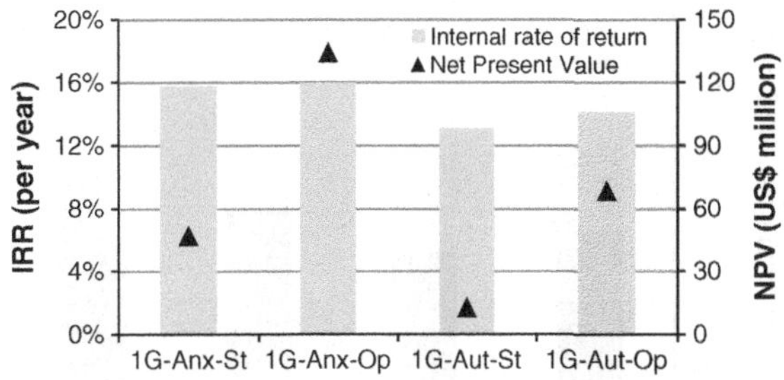

Fig. 7.1 Summary of results for first-generation annexed plants and autonomous distilleries (considering ethanol, electricity and sugar moving average prices over the last decade)

In order to comprehend the effect of price assumptions in the assessment, a new cash flow analysis was then performed considering ethanol, sugar and electricity prices of July 2014: US$0.60 per L, US$0.37 per kg and US$87 per MWh, respectively. Compared to average historical prices over the last decade, these refer to slightly higher prices for ethanol and electricity, but lower sugar price (see average historical prices in Sect. 6.1). Figure 7.2 shows that both 1G–Anx–Op and 1G–Aut–Op scenarios have similar economic results, with higher annual IRRs and NPVs: about 17.4 % and US$180 million. A similar effect of prices on standard scenarios was observed, resulting in similar IRR and NPVs. In fact, both annexed and autonomous scenarios are economically feasible, and as shown, their IRR and NPVs are very price sensitive.

Regarding ethanol production costs (see Fig. 7.3), the lowest ethanol production cost is associated with the 1G–Anx–Op scenario, mainly because of the cost allocation to sugar. Considering the relatively higher historical average sugar prices compared to its current value, a higher share of operating costs will be allocated to sugar production, thus decreasing ethanol production costs. The 1G–Aut–Op scenario is related to the second lowest ethanol production cost because of the

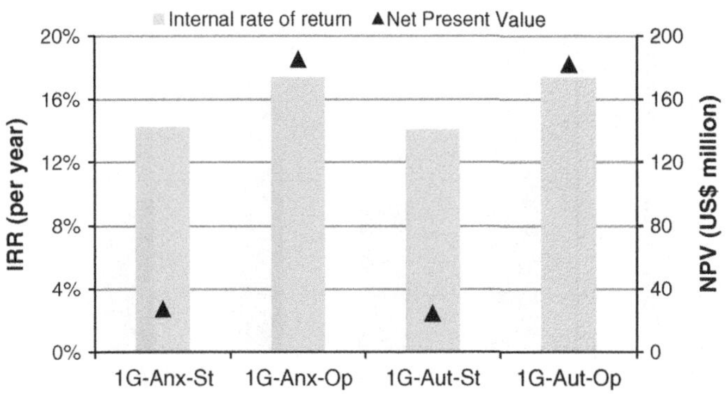

Fig. 7.2 Summary of results for 1G annexed plants and autonomous distilleries (considering ethanol, electricity and sugar average prices of July 2014)

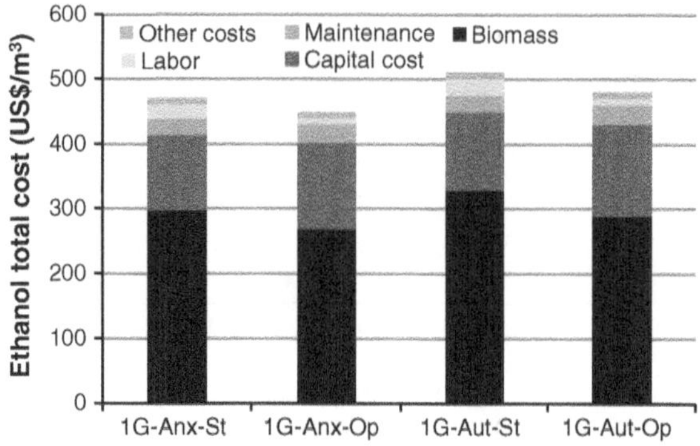

Fig. 7.3 First-generation ethanol costs of annexed plants and autonomous distilleries (considering ethanol, electricity and sugar moving average prices over the last decade)

scaling-up effect and the cost allocation to electricity. On the other hand, 1G–Anx–St and 1G–Aut–St are associated with higher ethanol production costs mainly because of lower ethanol production capacity over the year, as well as lower cost allocation to electricity.

The effect of price assumptions on the ethanol total cost was also assessed. When considering ethanol, sugar and electricity prices of July 2014, it is possible to observe that ethanol production costs are higher in the 1G–Anx–St scenario (see Fig. 7.4) when compared to the assessment with historical prices (see Fig. 7.3). This outcome was expected because, currently, lower sugar prices tend to increase ethanol cost because less operating costs are allocated to sugar. In the case of 1G–Anx–Op, the lower sugar price was compensated by the higher electricity price which, in turn, increased cost allocation to electricity production; the net effect was a slight decrease in the ethanol cost from US$0.450 to US$0.441 per L. In the 1G–Aut–St scenario, ethanol production cost remained the same (US$0.511 per L)

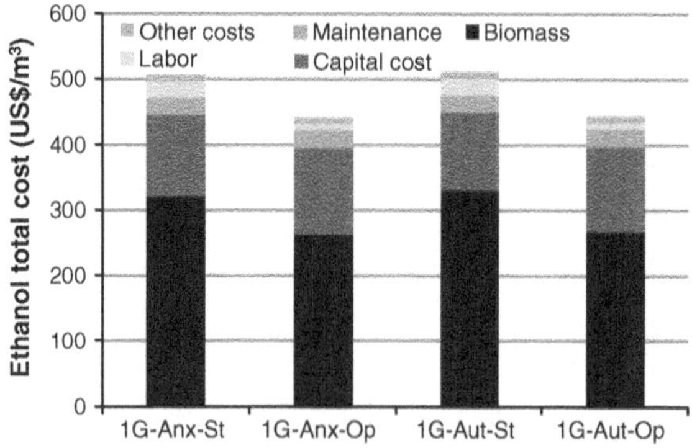

Fig. 7.4 First-generation ethanol costs of annexed and autonomous distilleries: standard and optimized configurations (considering ethanol, electricity and sugar average prices of July 2014)

because there are no coproducts; therefore, cost allocation does not interfere on ethanol production cost. In the case of 1G–Aut–Op, however, the increase in electricity price leads to lower ethanol total cost (decreased from US$0.480 to 0.443 per L) due to allocation of operating and capital costs to electricity.

7.1.3 Environmental Assessment

Environmental impact assessment for ethanol production in annexed plants and autonomous distilleries, considering standard and optimized scenarios, is evaluated using a cradle to gate Life Cycle Assessment. Industrial datasets for the evaluated scenarios are presented in Table 7.9, using one tonne of sugarcane processed as reference flow. The main differences on industrial datasets, besides products, are inputs and emissions specifically related to sugar, ethanol or CHP processes, which

Table 7.9 Industrial datasets for standard and optimized annexed plants and autonomous distilleries, per tonne of sugarcane processed

		Scenarios			
		1G–Anx–St	1G–Anx–Op	1G–Aut–St	1G–Aut–Op
Products					
Anhydrous ethanol	kg	42.1	42.3	67.0	67.2
Sugar	kg	51.3	51.4	–	–
Electricity	kWh	–	185.8	–	185.8
Bagasse surplus (dry mass)	kg	27.8	–	24.1	–
Allocation[a]					
Anhydrous ethanol	kg	57.0 %	48.6 %	97.9 %	82.3 %
Sugar	kg	40.8 %	34.7 %	–	–
Electricity	kWh	–	16.6 %	–	17.7 %
Bagasse surplus	kg	2.3 %	–	2.1 %	–
Inputs					
Sugarcane	kg	1000.0	1000.0	1000.0	1000.0
Straw (dry mass)	kg	–	45.8	–	45.8
Water	kg	1500.0	1500.0	1500.0	1500.0
Quicklime	kg	0.85	0.82	0.64	0.61
Sulfuric acid	kg	0.264	0.264	0.420	0.420
Phosphoric acid	g	236.2	176.5	231.3	172.4
Flocculant polymer	g	2.656	2.555	2.656	2.555
Antibiotics	g	0.696	0.698	1.105	1.110
Lubricating oil	g	13.0	13.0	13.0	13.0

(continued)

Table 7.9 (continued)

		Scenarios			
		1G–Anx–St	1G–Anx–Op	1G–Aut–St	1G–Aut–Op
Cyclohexane	kg	0.04	–	0.07	–
Zeolite	g	–	17.227	–	27.380
Steel	kg	1.30E−01	1.77E−01	1.25E−01	1.74E−01
Chromium steel	kg	7.59E−03	1.03E−02	7.31E−03	1.02E−02
Concrete	m^3	1.74E−04	2.37E−04	1.68E−04	2.34E−04
Building, hall	m^2	3.64E−05	4.97E−05	3.51E−05	4.88E−05
Emissions to air					
Carbon dioxide, biogenic[b]	kg	209.4	339.2	215.7	339.2
Carbon monoxide, biogenic[b]	g	132.1	214.6	136.1	214.6
Nitrogen oxides[b]	g	143.5	233.0	147.8	233.0
Dinitrogen monoxide[b]	g	8.61	13.98	8.87	13.98
Sulfur oxides[b]	g	0.50	0.82	0.52	0.82
Methane, biogenic[b]	g	0.23	0.37	0.23	0.37
VOC compounds[b]	g	0.02	0.03	0.02	0.03
Particulates, > 2.5 μm[b]	g	0.02	0.03	0.02	0.03
Particulates, < 2.5 μm[b]	g	0.01	0.02	0.01	0.02
Carbon dioxide, biogenic[c]	kg	41.9	42.1	66.4	66.6
Ethanol[d]	g	81.3	72.7	129.7	116.1
Carbon dioxide, biogenic[d]	g	787.6	704.2	1256.0	1124.0
Particulates, unspecified[e]	g	2.0	2.0	–	–
Organic acids[e]	g	1.34E−03	1.31E−03	–	–
Particulates, fine mineral fibers[e]	g	5.15E−03	5.10E−03	–	–
Phosphoric acid[e]	g	6.04E−05	5.31E−05	–	–
Coproducts[f]					
Vinasse	kg TC^{-1}	509.2	439.9	806.2	696.8
Filter cake mud (dry mass)	kg TC^{-1}	19.0	10.7	18.3	10.0
Ashes (dry mass)	kg TC^{-1}	5.8	9.4	6.0	9.4

[a]Economic allocation
[b]Emissions from bagasse burning
[c]Emissions from fermentation
[d]Emissions from distillation
[e]Emissions from sugar drying process
[f]Returned to the field

depend on production volume per tonne of sugarcane. Straw is used only in the optimized scenarios. Allocation factors are calculated based on the production volume and products prices, using economic criteria. Biomass production datasets are similar to those presented in Chap. 3, which were adapted to represent each scenario. Straw recovery for optimized scenarios and different amounts of vinasse, filter cake mud and ashes are taken into account, causing small modifications in relevant variables.

The comparative environmental impacts for ethanol production in annexed plants and autonomous distilleries, considering standard and optimized scenarios, were calculated using the ReCiPe midpoint method, hierarchist version v 1.08. Results for selected impact categories per unit of mass of ethanol are presented in Fig. 7.5. These scores give the relative environmental impacts of ethanol production including sugarcane production and transportation processes, industrial conversion in the biorefinery, and all the production of inputs used in all the life cycle stages. Ethanol commercialization and use were not included in this comparative assessment, since these stages will not differ among the evaluated scenarios.

Results show that, for all impact categories, the optimized scenarios for both annexed and autonomous plants present better environmental performance than the respective standard scenarios. These figures show that optimized plants have a great potential for decreasing environmental impacts of ethanol production in a sugarcane biorefinery, mainly due to surplus electricity commercialization. This extra product in the optimized biorefinery portfolio reduces allocation factors for ethanol, and consequently, final scores for the biofuel are lower offsetting inclusion of straw recovery inputs and higher process emissions per tonne of sugarcane in optimized scenarios (Table 7.9). Comparing annexed and optimized plants, the former have lower environmental impacts for ethanol production. Sugar production in annexed plants has similar effect to that observed for electricity in optimized biorefineries, reducing allocation factor for ethanol.

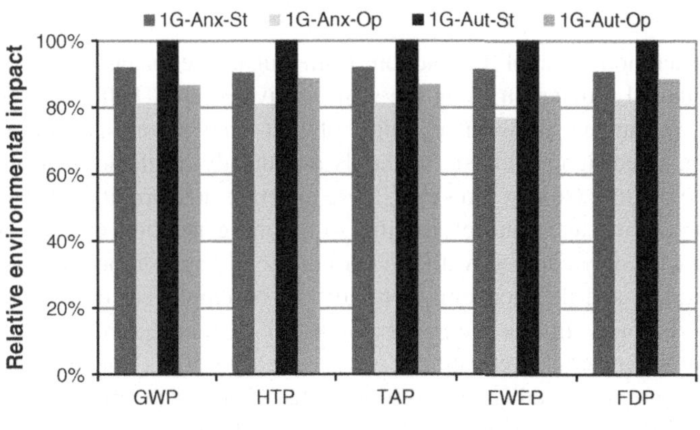

Fig. 7.5 Comparative environmental scores for ethanol production in standard and optimized scenarios of annexed plants and autonomous distilleries (**Note** *GWP* climate change; *HTP* human toxicity; *TAP* terrestrial acidification; *FWEP* freshwater eutrophication; *FDP* fossil depletion)

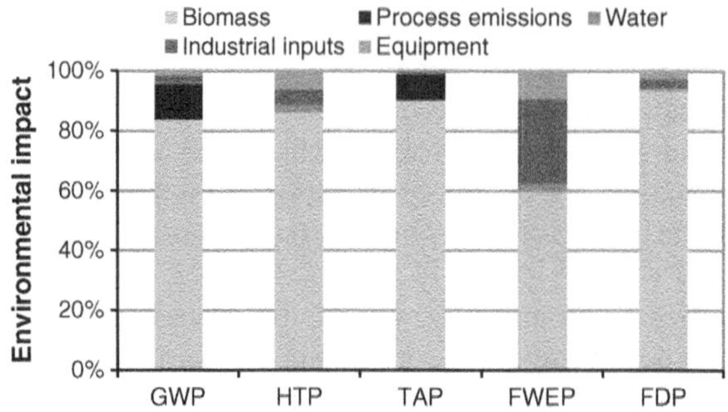

Fig. 7.6 Breakdown of the environmental impacts of ethanol production for optimized autonomous distillery (**Note** *GWP* climate change; *HTP* human toxicity; *TAP* terrestrial acidification; *FWEP* freshwater eutrophication; *FDP* fossil depletion)

Breakdown of the environmental impacts of ethanol production in the optimized autonomous distillery is presented in Fig. 7.6. Results indicate that agricultural phase has a very high impact in the ethanol production chain. Direct process emissions are observed only in GWP and TAP categories, due to emissions from bagasse and straw burning in industrial boilers. In GWP, impact is related to N_2O, CH_4 and CO emissions, since CO_2 from biomass is biogenic and has no characterization factor; in TAP, impact is related to NO_X and SO_X, gases that form nitrates and sulfates before deposition in soil surface. Industrial inputs have an important contribution to FWEP. The main contributor is phosphoric acid (used in juice treatment), since its production process releases considerable amounts of phosphate (one of the most important causes of eutrophication) to freshwater.

7.2 Case 2: Second-Generation Sugarcane Facilities

Second-generation ethanol production from sugarcane bagasse and straw can increase ethanol production using the same cultivated area. Differently from other feedstock, sugarcane allows integration of first- and second-generation ethanol production, since lignocellulosic materials are already available at plant site (bagasse), or close to it (straw). However, these materials are currently used as fuels for the production of steam and electricity, so second-generation ethanol production from sugarcane may compete with the current use of bagasse (and straw) as a fuel.

Biochemical and thermochemical routes can be employed to convert lignocellulose into biofuels. In this section, examples of biochemical route, comprised of lignocellulosic material pretreatment, enzymatic hydrolysis and fermentation are illustrated. Current and advanced conversion technologies are discussed, as well as the integration of first- and second-generation ethanol production in the same processing facility and its comparison with a stand-alone second-generation process facility.

7.2.1 *Current and Advanced Second-Generation Technologies*

7.2.1.1 Scenario Description

Since second-generation ethanol production is not a mature process, a comparison between current and advanced technologies, expected to become industrially available in the long term, is provided. The process flow sheet presented in Chap. 4 is same for both technologies, whereas process operating conditions and yields are differentiated as shown in Table 7.10. In the advanced technology, more optimistic figures were assumed, such as higher yields and recoveries as well as lower residence times, resulting from the improvement of second-generation process with

Table 7.10 Operating conditions and yields for current and advanced second-generation ethanol production technologies (Milanez et al. 2015)

Parameter	Current	Advanced
Steam explosion pretreatment		
Temperature (°C)	190	210
Residence time (min)	15	5
Solid content	Defined by the steam required to achieve reactor temperature	Defined by the steam required to achieve reactor temperature
Cellulose solubilization (%)	5.0	5.5
Xylan conversion to xylose (%)	30	60
Xylan conversion to xylose oligomers (%)	30	20
Xylan degradation to furfural (%)	10	10
Diffuser		
Proportion of water added (% of cellulignin fibers)	250	180
Insoluble solid retention (%)	99.5	99.5
Soluble solids recover in the liquor (%)	98.0	98.0
Cellulignin moisture (%)	50	50
Enzymatic Hydrolysis		
Temperature (°C)	50	65
Residence time (h)	48	36
Solid content (%)	15	25
Cellulose conversion to glucose (%)	60	80
Xylan conversion to xylose (%)	60	80
Filter		

(continued)

Table 7.10 (continued)

Parameter	Current	Advanced
Solid retention (%)	95	99
Soluble solids recover in the liquor (%)	92	99
Solid residues moisture (%)	55	50
C6/C12 fermentation		
Operational conditions	Same as 1G	Same as 1G
C6/C12 conversion to ethanol (%)	88	90
Ethanol content in C6/C12 wine (g L^{-1})	80	80
Deoligomerization and C5 fermentation		
Temperature (°C)	33	33
Residence time (h)	48	24
Xylose oligomers conversion to xylose (%)	80	90
C6 conversion to ethanol (%)	90	90
C5 conversion to ethanol (%)	80	85
Cell recycling (%)	80	95
Ethanol content in C5 wine (g L^{-1})	70	70

time. These figures were based on the study of Milanez et al. (2015), which present a comprehensive assessment of second-generation technologies with data from an extensive discussion with many biofuel sector stakeholders.

For the 1G2G scenarios, the considered 1G plant has the same configuration and operational conditions as the 1G optimized autonomous distillery previously described. However, in the 1G2G scenarios, steam is produced only to meet process requirement; then, back-pressure turbines are employed. The 1G sector of the plant operates for 200 days, while cogeneration and 2G sectors operate for 330 days. During harvest season (200 days), 2G sector runs with bagasse (and straw) coming directly from 1G plant; during the other 130 days of operation, 2G plant operates with bagasse and straw mixture that was stored during harvesting season.

The mixture of bagasse and straw sent to the 2G sector undergoes a steam explosion pretreatment with steam at 12.6 bar (current) or 19.1 bar (advanced). During the pretreatment, a fraction of xylan is converted to xylose and xylose oligomers. A diffuser is used to separate a liquid portion of the pretreated material, known as C5 liquor. The C5 liquor undergoes a neutralization treatment for pH correction prior to deoligomerization and fermentation of C5 liquor to ethanol. The solid fraction of the pretreated material is sent to enzymatic hydrolysis reactor.

After enzymatic hydrolysis, a solid–liquid separation takes place using a filter. The solid residues are used as fuel in the cogeneration sector. During harvesting season, when the 1G plant is also running, the liquid fraction (C6 liquor) is mixed with the sugarcane juice for fermentation to benefit from the dilution of possible inhibitors present in the C6 liquor. In the off-season, C6 fermentation reactor operates with C6 liquor only.

Wine obtained from the two fermentations (C5 and C6) is mixed and sent to distillation columns. Similar to 1G optimized autonomous distillery, ethanol dehydration is carried out using molecular sieves.

7.2.1.2 Techno-economic Analysis

Outputs for ethanol and electricity production for the integrated 1G2G scenarios, considering current (1G2G–Cur) and advanced (1G2G–Adv) technologies, are presented in Table 7.11.

A significant increase in 2G ethanol production (around 78 %) is observed for 1G2G–Adv scenario. This can be attributed mainly to the enhancements in the monomeric sugars (C5 and C6) recovery in pretreatment and hydrolysis units and by the increase in fermentation yields, as shown in Table 7.10. These factors also explain the reduction in electricity generation; since a larger part of the solid fraction of lignocellulosic material is solubilized, a smaller amount of solid residues is destined to cogeneration.

When compared to the first-generation process from Case 1, scenario 1G–Aut–Op, the 1G2G integration proves to be promising, promoting an increase of 26 and 46 % in ethanol production for 1G2G–Cur and 1G2G–Adv, respectively, in relation to scenario 1G–Aut–Op.

The economic analysis was performed to assess integrated 1G2G biorefineries considering capital expenditures summarized in Table 7.12. The 1G2G plant was assumed to be conceptually divided into two main sectors: (a) 1G and interface areas and (b) 2G areas. The 1G + interface areas include engineering, insurance, administrative and infrastructure costs as well as traditional 1G processing areas such as sugarcane reception and handling, juice extraction, evaporation and shared industrial infrastructure. The equipment processing capacities in these specific areas (e.g., distillation and generation and distribution of steam and electricity) are designed according to the amount of processed material. The capital expenses related to the 2G areas are associated with pretreatment, enzymatic hydrolysis, C5

Table 7.11 Outputs for the current and advanced 1G2G scenarios

Scenarios	1G2G–Cur	1G2G–Adv
Total anhydrous ethanol production (L TC^{-1})	107.5	124.8
1G anhydrous ethanol production (L TC^{-1})	85.4	85.4
2G anhydrous ethanol production (L TC^{-1})	22.1	39.4
Electricity (kWh TC^{-1})	77.4	61.6

Table 7.12 Summary of capital expenditures associated with the main industrial areas

CAPEX (US$ million)	1G2G–Cur	1G2G–Adv
1G + interface areas		
Administrative infrastructure, engineering and utilities	86.6	82.5
Sugarcane reception, juice extraction and evaporation	70.5	67.3
Ethanol production (fermentation, distillation/dehydration)	93.6	102.4
Combined heat and power (steam and electricity generation)	109.2	103.2
2G areas		
Pretreatment and C5 liquor separation/evaporation	75.9	64.6
Separated C5 fermentation	11.2	9.9
Enzymatic hydrolysis and C6 liquor separation	34.7	20.5
Yeast propagation	0.7	0.6
Total	482.4	451.0

fermentation and solid–liquid separation operations. The method used to estimate the capital expenditures in this study was intended to capture the differences between the different technological scenarios.

The sugarcane and straw production costs modeled in CanaSoft for the 1G2G–Cur scenario were US$28.50 and US$36.38 (dry basis) per tonne, respectively. In the case of 1G2G–Adv scenario, sugarcane production cost was US$28.76 per tonne, which is slightly higher mainly due to extra vinasse to be returned to the field. On the other hand, straw production cost remains the same for both scenarios.

As shown in Table 7.13, the 1G2G–Adv scenario presented the highest internal rate of return, roughly 18.5 % per year. Considering a minimum acceptable rate of return of 12 % per year, this scenario is associated with a positive NPV of approximately US$240 million. This result was expected because of the economies of scale and second-generation technological improvements that make higher ethanol yield possible in the advanced scenario. Because of that, 1G2G–Adv scenario is related to lower global (1G2G) and 2G ethanol production costs: US$0.418 and US$0.283 per L, respectively. Considering the results from the cash flow analysis and ethanol total cost, it is clear that the 1G2G–Adv scenario would be a feasible alternative because promising technologies overcome some technological gaps that are observed in current second-generation production such as hydrolysis yields.

Table 7.13 Techno-economic analysis of annexed distilleries

Parameter	1G2G–Cur	1G2G–Adv
IRR (% per year)	13.6	18.5
NPV (US$ mi)	57	240
1G2G ethanol cost (US$ m^{-3})[a]	488	418
2G ethanol cost (US$ m^{-3})[a]	516	283

[a]Total cost = operating costs + cost of investment

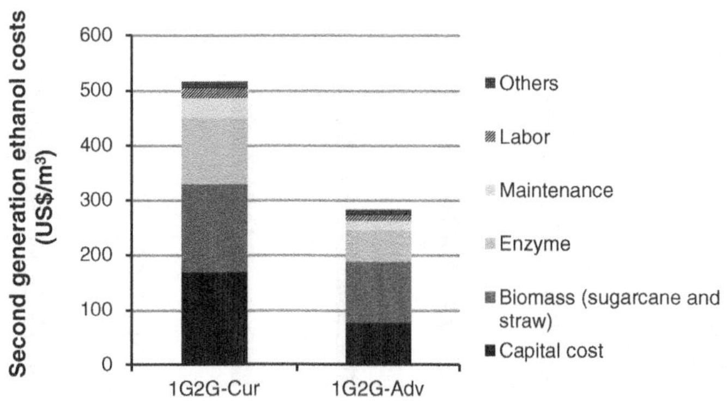

Fig. 7.7 Breakdown of second-generation ethanol production costs

Although techno-economic results of 1G2G–Cur scenario are below the potential of the 1G2G–Adv, the 1G2G–Cur scenario is also economically feasible. The comparison between current and advanced scenarios for second-generation ethanol production would inevitably result in a better economic outcome for the advanced scenario due to the assumptions on technology development of second-generation ethanol production in a near future. However, Table 7.13 is still useful to demonstrate the potential economic results of such technological improvements that would significantly increase both the IRR and NPV as well as decrease the second-generation ethanol total cost to levels below the ones from current first-generation ethanol cost.

The second-generation ethanol production cost breakdown is presented in Fig. 7.7 and highlights main differences between the cost components in 1G2G-Cur and 1G2G–Adv scenarios. It is possible to observe capital cost, biomass and enzyme as the main cost components for both scenarios. The 1G2G–Adv is associated with lower 2G production costs mainly because of the higher second-generation ethanol yield. Moreover, the expectations on technological improvements related to industrial equipment and hydrolysis yields may gradually decrease the contributions associated with capital and enzyme costs in a near future.

7.2.1.3 Environmental Assessment

In Fig. 7.8, the relative environmental impacts from the ethanol production life cycle are compared. Current and advanced second-generation technologies are compared with optimized first-generation (1G–Aut–Op). In this assessment,

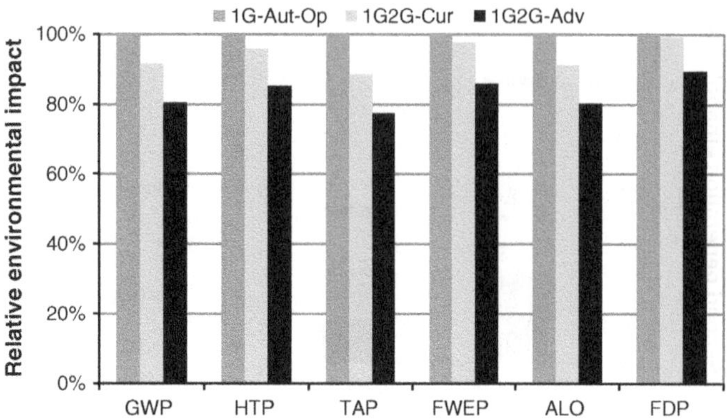

Fig. 7.8 Comparative environmental scores for ethanol production with current and 1G2G–Adv technologies (**Note** *GWP* climate change; *HTP* human toxicity; *TAP* terrestrial acidification; *FWEP* freshwater eutrophication; *ALO* agricultural land occupation; *FDP* fossil depletion)

economic allocation between product and coproducts was used. Changes on biomass production life cycle inventories due to different amounts of industrial residues (vinasse, filter cake and ashes) were considered as well.

Results show that integrating first- and second-generation processes using both current and advanced technologies for second-generation ethanol production presents lower environmental impacts than optimized first-generation ethanol for all the evaluated categories. Impacts are roughly 10 % lower for current and twice that for advanced technology in comparison with the first-generation ethanol production. For terrestrial acidification category (TAP), results for 1G2G ethanol were even slightly better due to lower emissions in the boilers, since only the required amount of lignocellulosic material (LCM) is burnt to supply process steam demand. Fossil depletion (FDP) presented results not as good as the other categories due to ammonia used in the 2G process. Advanced second-generation ethanol presented the lowest environmental impacts, mainly due to high ethanol yields per tonne of LCM, indicating that the technological improvements in second-generation process have a great potential to improve environmental performance of ethanol.

7.2.2 *Integrated Versus Stand-Alone Second-Generation Ethanol Production from Sugarcane*

7.2.2.1 Scenario Description

In Brazil, integration of first- and second-generation processes using sugarcane lignocellulosic materials (bagasse and straw) presents several advantages: (i) Feedstock is already at plant site (bagasse) or close to it (straw); (ii) part of infrastructure can be shared between 1G and 2G units, e.g., concentration, distillation, cogeneration and storage; (iii) additional possibilities of thermal integration; (iv) reduction of inhibitors concentration with addition of sugarcane juice into hydrolysate. However, the possibility of a stand-alone 2G plant cannot be disregarded, especially if physical separation of 1G and 2G ethanol is required in order to benefit from financial incentives.

An assessment was carried out to evaluate a stand-alone 2G plant, using surplus lignocellulosic material (bagasse and straw) from an optimized 1G autonomous distillery that burns only the amount of biomass required to supply its process steam demand. Process configuration and parameters for advanced 2G technology (see Table 7.10) were assumed based on Milanez et al. (2015). Similar to optimized 1G plants, efficient 65 bar boilers, ethanol dehydration with molecular sieves and 20 % of reduction on steam demand (2.5 bar) were assumed. Back-pressure turbines are employed, because steam is produced only to meet process requirement, using part of the lignocellulosic material and solid residues. The entire stand-alone 2G plant operates for 330 days.

7.2.2.2 Techno-economic Analysis

Outputs for advanced stand-alone 2G plant (2G–Adv) as well as optimized autonomous 1G distillery with surplus LCM (1G–LCM) and advanced integrated 1G2G plant (1G2G–Adv) are shown in Table 7.14.

When comparing 1G2G–Adv scenario and the combination of 1G–LCM and 2G–Adv scenarios, similar ethanol and electricity productions are observed, since same yields and optimization features were assumed for both advanced stand-alone 2G and integrated 1G2G scenarios.

The stand-alone 2G plant considering advanced technology (2G-Adv) is not physically integrated to a first-generation plant. However, it was assumed that it consumes lignocellulosic material from an optimized first-generation autonomous distillery whose bagasse and straw surplus is available in the market (scenario 1G–LCM). Considering that such market is not consolidated yet, the LCM market value was based on a fraction (60 %) of the potential revenues from burning LCM to produce bioelectricity. A 40 % discount was considered to account for the lack of

Table 7.14 Comparison between stand-alone 2G and integrated 1G2G plants, considering advanced technologies

Scenarios	1G2G–Adv	1G–LCM	2G–Adv
LCM for pretreatment—dry basis (10^6 kg year^{-1})	462.7	–	462.5
Total anhydrous ethanol production (10^6 L year^{-1})	499.1	341.4	157.7
1G anhydrous ethanol production (10^6 L year^{-1})	341.4	341.4	–
2G anhydrous ethanol production (10^6 L year^{-1})	157.7	–	157.7
Electricity (GWh year^{-1})	246.3	126.3	115.8

Table 7.15 Summary of capital expenditures for the main industrial areas of advanced 2G stand-alone plant

CAPEX (US$ million)	2G-Adv
1G areas	
Administrative infrastructure, engineering and utilities	33.4
Ethanol production (distillation/dehydration)	17.1
Combined heat and power (steam and electricity generation)	51.0
2G areas	
Pretreatment and C5 liquor separation/evaporation	64.6
Separated C5 fermentation	9.9
Enzymatic hydrolysis and C6 liquor separation	20.5
Separated C6 evaporation/fermentation	7.0
Yeast propagation	0.6
Total	204.1

investment in extra cogeneration equipment that would be necessary for producing this "extra" electricity. As a result, the opportunity cost of US$44.80 per tonne of dry lignocellulosic material was obtained.

Capital expenditures for 2G-Adv scenario are summarized in Table 7.15. In such scenario, as the LCM is directly sent to the 2G process, sugarcane preparation and juice extraction are not necessary. Despite that, the so-called 1G areas are still required, e.g., distillation/dehydration and CHP areas. The CHP was designed to only provide the amount of steam necessary to supply the process needs, not focusing on maximizing the electricity surplus. The 2G ethanol production technology is similar to the one in other 2G scenarios, except for the sucrose-rich stream to be mixed with the C6 liquor prior to fermentation.

The main results of the techno-economic analysis are presented in Table 7.16 and highlight internal rates of return, NPVs and ethanol production costs. It is clear that 1G2G–Adv scenario is related to better economic impacts when compared to the 2G–Adv. However, both alternatives are economically feasible and lead to positive results in terms of value added and profitability.

Table 7.16 Comparison of techno-economic results considering 2G–Adv and 1G2G–Adv scenarios

Parameter	2G-Adv	1G2G-Adv
IRR (% per year)	16.2	18.5
NPV (US$ mi)	67	240
1G2G ethanol cost (US$ m^{-3})[a]	–	418
1G ethanol cost (US$ m^{-3})[a]	–	480
2G ethanol cost (US$ m^{-3})[a]	411	283

[a]Total cost = operating costs + cost of investment

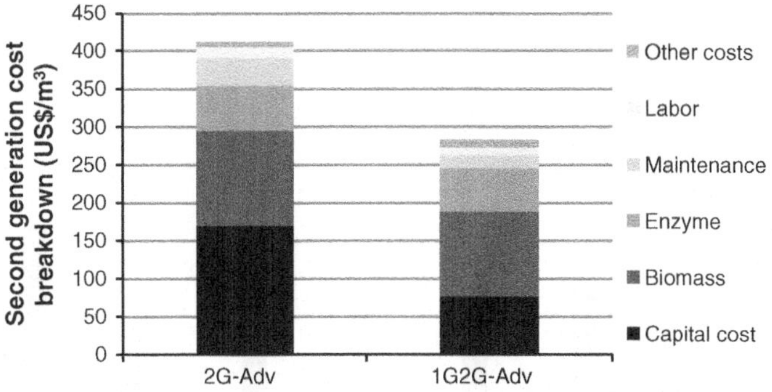

Fig. 7.9 Comparison of advanced stand-alone 2G and integrated 1G2G ethanol production costs

Ethanol production cost in 1G2G–Adv scenario is quite similar to the one in 2G-Adv stand-alone scenario: US$0.418 and US$0.411 per L, respectively. In the case of 1G2G–Adv, however, a significant share of 1G2G production costs is allocated to the first-generation ethanol (see the allocation procedure in Sect. 6.1). Second-generation ethanol cost is lower in the 1G2G–Adv scenario mainly due to the allocation procedure.

It is also important to highlight the main differences between 2G ethanol cost elements for both 2G-Adv and 1G2G–Adv scenarios. As Fig. 7.9 shows, they are related to a similar hierarchy of cost components. Capital cost, biomass and enzyme costs also play a key role on the second-generation ethanol, as they correspond to roughly 85 % of its total production costs in both scenarios. Although the magnitude of investment is significantly higher, maintenance and capital costs are significantly lower in 1G2G–Adv scenario when compared to the 2G–Adv. It is mainly because the cost allocation to the first-generation ethanol production contributes to decrease the second-generation ethanol cost, as previously mentioned.

7.2.2.3 Environmental Assessment

In 1G–LCM scenario, surplus LCM is a coproduct. Therefore, it receives an allocation factor based on its opportunity cost. Surplus LCM is an input to 2G stand-alone (2G–Adv) scenario. Similar to previous environmental analyses, all the product life cycle production phase is included and economic allocation was used to split impacts between coproducts. Environmental impacts of integrated and stand-alone 2G ethanol production are compared in Fig. 7.10. The optimized autonomous 1G distillery with surplus LCM (1G–LCM) is also included in this comparison. The environmental benefits of 2G ethanol are shown in an isolated manner, illustrating its great advantage when compared to the optimized first-generation ethanol production scenario. As discussed in techno-economic analysis, similar ethanol and electricity outputs are observed for 1G2G–Adv scenario and the combination of 1G–LCM and 2G–Adv scenarios. Consequently,

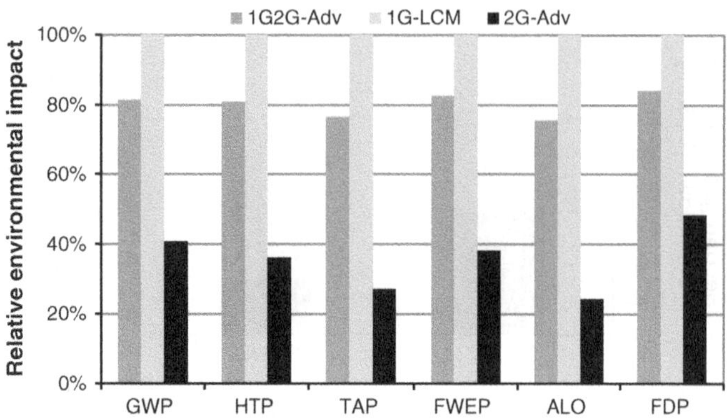

Fig. 7.10 Comparative environmental scores for ethanol production with integrated and stand-alone 2G (**Note** *GWP* climate change; *HTP* human toxicity; *TAP* terrestrial acidification; *FWEP* freshwater eutrophication; *ALO* agricultural land occupation; *FDP* fossil depletion)

scores for integrated 1G2G process are very similar to weighted scores of ethanol from stand-alone 2G and 1G. Indeed, integrated 1G2G process presents a rather small advantage because it shares some equipment (concentration, fermentation, distillation and cogeneration), thus requiring less building material per unit of ethanol produced than independent technologies.

Considering only the 2G-Adv scenario, from Fig. 7.11, it can be observed that LCM is the principal responsible for environmental impacts of 2G ethanol. Inputs required exclusively for 2G unit operations—ammonia and enzymes—have the main scores among the inputs related with the industrial process.

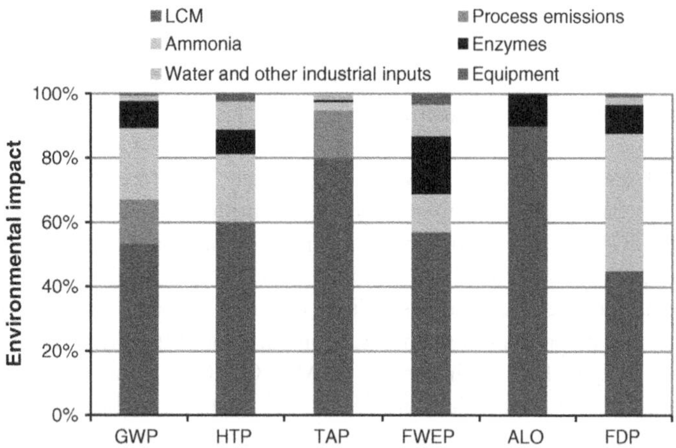

Fig. 7.11 Breakdown of the environmental impacts of stand-alone 2G ethanol production in 2G-Adv scenario (**Note** *GWP* climate change; *HTP* human toxicity; *TAP* terrestrial acidification; *FWEP* freshwater eutrophication; *ALO* agricultural land occupation; *FDP* fossil depletion)

7.3 Case 3: First-Generation Ethanol—Harvesting Extension

Sugarcane cannot be stored for more than a few hours and mills usually operate only during the harvesting period which, in the largest sugarcane producing region in Brazil, happens from April to November. Even though electricity production all year-round can already be found in some Brazilian mills, ethanol production in the off-season is not yet a reality. Longer operational periods may lead to more efficient use of existing production capacity, thus decreasing the impact of investments in equipment and reducing ethanol production costs.

In this case, three alternatives for extended operation through first- and second-generation ethanol production are compared.

7.3.1 Scenario Description

Alternatives for extended operation through the sugarcane off-season period include the use of sweet sorghum and corn as complementary feedstock and second-generation ethanol from sugarcane bagasse and straw.

7.3.1.1 Extended Operation with Sweet Sorghum

Sweet sorghum has been proposed as a promising alternative to extend the operation period in sugarcane mills, employing agricultural machinery and practices very similar to those used for sugarcane, therefore requiring only minor adjustments in industrial process (May 2012; Cavalett et al. 2013).

An optimized autonomous distillery with sweet sorghum processing in the sugarcane off-season (scenario 1G–Sorghum) was considered. Agricultural yield for sweet sorghum considered for this scenario was 60 t ha^{-1}, based on optimized agricultural varieties (May 2012). Sweet sorghum composition is presented in Table 7.17.

Table 7.17 Composition of sweet sorghum based on variety of BRS511 (Parrella and Schaffert 2012)

Parameter	Value
Fibers (%)	11.1
Soluble solids (°Brix)	14.4
Pol (% sucrose)	10.6
% reducing sugars	1.0
Water content (%)	74.5

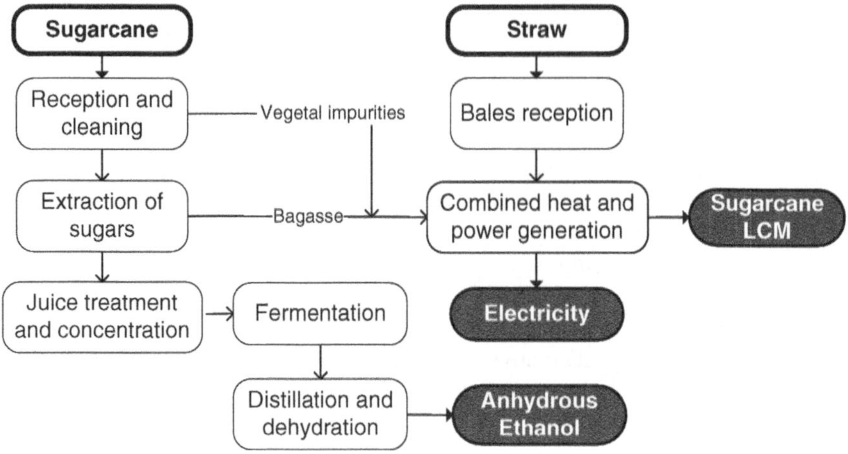

Fig. 7.12 Block flow diagram for optimized autonomous distillery with sugarcane processing

Block flow diagram showing season operation of an optimized autonomous distillery is depicted in Fig. 7.12. Off-season operation with sweet sorghum is shown in Fig. 7.13.

Process steps and parameters for optimized autonomous distillery processing sugarcane are correspondent to scenario 1G–Aut–Op (described in Case 1, Sect. 7.1). Although the same process steps are used in sweet sorghum processing, different yields are obtained. Operational characteristics for season and off-season periods are summarized in Table 7.18.

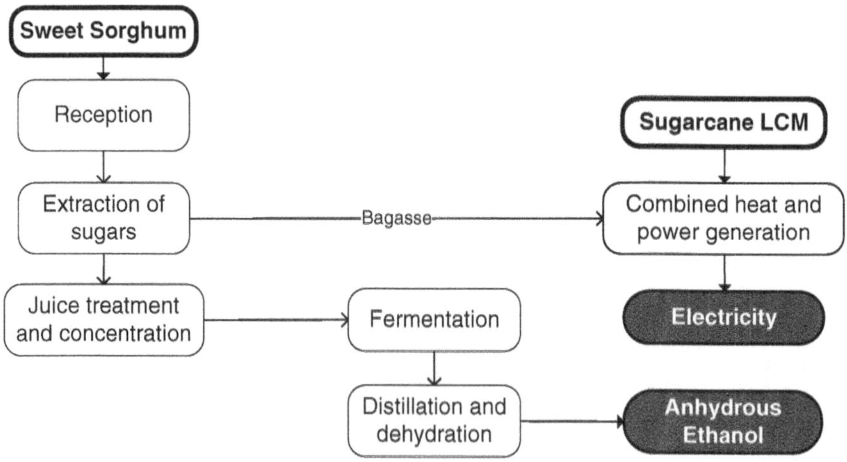

Fig. 7.13 Block flow diagram for optimized autonomous distillery with sweet sorghum processing

Table 7.18 Operational characteristics in sugarcane season and off-season with sweet sorghum as complementary feedstock

Parameter	Season	Off-season
Feedstock	Sugarcane	Sweet sorghum
Type of distillery	Optimized distillery	Optimized distillery
Products	Ethanol and electricity	Ethanol and electricity
Period of operation (days)	200	45
Processing capacity (10^6 t feedstock year^{-1})	4	0.9
Straw recovery (%)	50 (considering vegetal impurities and straw bales)	–[a]
Cleaning system	Dry cleaning	None
Efficiency of sugar extraction	96 %	90 %[b]
Sucrose content in the filter cake[c]	1 %	3 %[b]
Fermentation efficiency	90 %	88 %[d]
Wine ethanol content (g L^{-1})	80	not fixed[e]
Distillation efficiency	99 %	99 %[d]
Boiler pressure (bar)/efficiency (%, LHV basis)	65/87.7	65/87.7
Type of turbines/isentropic efficiency (%)	Extraction-condensing/85	Extraction-condensing/85
Assumed reduction in 2.5 bar steam consumption due to heat integration (%)	20	20
Electric demand (kWh/t feedstock)	30 + 1.6 (dry-cleaning system)	30.0

[a]It is considered that sorghum has no crop residues, since all the biomass (stalks, leaves and panicle) is collected in the harvest process
[b]Values based on Pacheco (2012)
[c]Represents the losses on juice treatment
[d]Values obtained from Junqueira (2015)
[e]Evaporation of sweet sorghum juice was designed to evaporate the same amount of water per hour as in sugarcane processing, since same equipment is employed

In the off-season period, the sweet sorghum bagasse is used as fuel in the CHP to supply steam and electricity to the process. In addition, part of the sugarcane surplus bagasse and straw (LCM) is stored during the harvest season and used in the off-season in such way that it is considered that the cogeneration system operates at full capacity all year-round.

The operation period of 45 days is based on the time available for sweet sorghum harvesting management during the sugarcane off-season. Sweet sorghum varieties are highly photoperiod-sensitive, and their production cycles need to start in the months of October/November to be harvested in February/March, limiting operational days for industrial processing. In this case, sugarcane replanting areas are not enough to produce the total amount of sweet sorghum required for this configuration, and additional areas for sweet sorghum production need to be considered. In both cases, same sweet sorghum production system is considered for the modeling, according to information in Chap. 3. Sweet sorghum milling capacity was fixed equal to that of sugarcane in the harvest season (approximately 834 t/h).

7.3.1.2 Extended Operation with Corn

The increase in the availability of corn at low cost due to the corn–soybean rotation in the Center-West region of Brazil has motivated the integration of this feedstock into sugarcane mills. Moreover, the possibility of using sugarcane lignocellulosic material, as energy source for corn processing, improves the environmental impacts of corn ethanol production, when compared with the values obtained for typical corn ethanol processes in the USA (Milanez et al. 2014).

In this scenario (1G–corn), corn is processed during sugarcane off-season to produce ethanol, electricity and fractionated coproducts. Agricultural yield of 6 t/ha was assumed (Milanez et al. 2014), with an average distance of 100 km between cornfields and sugarcane mill. Similar to the previous scenario, an optimized autonomous distillery processes sugarcane and stores surplus sugarcane LCM throughout the harvest season so the CHP operates at full capacity all year-round, supplying the process demand and generating surplus electricity. In the off-season, fermentation, distillation, dehydration and CHP sections of the sugarcane mill are employed for corn processing. Additional process steps (milling, liquefaction and separation) are included in the industrial facility in order to make fermentable sugars available and generate fractionated products. The block flow diagram for season operation is the same as Fig. 7.12. Off-season operation, represented by a corn dry-milling process, is shown in Fig. 7.14.

Average composition of corn used in the assessment is presented in Table 7.19. It was defined based on values provided in specialized literature (Corn Refiners Association 2006) and corn ethanol producers in Brazil. The operational features of the sugarcane plant are the same presented in Table 7.18. A summary of characteristics for off-season operation with corn is shown in Table 7.20. For corn processing, operating conditions and efficiencies, as well as steam and electricity consumption, were also calculated based on values provided by corn ethanol producers.

An operation period in the sugarcane off-season of 130 days was considered, totaling a year-round operation of 330 days. The amount of corn processed per hour was defined so that hourly corn ethanol production would be the same as sugarcane ethanol production, using the same distillation and dehydration systems in the existing infrastructure. Nevertheless, additional fermentors were necessary due to the longer fermentation time (40 h) for corn processing. Since straw is recovered from the field, the amount of surplus lignocellulosic material stored is more than sufficient to supply steam and electricity to corn processing.

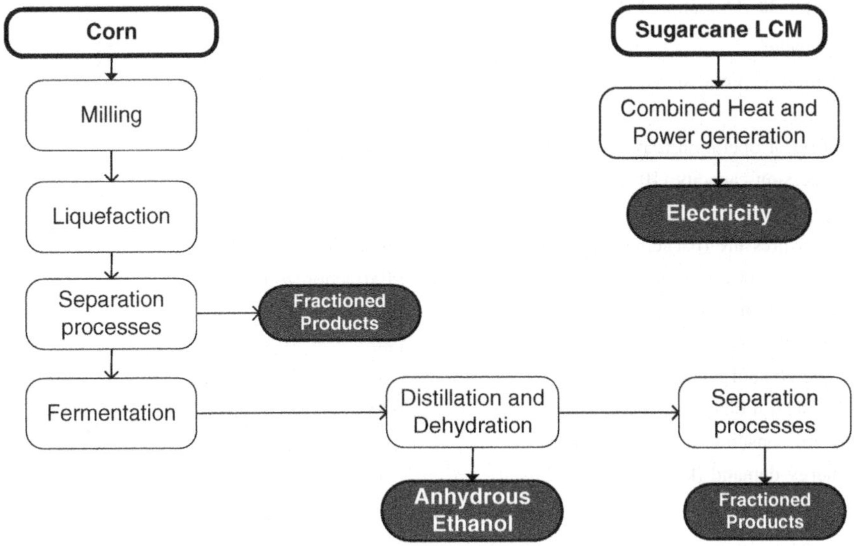

Fig. 7.14 Block flow diagram for corn processing

Table 7.19 Average corn composition

Parameter	Wet basis (%)	Dry basis (%)
Fibers	13	15
Starch	61	70
Water	13	–
Protein	8	9
Lipids	3	4
Minerals	1	1
Impurities	1	1

7.3.1.3 Extended Operation with Sugarcane Bagasse and Straw for Second-Generation Ethanol Production

Ethanol production from stored sugarcane lignocellulosic material (bagasse and straw) through biochemical route is another possible alternative to extend operation of sugarcane mills in the sugarcane off-season. The second-generation process was described in previous cases and in Chap. 4.

This scenario considers same process configuration and parameters of 1G2G–Adv scenario (Case 2). Unlike previous scenarios with sweet sorghum and corn, 2G process is carried out all year-round (330 days) to allow a better use of equipment, since other process steps with considerable additional investment—e.g., pretreatment and hydrolysis—are required to produce fermentable sugars from

Table 7.20 Operational characteristics in corn processing plant

Parameter	Corn processing plant
Method of processing	Corn dry-milling process
Products	Ethanol, electricity and fractioned products
Period of operation (days)	130
Processing capacity (10^6 t corn year^{-1})	0.55
Dehydration method	Molecular sieves
Boiler pressure (bar)/efficiency (%, LHV basis)	65/87.7
Type of turbines/isentropic efficiency (%)	Extraction-condensing/85
Fermentation efficiency (%)	93
Fermentation time (h)	40
Wine ethanol content (wt%)	16
Distillation efficiency (%)	99
Steam consumption (t/t corn)	1.4
Energy demand (kWh/t corn)	106

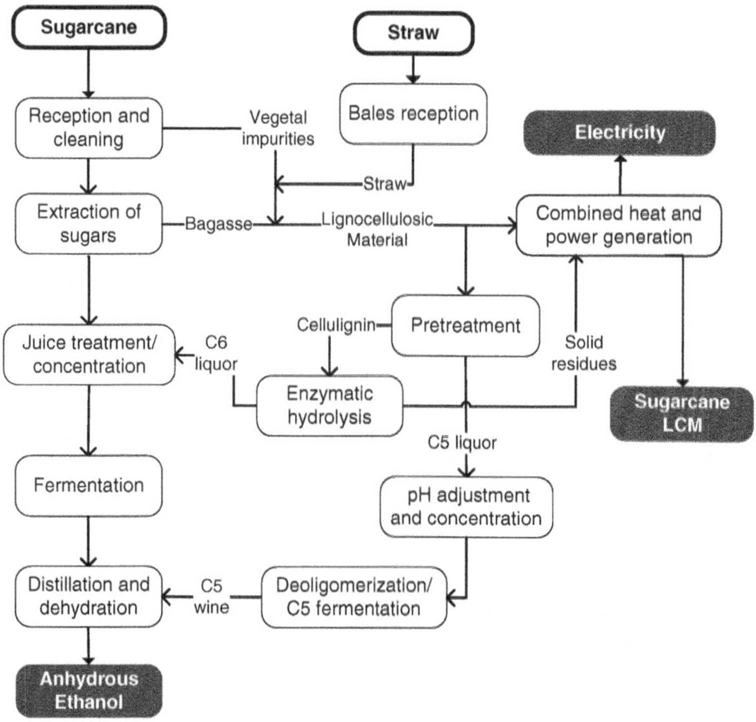

Fig. 7.15 Block flow diagram for season period: sugarcane processing and 1G2G ethanol production

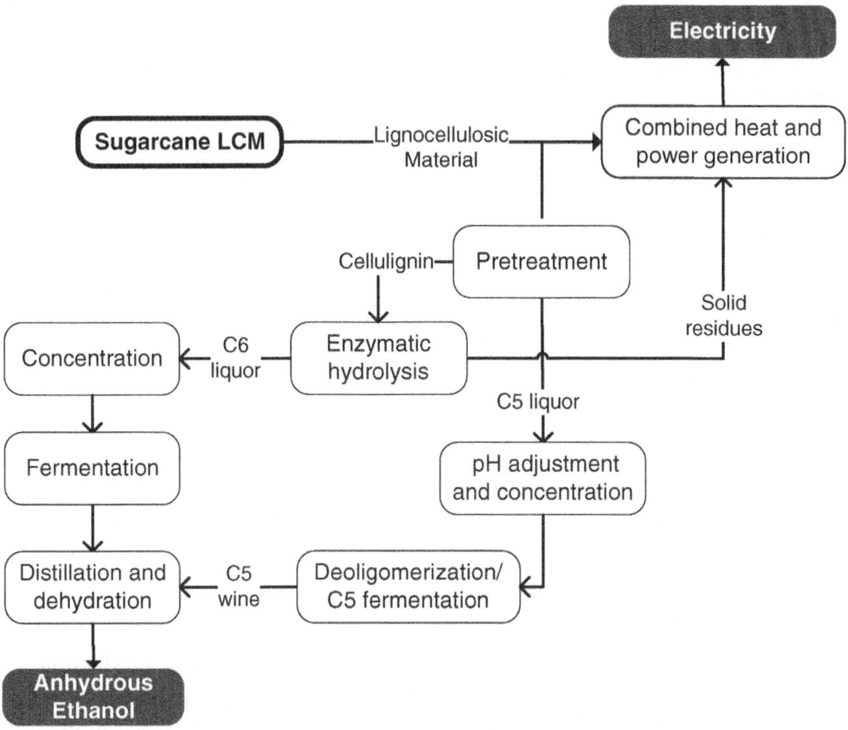

Fig. 7.16 Block flow diagram for off-season period: 2G ethanol production from sugarcane LCM

lignocellulosic material. In the season, with integration of 1G and 2G processes, C6 liquor and sugarcane juice are fermented together and the resulting wine is mixed with C5 wine for subsequent distillation. In the off-season, C6 liquor is fermented without sugar complementation. Block flow diagrams for season and off-season periods are shown in Figs. 7.15 and 7.16, respectively.

Part of the lignocellulosic material is stored for off-season operation (130 days), while the remaining is sent to 2G process in such a way that the feed in pretreatment stage is the same all year-round (58.4 t/h, dry basis). Therefore, CHP does not operate at full capacity in the off-season, since steam consumption is much lower in this period compared to sugarcane harvest season, when 1G process also takes place.

7.3.2 Techno-economic Analysis

The three alternatives proposed for extended operation (sweet sorghum, corn and second-generation ethanol) were assessed and compared to optimized 1G autonomous plant operating only in the season. Technical results for each scenario are presented in Table 7.21.

Table 7.21 Production of ethanol, electricity and fractionated products for the evaluated scenarios

Scenarios	1G–Aut–Op	1G–Sorghum	1G–Corn	1G2G–Adv
Feedstock/specific ethanol production (L t^{-1})	Sugarcane/85.4	Sweet sorghum/61.2	Corn/402.8	LCM (dry basis)/ 340.7
1G anhydrous ethanol production—season (10^6 L $year^{-1}$)	341.4	341.4	341.4	341.4
1G anhydrous ethanol production—off-season (10^6 L $year^{-1}$)	–	55.1	220.7	–
2G anhydrous ethanol production—total (10^6 L $year^{-1}$)	–	–	–	157.7
Overall anhydrous ethanol production (10^6 L $year^{-1}$)	341.4	396.5	562.2	499.1
Electricity (GWh $year^{-1}$)	743.4	814.7	596.7	246.3
Fractionated products (10^6 t $year^{-1}$)	–	–	0.14	–

Compared to the base case (1G–Aut–Op), incremental ethanol production per year was 16, 65 and 46 % for scenarios using sweet sorghum, corn and 2G process, respectively. The highest production in the 1G–Corn scenario is related to the high specific ethanol production of corn and the definition of the plant capacity (i.e., same hourly ethanol production of sugarcane processing).

In terms of surplus electricity, 1G–Sorghum presented the highest output mostly due to the additional fuel (sweet sorghum bagasse). On the other hand, surplus electricity in the 1G2G–Adv scenario was the lowest among the evaluated scenarios, since the lignocellulosic material is primarily sent for 2G processing rather than to CHP system.

The economic analysis was performed to assess off-season scenarios considering the capital expenditures summarized in Table 7.22. During the off-season period, the industrial facility dedicated to sugarcane can be easily adapted to process sweet sorghum with no additional investment. In the case of corn, however, additional investment on equipment—roughly US$150 million—is required. Additional operating expenditures—such as amylase for corn starch hydrolysis and labor costs—

Table 7.22 Summary of capital expenditures for the scenarios with sweet sorghum and corn processing in the off-season

CAPEX (US$ million)	1G–Sorghum	1G–Corn
Season investment	403.7	353.9
Off-season investment	0	151.1
Total	403.7	505.0

were considered in the cash flow analysis for both off-season scenarios. The lower investment in the sugarcane processing plant of the scenario with sweet sorghum in the off-season is due to the reduced amount of steam and electricity produced in the CHP area, since part of sugarcane LCM is stored to be burnt along with sweet sorghum bagasse.

This assessment includes other products than ethanol and electricity which are sold to the market, the coproducts from corn processing. The coproducts from corn are basically feed products, and their selling prices are presented in Table 7.23.

As discussed in Chap. 3, CanaSoft has been adapted to model sweet sorghum and corn production costs. Results obtained in the simulations of agricultural production systems were US$26.34 per tonne of sorghum and US$155.91 per tonne of corn. These figures were also used as inputs to the techno-economic analysis.

Results of cash flow analysis are compiled in Fig. 7.17, which also includes results of the 1G2G–Adv scenario (previously described in Sect. 7.2). Conceptually, this scenario was included in the assessment because surplus bagasse and straw produced over the sugarcane season represent a source of biomass, which makes the production of second-generation ethanol and electricity possible also in the off-season.

It is clear that scenarios with extended season are related to better economic results when compared to the "season-only" scenario (1G–Aut–Op). The use of sweet sorghum during the off-season (1G–Sorghum) increased both theIRR and

Table 7.23 Summary of selling prices considered in this section

Selling prices	Values
Anhydrous ethanol (US$ L^{-1})	0.58
Electricity (US$ MWh^{-1})	58
Coproducts from corn (US$ kg^{-1})	0.33

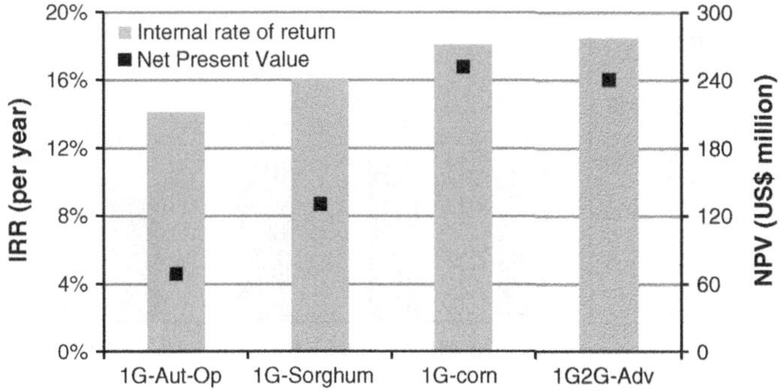

Fig. 7.17 Comparison of season (1G–Aut–Op) and off-season scenarios (1G–Sorghum, 1G–Corn and 1G2G–Adv)

NPV when compared to the 1G–Aut–Op: from 14 to 16 % per year and from US $68 to 130 million, respectively. In the case of corn (1G–Corn), IRR increased to 18.1 % per year, and the NPV to US$251 million. These positive economic impacts of operation during the off-season period were expected mainly because it solves the issue of idle equipment. In these scenarios, potential producing capacity of the plant is fully explored, and therefore, economic feasibility is enhanced due to the cash inflows associated with the extra revenues obtained in the off-season period.

Among all scenarios, the 1G2G–Adv leads to the highest internal rate of return, roughly 18.5 % and NPV around US$240 million. This scenario, 1G2G–Adv, takes advantage of all year-round operation using straw and bagasse. The use of corn during the sugarcane off-season is related to the highest NPV because of its investment and its net revenues that are slightly higher than those observed in the 1G2G–Adv scenario. In general, all off-season (1G–Sorghum, 1G–Corn and 1G2G–Adv) scenarios are economically feasible and contribute to improve the economic results of operating only during sugarcane season.

Figure 7.18 shows the impact of different off-season scenarios on the ethanol total cost. In general, costs are reduced as much as additional biomass feedstock is added in the off-season operation of the plant. This outcome was expected because additional biomass implies on additional ethanol production during the off-season which, in turn, reduces operating costs, mainly fixed costs per liter of ethanol. Besides the reduction on idle time of equipment, additional products—such as extra ethanol and coproducts from corn—reduce the ethanol cost because operating and capital costs are proportionally allocated to these additional products.

For all scenarios, biomass and capital costs represent 85 % up to 90 % of the total costs. Once again, 1G2G–Adv scenario is related to the lowest ethanol total cost when considering the average production of both first- and second-generation ethanol over the year—approximately US$0.42 per L. It is mainly because biomass cost is clearly lower when compared to the other scenarios. This effect was expected

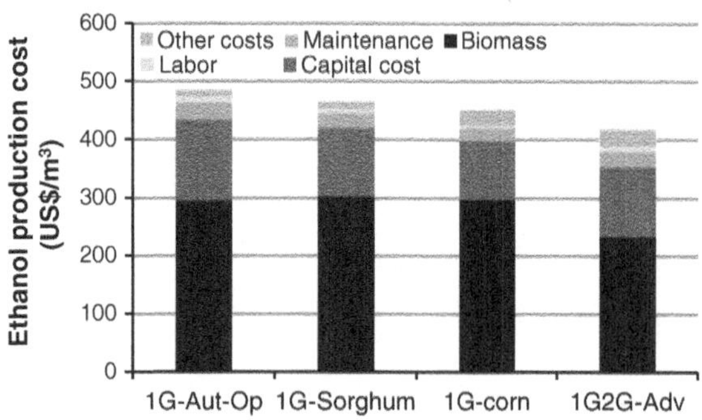

Fig. 7.18 Ethanol production costs considering different season and off-season configurations

because LCM, especially bagasse, is diverted to second-generation ethanol production in the 1G2G–Adv and does not represent an additional biomass cost during the off-season operation.

In the case of 1G–Corn and 1G–Sorghum scenarios, ethanol production cost values were slightly higher than the value observed in the 1G2G–Adv, US$0.45 and 0.46 per L, respectively. As Fig. 7.18 shows, these values were higher mainly due to the costs associated with additional corn and sorghum production cost.

7.3.3 Environmental Assessment

Figure 7.19 shows the comparative cradle to gate Life Cycle Assessment results per unit of mass of ethanol produced in the different harvest extension scenarios. The lowest impacts were obtained for 1G2G–Adv scenario, which increases 46.2 % ethanol production using the same amount of biomass from 1G–Aut–Op scenario, despite of additional industrial impacts for 2G process and reduction on surplus electricity.

Results also show that integrating other feedstocks for ethanol production during the sugarcane off-season leads to higher environmental impacts when compared to the optimized first-generation using only sugarcane. This fact occurs mainly due to lower ethanol yields per hectare obtained with sorghum and corn, associated with higher use of inputs in agricultural stages. Environmental impacts of ethanol from corn are more sensitive for impact categories specially related to fertilizer and agrochemicals used in agricultural stage (HTP, TAP and FWEP). Beyond that,

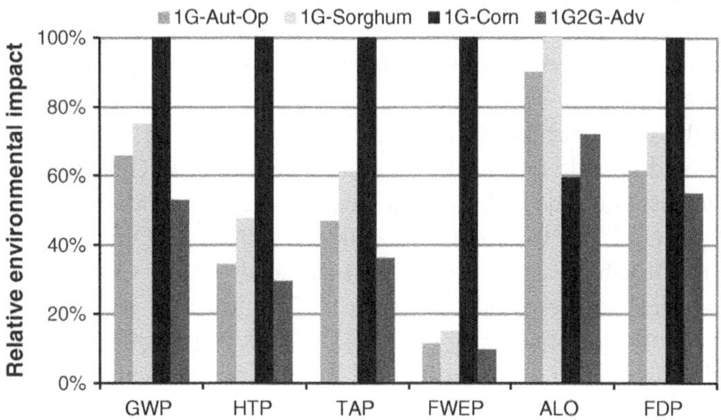

Fig. 7.19 Comparative environmental impact scores for ethanol production in harvest extension scenarios (**Note** *GWP* climate change; *HTP* human toxicity; *TAP* terrestrial acidification; *FWEP* freshwater eutrophication; *FDP* fossil depletion)

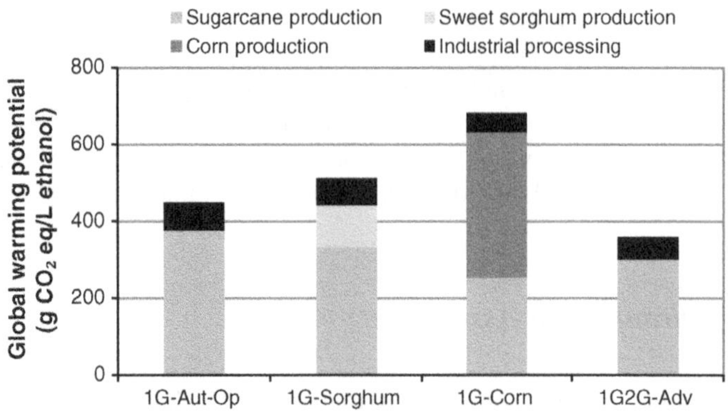

Fig. 7.20 Breakdown of global warming potential shares for ethanol production in harvest extension scenarios

while increment on ethanol production in 1G–Sorghum scenario is only 16.1 %, corn increases ethanol production to 64.7 % when compared to 1G–Aut–Op scenario, making its contribution to final scores more important.

High impact of biomass production processes in global warming potential can be observed in Fig. 7.20. The absolute environmental impacts per liter of ethanol are given in this figure, including agricultural production, transportation of biomass and the industrial conversion to ethanol. It is important to note that agricultural stages are responsible for around 85 % of total impacts in all the scenarios. Besides the fact that ethanol from corn represents only 39.4 % of total ethanol production in scenario 1G–Corn, corn production process is responsible for 55.4 % of total environmental impacts. Similar results are observed for sweet sorghum, stressing the important contribution of the (agricultural production system of) off-season processing biomass in the environmental impact of ethanol in the 1G–Sorghum and 1G–Corn scenarios.

Despite the difference in the results, ethanol obtained in all season extension scenarios would be able to promote significant reduction in GHG emissions when compared to gasoline. The highest calculated value, obtained for 1G–Corn scenario, was 682.1 g CO_2eq per L of anhydrous ethanol (30.58 g CO_2eq/MJ). Even when emissions for distribution and use are also included (32.7 g CO_2eq/MJ) (Milanez et al. 2014), final results would be at least 65 % lower in comparison with the reference value for gasoline (93 g CO_2eq/MJ) (EPA 2010), indicating that ethanol could be classified as an advanced biofuel for all assessed scenarios.

7.4 Case 4: The Biorefinery Concept—The Case of *n*-Butanol Production

Different biorefinery routes for production of biofuels and chemicals can be envisioned today. Among liquid biofuels for transportation, ethanol has been extensively studied and it is used in large scale as fuel for road transport in countries such as Brazil and USA. Other products may be used as fuels as well; *n*-butanol has received a lot of attention, since it has higher energy content than ethanol and may be used in current gasoline driven engines with practically no changes in engines and cars (Riittonen et al. 2012).

Butanol production from renewable resources usually considers the ABE (acetone–butanol–ethanol) fermentation of sugars, such as sucrose derived from sugarcane. This process has been extensively studied but faces technological challenges such as low productivity and low *n*-butanol concentration in the reactor due to product inhibition.

Another potential route for the production of *n*-butanol from sugarcane is the alcoholchemistry (or ethanolchemistry), in which ethanol is used as feedstock in catalyzed chemical reactions. Using ethanol as feedstock for *n*-butanol production would require less retrofitting than replacing ethanol by *n*-butanol fermentation (ACS 2013); thus, existing ethanol facilities could produce *n*-butanol by using catalysts to convert ethanol.

7.4.1 Scenario Description

In face of the objective of this study, two butanol-producing scenarios have been developed assuming as base cases optimized autonomous and annexed first-generation scenarios (1G–Aut–Op and 1G–Anx–Op from Sect. 7.1). For all scenarios investigated in this section, 200 days of operation and a sugarcane processing capacity of 4 Mt per year were considered.

7.4.1.1 Sugarchemical *n*-Butanol Production (1G–But–ABE/Base Case for Comparison: 1G–Anx–Op)

An optimized autonomous first-generation configuration is considered in which 50 % of the sugarcane is used for the production of *n*-butanol through ABE fermentation, as shown in Fig. 7.21.

The juice containing sugars is diluted to 50–60 g/L in order to avoid microbial butanol inhibition (Roffler et al. 1987). The diluted feed stream is then continuously sterilized (100 °C) and sent to fermentation. The fermentation process using *Clostridium* cells consists of a continuous cell production stage (seed fermentors) and a second batchwise fermentation stage where mainly acetone, *n*-butanol and

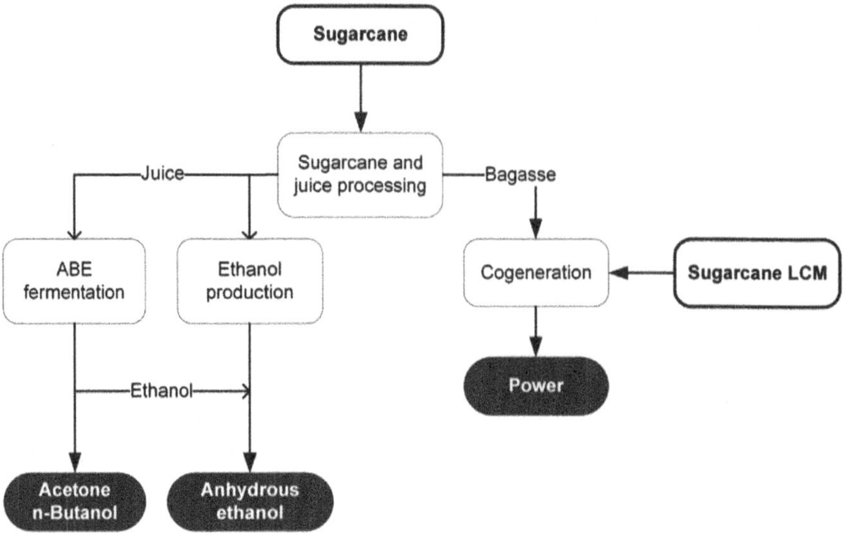

Fig. 7.21 Simplified block flow diagram of the sugarcane biorefinery producing *n*-butanol through ABE fermentation

ethanol (ABE) are produced (Afschar et al. 1990). As such, the liquor stream is split in two streams, and the smaller is fed to the seed fermentors after being further diluted to 20 g/L and sterilized at 130 °C (Mariano et al. 2013a, b).

The separation of the fermentation products (acetone, *n*-butanol and ethanol) takes place in a series of five continuous distillation columns, with the last two being responsible for the separation of butanol from water (Roffler et al. 1987). The water stream is recycled to the fermentation unit for juice dilution, and the ethanol stream is sent to the ethanol distillation unit for water removal. A detailed description of the distillation unit can be found in Mariano and Maciel Filho (2012).

Main performance parameters considered for ABE fermentation are presented in Table 7.24. These parameters were retrieved from Mariano et al. (2013a, b) and are based on experimental data obtained from studies on fermentation of hydrolysates of different biomass feedstocks to ABE (Ezeji and Blaschek 2008; Qureshi et al. 2010; Liu et al. 2010).

7.4.1.2 Alcoholchemical *n*-Butanol Production (1G–But–ALC/Base Case for Comparison: 1G–Aut–Op)

An optimized autonomous first-generation configuration is considered in which all anhydrous ethanol produced is used as feedstock for the catalysis to *n*-butanol and coproducts, as illustrated in Fig. 7.22.

A hydroxyapatite-based catalyst from Tsuchida et al. (2008) was considered for the conversion: hydroxyapatite (HAP-4 with Ca/P ratio of 1.67)—20 % ethanol

Table 7.24 Process parameters of ABE fermentation (%)

Parameter	Value
Fermentation	
Total conversion of sucrose	100.0
C6 to ethanol	2.3
C6 to n-butanol	70.2
C6 to acetone	22.0
C6 to butyric acid	0.5
C6 to cells	2.0
Recovery	
Ethanol	54.8
n-butanol	92.1
Acetone	98.8
Purity	
Ethanol[a]	77.1
n-butanol	99.0
Acetone	75.4

[a]Ethanol is mixed with the anhydrous ethanol stream

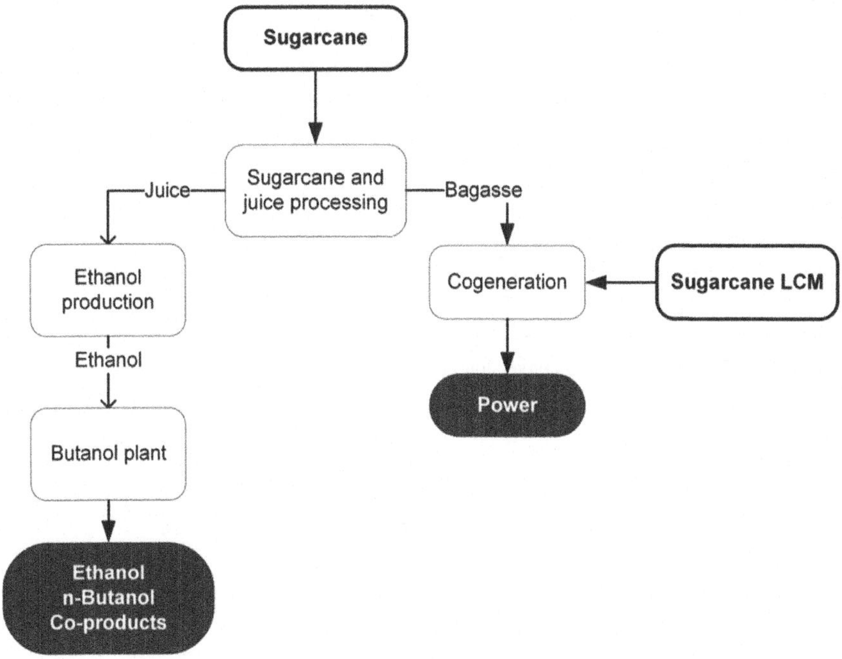

Fig. 7.22 Simplified block flow diagram of the sugarcane biorefinery producing *n*-butanol through ethanol catalysis

conversion with 69.8 % *n*-butanol selectivity at 298 °C (which requires the use of high pressure—40 bar—steam), with a reaction time of 1.78 s. 2-ethylbutanol, n-hexanol and unsaturated C4 alcohols are the most important coproducts, with selectivity of 7.0, 5.8 and 3.5 %, respectively, but 2-ethylhexanol and *n*-octanol are produced as well. Because of the low reaction time (1.78 s) with the hydroxyapatite catalyst, an arrangement of catalytic reactors in a series–parallel configuration described in a previous work by Dias et al. (2014) and Pereira et al. (2014) was used. In this configuration, anhydrous ethanol is split into six streams, and each one is fed to a different reactor. Products obtained in the first reactor are mixed with fresh ethanol feed and both are fed in the second reactor, and so on until the sixth reactor, which has therefore the largest capacity. The same selectivities and conversion provided in the literature are assumed for each reactor. Reaction products are cooled and sent to the purification system, comprised of absorption and distillation columns and adsorption for products purification. Distillation columns recover unreacted ethanol, which is dehydrated using molecular sieves, and *n*-butanol; 2-ethylbutanol is separated from n-hexanol using selective adsorption, producing streams with >98 wt% purity. Small amounts of 2-ethylhexanol and *n*-octanol are also produced and separated using selective adsorption as well, producing streams with >96 wt% purity.

Main performance parameters considered for the ethanol catalysis plant are presented in Table 7.25.

Table 7.25 Main process parameters of ethanol catalysis (%)

Parameter	Value
Reaction	
Total ethanol conversion	20.0
Ethanol–*n*-butanol conversion (2:1)	87.4
Ethanol + *n*-butanol–2-ethylbutanol conversion (1:1)	3.2
n-butanol–2-ethylhexanol conversion (2:1)	0.1
n-butanol–*n*-octanol conversion (2:1)	0.1
Ethanol + *n*-butanol–*n*-hexanol conversion (1:1)	2.5
Recovery	
Ethanol	97.1
n-butanol	95.3
n-hexanol	98.5
2-ethylbutanol	87.9
2-ethylhexanol	96.4
n-octanol	99.0

7.4.2 Techno-economic Analysis

Process simulation results for the production of anhydrous ethanol, sugar, electricity, *n*-butanol and coproducts are presented in Table 7.26.

Results show that reduced productions of anhydrous ethanol are obtained for *n*-butanol producing scenarios in comparison with the base cases due to the use of ethanol (1G–But–ALC) and sugarcane juice (1G–But–ABE) as feedstocks in the butanol plant. It is also important to highlight that higher steam demand in the butanol plants (more evident for the alcoholchemical scenario) results in less surplus electricity being produced in comparison with the base scenarios.

This assessment involves several products which are sold mainly to the chemical market, other than the usual products (ethanol, sugar and electricity). Their selling prices are presented in Table 7.27 as well as the probability distributions assigned in the subsequent risk analysis.

The results obtained in the financial analysis of butanol producing scenarios as well as of the base cases are shown in Table 7.28.

Obtained results show that although generating higher value products such as *n*-butanol, acetone and coproducts, and presenting higher capital investment (required for the butanol plants), 1G–But–ABE and 1G–But–ALC end up with lower economic return than the base scenarios (1G–Anx–Op and 1G–Aut–Op, respectively). The production of *n*-butanol by ABE fermentation using 50 % of the sugarcane juice presents similar financial results and higher revenues per amount of processed feedstock, if compared to 1G–Aut–Op, which could indicate an attractive option in case of interest in diversified and flexible production.

Table 7.26 Production of ethanol, electricity, *n*-butanol and coproducts

Scenarios	1G–Anx–Op	1G–Aut–Op	1G–But–ABE	1G–But–ALC
Anhydrous ethanol (L TC^{-1})	53.7	85.4	42.6	40.4
Sugar (kg TC^{-1})	51.4	–	–	–
Surplus electricity (kWh TC^{-1})	185.8	185.8	162.5	140.8
n-butanol (kg TC^{-1})	–	–	19.7	19.3
n-hexanol (kg TC^{-1})	–	–	–	2.2
2-ethylbutanol (kg TC^{-1})	–	–	–	2.4
2-ethylhexanol (kg TC^{-1})	–	–	–	0.7
n-octanol (kg TC^{-1})	–	–	–	0.4
Acetone (kg TC^{-1})	–	–	6.8	–

Table 7.27 Point values and probability distributions assigned to the selling prices of *n*-butanol and coproducts

Selling prices	Values	Distribution[a]
n-butanol (US$ kg^{-1})[b]	0.97	(0.74; 0.97; 1.32)
2-ethylbutanol (US$ kg^{-1})[c]	1.23	(1.15; 1.23; 1.32)
2-ethylhexanol (US$ kg^{-1})[d]	1.38	(1.33; 1.38; 1.43)
n-hexanol (US$ kg^{-1})[e]	1.38	(1.33; 1.38; 1.43)
n-octanol (US$ kg^{-1})[f]	1.75	(1.00; 1.75; 2.00)
Acetone (US$ kg^{-1})[g]	1.40	(1.00; 1.40; 1.50)

[a]Triangular distribution (minimum; most likely; maximum)
[b]Minimum price assumed to be the same as anhydrous ethanol's price per mass; most likely price assumed to be the same as anhydrous ethanol's price per energy; and maximum price considering *n*-butanol as chemical (Orbichem 2013a)
[c]Nominally assumed to be the same as *n*-butanol's price as chemical (Orbichem 2013a)
[d]Orbichem (2013b)
[e]Nominally assumed to be the same as 2-ethylhexanol's price
[f]CNCIC (2014)
[g]Orbichem (2013c)

Table 7.28 Financial analysis results

Parameter	1G–Anx–Op	1G–Aut–Op	1G–But–ABE	1G–But–ALC
CAPEX (MM US$)	424.8	417.8	453.2	494.0
IRR (% per year)	16.0	14.1	14.2	12.1
NPV (US$ mi)	134	68	77	4
Revenues (US$ per TC)	64.3	60.5	62.8	58.1

7.4.3 *Risk Analysis*

Risk analysis was performed to evaluate to what extent would the selling prices of *n*-butanol and coproducts obtained in 1G–But–ABE and 1G–But–ALC influence the economic return (IRR) of such scenarios.

Figure 7.23 shows the probability distribution for the IRR of both *n*-butanol producing scenarios.

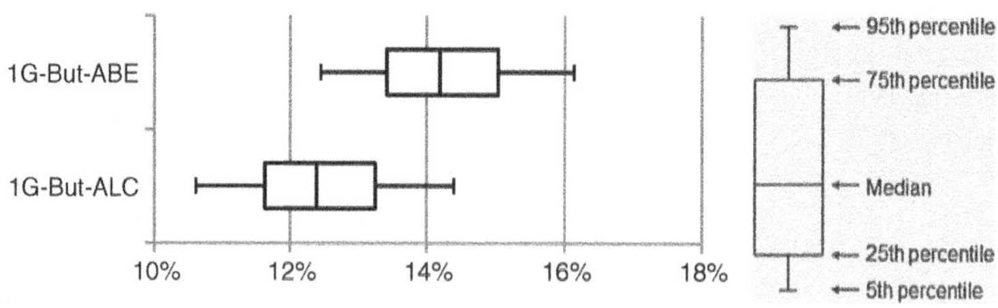

Fig. 7.23 IRR values considering *n*-butanol and coproducts selling prices uncertainties

Considering a 90 % confidence interval (i.e., 5th percentile and the 95th percentile), the probability that the IRR of 1G–But–ALC is greater than the hurdle rate (MARR) is 58.6 %, whereas for 1G–But–ABE it is 100 %.

The probabilities that the IRR of 1G–But–ALC and 1G–But–ABE is greater than their respective base scenarios 1G–Aut–Op (14.7 %) and 1G–Anx–Op (16.0 %) are 5.1 % and 0.2 %, respectively, meaning that adding a butanol plant to ethanol producing facilities is not expected to be an attractive configuration in a comparative manner, considering the parameters adopted in the analysis.

7.5 Case 5: Thermochemical 2G Ethanol Production

Ethanol production in Brazil has been historically performed via biochemical routes, namely through fermentation of sugarcane juice (1G biorefinery) and more recently through fermentation of sugars obtained from the hydrolysis of sugarcane lignocellulosic material (LCM) in 2G biochemical biorefineries. In this section, the possibility of employing a thermochemical route for the production of ethanol from sugarcane LCM in Brazil in an integrated biorefinery context is assessed taking into account the process design and parameters presented in Sect. 4.5.4. The performance of the thermochemical biorefinery is compared to that of conventional 1G and 1G2G biochemical biorefineries in terms of technical, economic and environmental impacts.

7.5.1 Scenario Description

For the simulation of the thermochemical conversion of sugarcane LCM, scenarios employing conventional sugarcane (CC) during the season (200 days) and energy cane (EC) during the off-season (130 days) were designed. Such configuration was chosen in order to increase the availability of LCM for gasification during the whole year, without recurring to storage of LCM obtained in the season for use in the off-season. Composition of ultimate and proximate analyses of LCM of both CC and EC is shown in Table 7.29, calculated based on the composition obtained in the VSB process simulations. Among biomasses suitable to be used as feedstock for the production of ethanol, such as sweet sorghum and corn, EC is the biomass considered in this assessment due to its great productivity potential (Milanez et al. 2015) and to the advantage of being processed in CC mills without major modifications in the plant. This variety of sugarcane presents lower sucrose content and higher fiber content than CC.

Three scenarios employing the same productive system—with CC during the season and EC during the off-season—were considered: 1G biorefinery (1G-Anx-Op), 1G2G biochemical biorefinery (1G2G-Bio) and 1G2G thermochemical biorefinery (1G2G-The). In all cases, the 1G plant produces both ethanol and

Table 7.29 Ultimate and proximate analyses' composition of LCM of CC and EC entering the gasifier

Composition	CC LCM	EC LCM
Ultimate Analysis (wt%, dry basis)		
C	46.69	47.81
H	5.72	5.84
O	44.03	44.85
N	0.32	0.27
S	0.05	0.04
Cl	0.04	0.02
Proximate Analysis (wt%, dry basis)		
Fixed carbon	16.79	19.77
Volatile matter	80.06	79.06
Ash	3.15	1.17
Moisture (%)	10	10

sugar from CC juice during the season and produces only ethanol from EC juice during the off-season. Each integrated facility processes a total amount of 4 million tonnes of CC and 0.34 million tonnes of straw during the season and 1.7 million tonnes of EC during the off-season. The amount of EC processed in the off-season was determined by considering the same hourly processing rate of fibers as in the season, since the same mills are used throughout the year. The 2G thermochemical unit gasifies the totality of the LCM, i.e., bagasse from sugar extraction and straw recovered from the field, obtained from both sugarcane varieties. The 2G thermochemical unit generates power and heat to the integrated 1G2G thermochemical biorefinery; thus, the CHP section present in 1G and 1G2G biochemical biorefineries is not necessary. A combined cycle gas turbine (CCGT) using unconverted syngas is employed in this scenario in order to increase the supply of electricity and steam to the integrated plant. This configuration was presented previously in Chap. 4.

Table 7.30 presents the main 2G biochemical parameters considered in the assessment, corresponding to a medium-term target (Milanez et al. 2015).

Table 7.30 Main parameters involved in the simulation of the 2G biochemical unit

Parameter	Value (%)
Pretreatment (steam explosion)	
Cellulose to glucose conversion	1
Xylan to xylose conversion	45
Enzymatic hydrolysis	
Cellulose to glucose conversion	70
Xylan to xylose conversion	70
Separate fermentations	
C6/C12 fermentation yield	90
C5 fermentation yield	80

Table 7.31 Main industrial outputs associated with the scenarios

Scenarios[a]	1G–Anx–Op	1G2G–Bio	1G2G–The
Ethanol (million L $year^{-1}$)	310	517	596
Sugar (million kg $year^{-1}$)	206	206	206
Higher alcohols (million L $year^{-1}$)	–	–	30
Electricity (GWh $year^{-1}$)[b]	1056	357	–

[a]Scenarios: *Anx* annexed plant, *Op* optimized configuration, *Bio* biochemical 2G ethanol, *The* thermochemical 2G ethanol

[b]Equivalent to an average power output to the grid of 134 MW (1G–Anx–Op) and 45 MW (1G2G–Bio)

7.5.2 Techno-economic Analysis

The techno-economic analysis was performed to assess biochemical and thermochemical routes integrated to first- and second-generation biorefineries considering the estimations on capital expenditures as summarized in Table 7.32.

7.5.2.1 Technical Results

The main industrial outputs from each scenario are presented in Table 7.31.

It is worthwhile to mention that the 1G–Anx–Op scenario presented here is described in Sect. 7.5.1, thus being different from the homonymous scenario in Sect. 7.1.2.

7.5.2.2 Economic Assessment

The capital expenditures on thermochemical second-generation equipment were based on a NREL report (Humbird et al. 2011), which depicts the investment breakdown of the thermochemical 2G plant. Values described in that report refer to 2007 and were updated to 2014. Capital expenditures for modifications required to represent operation in the Brazilian context (LCM dryer and CCGT, for example) were calculated based on the VSB database. A scaling size factor of 0.6 and an installation factor of 2.31 were considered in the calculations. The estimated CAPEX is summarized in Table 7.32.

The cash flow analysis was then performed to compare 1G–Anx–Op, 1G2G–Bio and 1G2G–The scenarios. Regarding biomass costs calculated using CanaSoft, production costs of CC, CC dry straw and EC were US\$20.73, US\$20.47 and US \$14.33 per tonne, respectively. It is important to highlight that the considered biomass production costs are significantly lower than those presented in Sects. 7.1 and 7.2, due to higher agricultural productivities for both CC and EC (Milanez et al. 2015).

Table 7.32 Summary of total capital expenditures associated with the scenarios

CAPEX (US$ million)	1G–Anx–Op	1G2G–Bio	1G2G–The
1G areas	426	464	230
2G areas	–	197	658
Total	426	661	888

Higher alcohols, produced in the 1G2G–The scenario, are considered to be sold to the fuel market. Therefore, the selling price is proportional to the lower heating value (LHV) of the product, considering hydrous ethanol as the reference price.

Results from techno-economic analysis are provided in Table 7.33. The 1G2G–The scenario presents an internal rate of return of roughly 18.6 % per year. Considering a minimum acceptable rate of return of 12 % per year, this scenario is associated with a positive NPV of approximately US$480 million. Although 2G technological improvements make higher ethanol yield possible in both 1G2G–Bio and 1G2G–The plants, the high CAPEX of 1G2G–The scenario impacts the IRR results. The 1G–Anx–Op scenario, on the other hand, is associated with a higher IRR (23.8 % per year). The combined 1G2G ethanol production costs is practically the same for 1G2G-Bio and 1G2G-The scenarios, around US$365 per cubic meter.

The second-generation ethanol production cost breakdown is presented in Fig. 7.24, which highlights the main differences among the cost components in 1G2G–Bio and 1G2G–The scenarios: Capital cost, biomass, enzyme (for biochemical route) and catalyst makeup (for thermochemical route) are the main cost components. As discussed in previous sections, these calculations are based on the allocation criteria described in Sect. 6.1. Although 2G ethanol total costs are quite similar for both 1G2G–Bio and 1G2G–The scenarios, a higher CAPEX associated with 1G2G–The implies on higher capital and maintenance costs. On the other hand, the biochemical route has a lower 2G ethanol yield; therefore, higher operating costs associated with biomass, labor and other inputs are observed when compared to the thermochemical route.

Table 7.33 Techno-economic analysis of annexed distilleries

Parameter	1G–Anx–Op	1G2G–Bio	1G2G–The
IRR (% per year)	23.8	20.4	18.6
NPV (US$ mi)	446	469	480
1G2G ethanol cost (US$/$m^3$)	346	364	365
2G ethanol cost (US$/$m^3$)	–	390	385

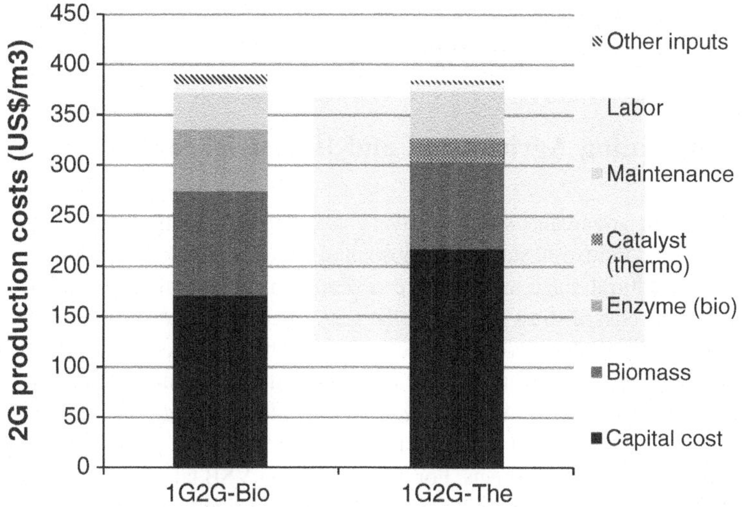

Fig. 7.24 Breakdown of second-generation ethanol production costs (US$/m^3)

7.5.3 Environmental Analysis

Figure 7.25 shows the relative environmental impact indicators from the life cycle of ethanol production in evaluated scenarios, including integrated processes for biomass production, transportation and conversion in the biorefineries. Economic allocation criterion was used to partitioning the environmental impacts between the products.

Results show that second-generation processes using bio- or thermochemical technology for 2G ethanol production present better environmental indicators than optimized first-generation ethanol for all the selected impact categories. Thermochemical route has potential to substantially decrease environmental

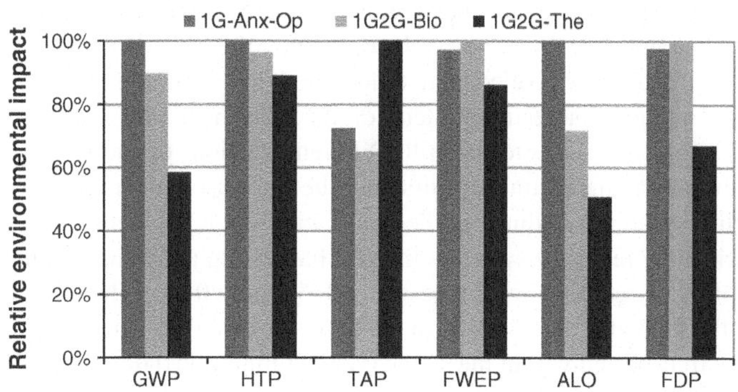

Fig. 7.25 Comparative environmental scores for ethanol production with biochemical and thermochemical 1G2G technologies (**Note** *GWP* climate change; *HTP* human toxicity; *FWEP* freshwater eutrophication; *ALO* agricultural land occupation; *FDP* fossil depletion)

impacts of ethanol, even when compared to biochemical route, mainly due to high ethanol yields per tonne of biomass.

7.6 Integrating Agricultural and Industrial Stages

Scenarios with sugarcane straw recovery may present conflicting results in the agricultural and industrial stages (Cardoso et al. 2013; Cardoso 2014) with increased costs in agricultural stage and positive revenues in industrial stage due to surplus electricity. The cost of straw recovery changes, for example, according to the amount of recovered straw, the recovery system, the machine efficiency and the transport distance. Similarly, the effects of straw in the mill must be evaluated, considering the quantity and quality of straw processed, such as moisture and mineral impurities.

Therefore, it is important to assess the effects of straw recovery systems on both agricultural and industrial aspects. For this reason, the VSB takes into account vertical integration of sugarcane biorefineries (including both agricultural and industrial stages) which enables analyzing the results in an aggregate form. In order to perform this study, the agricultural stage scenarios were modeled using CanaSoft, described in Chap. 3, while the industrial stage was simulated using Aspen Plus, described in Chap. 4.

7.6.1 Straw Recovery

7.6.1.1 Scenario Description

Agricultural Stage

The agricultural stage includes all the process steps, from sugarcane production to its transportation (in the form of stalks and straw) to the industry. According to the method described in Chap. 3, in the baling system, straw is left on the field for about 15 days after sugarcane harvesting to decrease its water content through natural drying, and after that straw is windrowed, collected and compacted in bales, which are subsequently loaded and transported to the mill separately from the stalks. In the integral harvesting system, on the other hand, the straw is transported along with sugarcane stalks to the mill. Different fractions of straw recovery can be obtained in integral harvesting by adjusting the blower of sugarcane harvester and, in the baling system, by adjusting the straw windrow equipment.

For each straw recovery system—integral harvesting and baling—three different fractions of straw recovery were assessed: 30, 50 and 70 % of the straw that would remain in the field after the harvest. In the base scenario, it is considered that part of the straw is transported within the stalks (vegetal impurities) due to inefficiencies in the mechanical harvesting process. Six scenarios with straw recovery were evaluated and compared to a scenario without straw recovery (base scenario). Therefore, seven scenarios were considered in this study (Table 7.34).

Table 7.34 Description of evaluated scenarios

Scenario	Description	
Base	Without straw recovery	
I30	Integral harvesting	30 % straw recovery
I50		50 % straw recovery
I70		70 % straw recovery
B30	Baling system	30 % straw recovery
B50		50 % straw recovery
B70		70 % straw recovery

The basic agricultural parameters considered for all scenarios in this study were as follows: 200 days season; production of 4 million tonnes of sugarcane stalks; mechanized planting using 20 tonnes of cane seedlings per hectare; mechanized sugarcane harvesting without previous burning (green harvesting); sugarcane stalks yield of 80 tonnes per hectare per year; production of 140 kg of straw (dry mass) per tonne of stalks (wet basis); transportation capacity of 184 m^3 per truck with variable transport density in integral harvesting system according to the straw recovery fractions; and average transport distance of 35 km.

In the integral harvesting, the absence of straw separation in the field reduces stalk losses (Hassuani et al. 2005) which, in turn, increases sugarcane yield per hectare and, consequently, decreases sugarcane cultivation area and transport distance, as shown in Table 7.35.

Industrial Stage

An optimized autonomous distillery producing first-generation ethanol and surplus electricity was considered for the industrial stage, as described in Sect. 7.1.1.

In the integral harvesting scenarios, straw is separated by an air separation system (dry-cleaning station). Additional straw passing through the mill along with

Table 7.35 Agricultural parameters

	Harvester losses (%)	Area of sugarcane cultivation (ha)	Transport distance (km)
Base	10.0	52,625	35.0
I30	7.3	51,011	34.5
I50	5.9	50,240	34.2
I70	5.0	49,690	34.0
B30	10.0	52,625	35.0
B50	10.0	52,625	35.0
B70	10.0	52,625	35.0

sugarcane stalks promotes reduction in the sugar separation efficiency and crushing capacity. For the straw recovery by baling system, a higher incorporation of mineral impurities was assumed as the fraction of straw recovery increases, mainly because more soil is collected along with the straw which is in direct contact with it. The main differences related to the industrial parameters considered for different scenarios are shown in Table 7.36.

Results for ethanol production and electricity surplus are shown in Table 7.37. As Table 7.37 shows, larger straw recovery fractions promote higher electricity surplus. The integral harvesting system resulted in lower ethanol production than the baling system due to higher sugar losses during extraction in the industrial plant. Moreover, the larger the amount of straw processed through the mill, the higher its energy consumption, due to the increased amount of material to be crushed. Hence, lower electricity surplus occurs in the integral harvesting system scenarios compared to baling system scenarios.

Table 7.36 Industrial parameters

Scenarios		Base	I30	I50	I70	B30	B50	B70
Sugarcane quality								
Vegetal impurities	$kg_{(dry\ mass)}\ TC^{-1}$	24.2	24.2	24.2	24.2	24.2	24.2	24.2
Vegetal impurities moisture	%	52.1	52.1	52.1	52.1	52.1	52.1	52.1
Mineral impurities	$kg\ TC^{-1}$	12.0	12.0	12.0	12.0	12.0	12.0	12.0
Integral straw								
Straw recovery	$kg_{(dry\ mass)}\ TC^{-1}$	–	40.0	65.5	90.5	–	–	–
Straw moisture	%	–	39.6	33.6	29.6	–	–	–
Additional mineral impurities	$kg\ TC^{-1}$	–	1.87	3.05	4.37	–	–	–
Baled straw								
Straw recovery	$kg_{(dry\ mass)}\ TC^{-1}$	–	–	–	–	41.5	69.2	96.8
Straw moisture	%	–	–	–	–	13	13	13
Mineral impurities	%	–	–	–	–	2.0	4.0	8.0
Dry-cleaning station efficiencies								
Vegetal impurities separation	%	–	55.4	56.3	57.6	–	–	–
Mineral impurities separation	%	–	65.0	65.0	65.0	–	–	–
Stalk losses	%	–	0.8	1.0	1.2	–	–	–
Extraction								
Sugar extraction in mills	%	96.0	95.6	95.3	95.1	96.0	96.0	96.0

Table 7.37 Ethanol and surplus electricity production

Scenarios		Base	I30	I50	I70	B30	B50	B70
Anhydrous ethanol production	L TC^{-1}	85.36	85.24	85.14	85.06	85.36	85.36	85.36
Surplus electricity	kWh TC^{-1}	118.9	169.5	202.1	234.5	179.7	220.4	260.9

7.6.1.2 Techno-economic Analysis

In order to assess the impacts of straw recovery on the economic viability of the integrated sugarcane biorefinery (both agricultural and industrial phases), the biomass costs that were calculated using CanaSoft were then considered as operating costs in the industrial stage. Therefore, the overall effects of straw recovery systems can be fully comprehended considering a broader system boundary.

Agricultural Stage

The production costs for the agricultural scenarios are presented in Table 7.38. Straw recovery costs are calculated as the additional costs per hectare of each scenario with straw recovery in comparison with the scenario without straw recovery. Reduction of stalk losses in the harvest and the criteria of division of additional costs between straw and extra stalks, according to their mass, contribute to decreased costs in the integral harvest. These factors associated with lower use of machines in integral harvest make stalks cost even lower.

In the integral harvesting system, a lower use of machinery associated with the reduction of losses in the harvest leads to lower sugarcane stalks costs. In the baling system, as much as the straw quantity collected increases, the cost per tonne of straw (dry basis) with bale decreases due to a more efficient use of agricultural machinery. For the integral harvesting, costs are practically unchanged, since the effect of load density reduction is balanced by decreased stalk losses.

In regard to transport distance, both systems show higher costs for longer distances. However, integral harvesting shows higher variations than those observed in the baling system, as presented in Fig. 7.26.

Table 7.38 Production costs of sugarcane and straw recovery

	Base	B30	B50	B70	I30	I50	I70
Total cost (US$ ha^{-1})	2,148	2,270	2,306	2,351	2,255	2,333	2,388
Stalks (US$ t^{-1})	28.26	28.26	28.26	28.26	27.83	27.71	27.59
Straw (db) (US$ t^{-1})	–	38.52	29.99	27.53	23.38	24.33	22.96

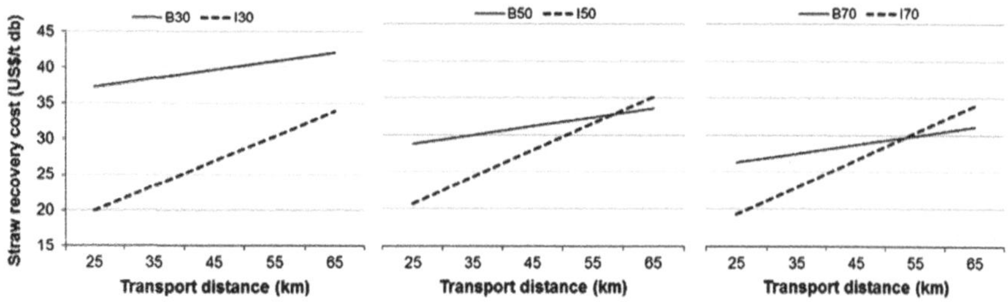

Fig. 7.26 Trends in the straw recovery costs for different fractions of straw recovery and transport distances

Vertically Integrated Scenarios

Techno-economic analysis was then performed to assess the impact of different biomass production costs on the economics of the industrial stage. As previously described, the base scenario is represented by an autonomous distillery with optimized technology processing 4 million tonnes of sugarcane stalks (similar to the 1G–Aut–Op scenario described in Sect. 7.1, except for the straw recovery fraction which is zero in the base case). Some industrial equipment, however, will be added to the base scenario in order to deal with the extra straw from the different recovery scenarios—integral harvest, for instance, requires a dry-cleaning station to separate straw from sugarcane stalks in the industry and baling system requires a bale shredder to make straw available for industrial processing, for example. The investment on the base scenario is roughly US$377.50 million, and the supplementary investment for each scenario is presented in Table 7.39. Moreover, increase in the processed material increases the electricity production capacity which, in turn, implies on higher capital expenditures on the cogeneration area (CHP). The CHP area, in turn, contributes to increased insurance, engineering and contingency costs. As expected, the supplementary investment increases as much as the straw recovery fraction gets higher. As Table 7.39 highlights, the investment associated with the integral recovery systems is slightly higher when compared to baling scenarios. Although these differences may not seem so significant, they are very important in techno-economic analyses.

Table 7.39 Summary of capital expenditures associated with the industrial scenarios

CAPEX (US$ million)	Base	I30	I50	I70	B30	B50	B70
Extra investment[a]	–	24.00	33.18	41.50	17.49	27.55	35.13
Total	377.50	401.50	410.69	419.00	395.00	405.06	412.64

[a]Supplementary investment associated with industrial equipment for straw processing and extra capacity in the cogeneration equipment

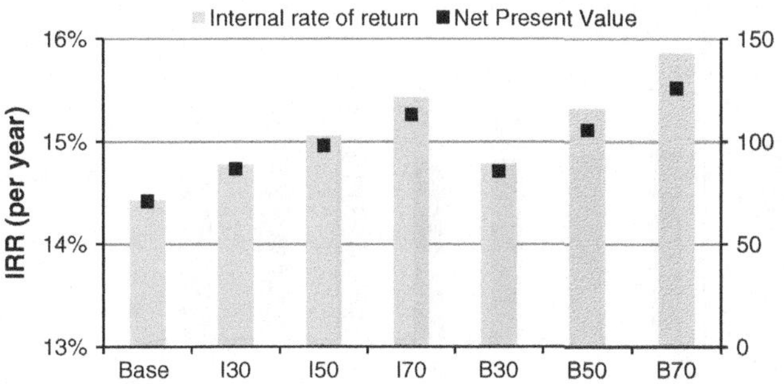

Fig. 7.27 Economic results of the vertically integrated scenarios considering different straw recovery systems

The cash flow analysis was then performed to compare the base case with the six straw recovery scenarios. The biomass production costs (stalks and straw) presented in the previous section were calculated according to the CanaSoft model, and their results were used as inputs to estimate biomass operating costs for the industrial stage.

Results from the economic analysis are described in Fig. 7.27. Both baling and integral straw recovery systems improved the economic results compared to the base case. All scenarios presented results that highlight their economic viability due to the positive NPVs and internal rates of return above the minimum acceptable rate of return (12 % per year). Compared to the base case—whose IRR and NPV were of 14.4 % per year and US$71 million, respectively—the economic results of straw recovery systems were better: IRR varied from 14.8 to 15.9 % per year and NPV from US$85 to 126 million.

In regard to the straw recovery technology, the integral recovery system presented slightly better results when considering a fraction of 30 % recovery, because the I30 scenario leads to higher NPV when compared to B30. For other straw recovery fractions such as 50 and 70 %, the B50 and B70 scenarios presented slightly higher internal rates of return and NPVs if compared with I50 and I70 scenarios. Considering that the higher the amount of straw transported from the field to the plant, the higher the transportation distance, it is clear that integral system is more advantageous for shorter distances, whereas baling is more economically feasible when considering longer transportation distances.

It is important to highlight that the economic results presented in this section can be very sensitive to the investment on cogeneration and electricity prices, since sugarcane straw recovered from the field is burnt mainly to produce bioelectricity. For this reason, the main impact of the straw recovery fraction on the industrial plant is the extra investment required by the expansion of the CHP area. Considering the relatively high industrial investment (around US$450 million), relatively small uncertainties on investment may significantly affect the economic results for the integrated straw recovery scenarios. Moreover, electricity market prices may also affect the economic viability of straw recovery in these scenarios

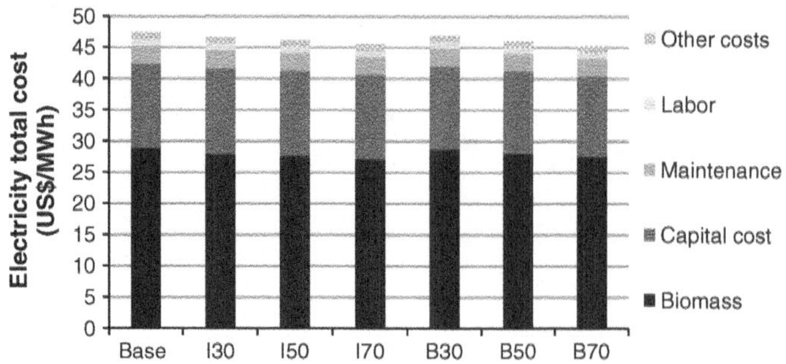

Fig. 7.28 Breakdown of electricity production costs (considering both operating and capital costs)

because electricity is the only product related to fluctuating output (ethanol production is only slightly different).

The electricity production cost breakdown is presented in Fig. 7.28, which highlights the main differences between the cost components for the industrial phase considering different straw recovery scenarios. It is important to highlight that electricity production leads to very similar costs and ranged from US$45 to 47 per MWh for all scenarios. The lowest cost is associated with the B70 scenario—US$45.24 per MWh; the highest cost is related to the base case—US $47.40 per MWh. These results are in accordance with the previous results that favored the highest straw recovery fraction using baling system technology. The electricity cost associated with I70 (highest integral recovery), however, presented a quite similar result: US$45.54 per MWh. Biomass was the main cost component in all scenarios and represented about 60 % of the total electricity costs. Capital costs were the second most important component and remained almost unchanged over the scenarios—approximately 30 % of the total cost.

With respect to the total ethanol cost (Fig. 7.29), it is possible to observe that results also favored B70 and I70 scenarios whose production costs were the lowest—roughly US$0.456 and 0.459 per L, respectively. This result was expected because of the higher cost allocation to electricity. As a result of increased electricity production, ethanol costs become slightly lower. Similar to electricity, biomass and capital were the main cost components and represented about 90 % of the ethanol total cost. The highest ethanol total costs were associated with base, B30 and I30 scenarios—US $0.478, 0.473 and 0.470 per L, respectively—mainly because of higher operating costs related to biomass.

7.6.2 Optimization of the Vertically Integrated Scenario

Through the methodology described in Sect. 6.6, reduced-order models (surrogate models) for optimizing the IRR (calculated with VSB) were developed as a function

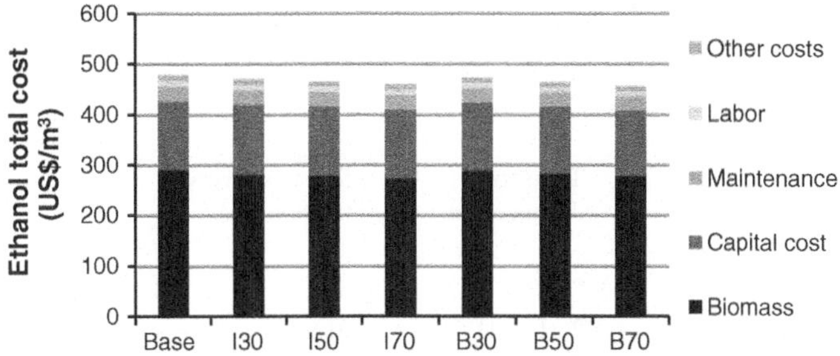

Fig. 7.29 Breakdown of the ethanol total cost (considering both operating and capital costs)

of sugarcane productivity (P), transport distances (D) and straw recovery fractions (R) considering baling system (BS) and integral harvesting (IS).

For the vertically integrated scenario, the IRR was calculated by VSB according to a central composite design simulations (Table 7.40). Then, from this design, a reduced-order model was obtained for each response: BS_{IRR} (IRR for the baling system) and IS_{IRR} (IRR for the integral harvesting system).

Equations 7.1 and 7.2 showed the codified model adjusted for BS_{IRR} and IS_{IRR} as function of X_1, X_2 and X_3.

$$\begin{aligned} BS_{IRR} = {} & 0.147803 + 0.013314X_1 - 0.000528X_1^2 - 0.004341X_2 \\ & + 0.001483X_2^2 + 0.005590X_3 + 0.000534X_3^2 \\ & + 0.000224X_1X_2 + 0.000791X_1X_3 \\ & + 0.001631X_2X_3 \end{aligned} \tag{7.1}$$

$$\begin{aligned} IS_{IRR} = {} & 0.1499 + 0.012195X_1 - 0.002337X_1^2 - 0.005690X_2 \\ & - 0.000979X_2^2 + 0.001625X_3 + 0.000843X_3^2 \\ & + 0.000202X_1X_2 - 0.000332X_1X_3 \\ & - 0.000259X_2X_3 \end{aligned} \tag{7.2}$$

The relation of physical factors (P, D and R) with X_1, X_2 and X_3 is as follows:

$$X_1 = (P - 80.0)/(8.9191) \tag{7.3}$$

$$X_2 = (D - 35.0)/(8.9191) \tag{7.4}$$

$$X_3 = (R - 50.0)/(11.8921) \tag{7.5}$$

For screening purposes, the responses obtained from statistical analysis were evaluated by using the Pareto chart (Fig. 7.30).

From the Pareto chart, the linear effects of $P_{(L)}$, $R_{(L)}$ and $D_{(L)}$ are statistically significant for BS_{IRR}, considering the vertical integration of sugarcane biorefineries. In the integral harvesting system, the linear effects of stalk productivity ($P_{(L)}$) and

Table 7.40 Central composite design with the real and coded values (between parentheses) for the responses BS_{IRR} and IS_{IRR}

Scenario	Independent variables			Response variables	
	Sugarcane productivity (t ha^{-1})	Transport distance (km)	Straw recovery fraction (%)	BS_{IRR} (%)	IS_{IRR} (%)
S1	71.1 (−1)	26.1 (−1)	38.1 (−1)	13.79	13.82
S2	71.1 (−1)	26.1 (−1)	61.9 (1)	14.50	14.26
S3	71.1 (−1)	43.9 (1)	38.1 (−1)	12.39	12.94
S4	71.1 (−1)	43.9 (1)	61.9 (1)	13.77	13.28
S5	88.9 (1)	26.1 (−1)	38.1 (−1)	16.08	16.43
S6	88.9 (1)	26.1 (−1)	61.9 (1)	17.13	16.74
S7	88.9 (1)	43.9 (1)	38.1 (−1)	14.79	15.64
S8	88.9 (1)	43.9 (1)	61.9 (1)	16.47	15.84
S9	65.0 (−1.68)	35.0 (0)	50.0 (0)	12.29	12.27
S10	95.0 (1.68)	35.0 (0)	50.0 (0)	17.15	16.02
S11	80.0 (0)	20.0 (−1.68)	50.0 (0)	15.84	15.79
S12	80.0 (0)	50.0 (1.68)	50.0 (0)	14.74	13.28
S13	80.0 (0)	35.0 (0)	30.0 (−1.68)	14.19	14.78
S14	80.0 (0)	35.0 (0)	70.0 (1.68)	15.86	15.32
S15	80.0 (0)	35.0 (0)	50.0 (0)	14.75	15.05

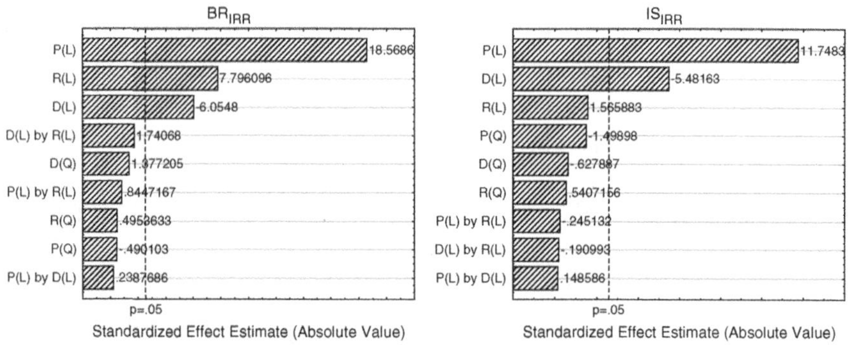

Fig. 7.30 Pareto chart of standardized effects. *P*(*L*) linear effect of stalk productivity; *D*(*L*) linear effect of transport distance; *R*(*L*) linear effect of straw recovery fraction; *P*(*Q*) quadratic effect of stalk productivity; *D*(*Q*) quadratic effect of transport distance; *R*(*Q*) quadratic effect of straw recovery rate

transport distance ($D_{(L)}$) showed statistically significant influence for internal rate of return (IS_{IRR}).

The productivity effects are important because of the high influence of this parameter on sugarcane stalks costs. As previously observed, higher transport distances increase the stalk and straw costs. Higher straw recovery fraction implies on lower agricultural costs for baling system. Due to large surplus electricity and no reduction on sugar extraction with baling system, IRR is favorably impacted.

A new statistical analysis was made, neglecting the non-significant effects according to Pareto charts, and a simplified model was generated for each recovery system. The models with only significant effects are presented in Eqs. 7.6 and 7.7 for BS_{IRR} and IS_{IRR}.

$$BS_{IRR} = 0.149158 + 0.013314X_1 - 0.004341X_2 + 0.005590X_3 \tag{7.6}$$

$$IS_{IRR} = 0.147649 + 0.012195X_1 - 0.005690X_2 \tag{7.7}$$

The analysis of variance (ANOVA) was performed to ensure that the generated simplified models had statistical significance. As shown in Table 7.41, the models accounted for a high percentage of the explained variance: R^2 = 97.0 % for BS_{IRR} and R^2 = 92.6 % for IS_{IRR}. The calculated regression F values (regression/residual) for IRR response for baling and integral systems are 33 and 19 times larger than the critical F value at 95 % confidence, respectively. ANOVA results indicate that the regressions are statistically significant for both responses.

Equations 7.6 and 7.7 were used to predict IRR for each system. Relative errors compared to VSB deterministic results (Table 7.40) are presented in Table 7.42.

Table 7.41 ANOVA for the simplified model

	Source of variation	Sum of squares	Degrees of freedom	Mean square	F-ratio
BS_{IRR}	Regression	0.003105	3	0.001035	118.5974
	Residual	0.000096	11	0.000008727	
	Total	0.003201	14		
	F-listed value (95 % of confidence)				$F_{3,11(95\%)} = 3.5874$
	Percentage of explained variance (R^2) = 97.0				
IS_{IRRs}	Source of variation	Sum of squares	Degrees of freedom	Mean square	F-ratio
	Regression	0.002474	2	0.001237	74.9697
	Residual	0.000198	12	0.0000165	
	Total	0.002672	14		
	F-listed value (95 % of confidence)				$F_{2,12(95\%)} = 3.8852$
	Percentage of explained variance (R^2) = 92.6				

Table 7.42 Internal rate of return predicted by CCD-simplified models and comparison with VSB results

Scenario	IRR (%)		Error (%)	
	BS	IS	BS	IS
S1	13.46	14.11	2.41	2.14
S2	14.58	14.11	0.53	1.06
S3	12.59	12.98	1.63	0.26
S4	13.71	12.98	0.48	2.31
S5	16.12	16.55	0.26	0.73
S6	17.24	16.55	0.64	1.14
S7	15.25	15.42	3.13	1.41
S8	16.37	15.42	0.59	2.71
S9	12.68	12.71	3.13	3.59
S10	17.16	16.82	0.04	4.95
S11	15.65	15.72	1.22	0.40
S12	14.19	13.81	3.75	3.98
S13	13.98	14.76	1.48	0.08
S14	15.86	14.76	0.00	3.62
S15	14.92	14.76	1.13	1.91

Simplified models showed good agreement as shown by the low relative error values.

The simplified models described in Eqs. 7.6 and 7.7 can assist the analysis of complex systems such as the integration of agricultural and industrial phases simulated on VSB in a simple way, allowing to proceed the optimization of the whole system. It highlights that optimization must be performed within the ranges of the variables that were used for the model fit.

Through the formulation of an optimization problem, the models in Eqs. 7.6 and 7.7 were used to determine the optimum conditions for IRR. An optimized IRR (18.8 %) was obtained for baling system, with low transport distance (20 km), maximum sugarcane stalk productivity (95 t ha^{-1}) and maximum recovery fraction (70 %). The same conditions lead to highest IRR (17.7 %) for integral system.

IRR lines obtained with CCD-simplified models for both recovery systems are shown in Fig. 7.31, for different recovery fractions and transport distances, and stalk productivity of 80 t ha^{-1}. Results followed the same tendency discussed in the integrated analysis performed in Sect. 7.6.1.

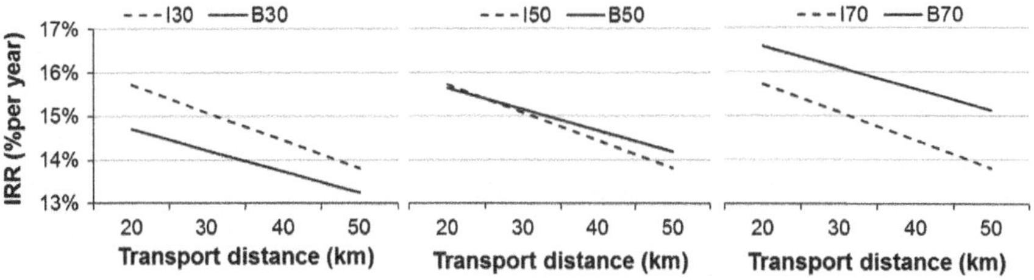

Fig. 7.31 Trends in IRR for vertical integration of sugarcane biorefineries considering different transport distances

7.7 Input–Output Analysis: Case Study of Ethanol Production Technologies

In this section, the integration of Input–Output Analysis and Life Cycle Assessment is explored to simulate the impacts of sugarcane and ethanol production on the economic activity of Brazilian sectors. This assessment captures direct and indirect effects of ethanol production and, consequently, covers all the sectors involved in the sugarcane ethanol supply chain.

7.7.1 Scenario Description

In this study, three ethanol production scenarios were created to demonstrate the effects of improvements on ethanol production technologies: 1G-basic, 1G-optimized and 1G2G. These scenarios correspond to vertically integrated models of sugarcane and ethanol production representing current (1G-basic and 1G-optimized) and future autonomous distilleries (1G2G) using biomass cultivated under different agricultural management practices. The 1G-basic represents a scenario producing ethanol under relatively outdated technologies both in industrial and agricultural stages. For instance, preharvest straw burning is practiced in approximately 30 % of the total harvested area. In the industrial phase, a plant processing 2 million tonnes of cane per year uses mechanical drives in the mill engines and there is no energetic integration. The 1G-optimized represents a scenario by which the best current agricultural technologies are observed, such as mechanical harvesting and straw recovery to produce bioelectricity. Also, improvements in the first-generation plant are observed, for instance, electric drives in mill engines, steam turbines and energetic integration that allow the distillery to sell surplus electricity to the grid. In the 1G2G scenario, a future biorefinery integrating first- and second-generation ethanol production technologies is represented. It considers future technologies of second-generation ethanol production such as pentose (C5) and hexose (C6) fermentation to ethanol and better agricultural technologies that increase sugarcane productivity to 100 tonnes of cane per

Table 7.43 Scenario description of the autonomous distilleries assessed in this study

Parameters	1G-basic	1G-optimized	1G2G
Agricultural			
Sugarcane yield (TC ha^{-1} $year^{-1}$)	80	80	100
Straw recovery fraction (% of total mass on the field)	–	50	50
Straw recovery system	–	Baling	Baling
Mechanical planting (% of total cultivated area)	60	100	100
Harvest mechanization (% of total harvested area)	70	100	100
Preharvest burning (% of total harvested area)	30	–	–
Biomass cost (US$ TC^{-1})	22.90	24.39	20.12
Industrial			
Processing capacity (10^6 TC $year^{-1}$)	2	4	4
1G ethanol yield (L TC^{-1})	85.03	85.36	85.36
2G ethanol yield (L TC^{-1})	–	–	39.41
Electricity surplus (kWh TC^{-1})	–	185.84	61.58

hectare per year. Details on agricultural and industrial parameters are described in Table 7.43.

The agricultural and industrial parameters modeled in the VSB were then used to compile the life cycle inventories related to the sugarcane ethanol production. Data from the life cycle inventories were then translated into a list of purchased inputs that gave rise to the technical coefficients of both agricultural and industrial scenarios considered in this study. These scenarios and their technical coefficients were inserted into the input–output table of direct requirements in order to simulate the impacts of different scenarios on the Brazilian economic sectors. The procedure of integrating the life cycle inventory and the input–output table characterizes a hybrid model whose details are described in a previous study (Watanabe et al. 2015).

7.7.2 *Impact of Ethanol Production on the Economic Activity of Sectors*

Three changes in ethanol final demand were simulated separately to calculate the economic effects of a US$1 billion increase in the production considering the different technological scenarios. It is important to highlight that a US$1 billion increase in 1G-basic ethanol demand results in a US$1 billion of output from 1G-base sector because this biorefinery only produces ethanol. The same shock to the 1G-optimized and 1G2G biorefineries, however, calls for the same ethanol production and electricity generation as a coproduct in both scenarios. In order to produce US$1 billion worth of 1G-optimized ethanol, there is an actual US$1.22 billion shock due to an additional US$0.22 billion worth of electricity output. In addition, in order to produce 1G2G ethanol, the actual shock in final demand would

Table 7.44 Direct and indirect economic effects for a $1 billion increase in ethanol demand on Brazilian commodity outputs

Commodity outputs, US$ billion	1G-basic			1G-optmized			1G2G		
	Total	Direct effect[a]	Indirect effect[b]	Total	Direct effect[a]	Indirect effect[b]	Total	Direct effect[a]	Indirect effect[b]
Ethanol	1.000	1.000	–	1.000	1.000	–	1.000	1.000	–
Sugarcane	0.468	0.447	0.021	0.534	0.416	0.118	0.299	0.282	0.017
Bioelectricity	n.a.	n.a.	n.a.	0.215	0.215	–	0.049	0.049	–
Inorganic chemicals	0.056	0.002	0.054	0.068	0.003	0.065	0.031	0.003	0.028
Diesel	0.045	–	0.045	0.057	0.057	–	0.035	–	0.035
Trade	0.038	0.007	0.031	0.045	0.007	0.038	0.033	0.009	0.024
Repair services	0.029	–	0.029	0.042	0.042	0.042	0.024	–	0.024
Oil and gas	0.025	–	0.025	0.030	–	0.030	0.019	–	0.019
Transportation	0.024	0.007	0.017	0.029	0.023	0.006	0.019	0.005	0.014
Others (111)	0.209	0.052	0.157	0.028	0.040	0.179	0.213	0.078	0.135
Total	1.895	1.515	0.380	2.267	1.700	0.567	1.722	1.426	0.296

[a]Economic activity due to the production of commodities directly purchased by biorefineries

[b]Economic activity due to the production of commodities required to produce commodities directly purchased by biorefineries

n.a. not applicable

be of US$1.05 billion due to the additional US$0.05 billion worth of electricity output.

Table 7.44 shows the impact of changes in ethanol demand on the Brazilian commodity outputs considering the three different scenarios. The top 10 commodity outputs are related to ethanol, sugarcane, bioelectricity, inorganic chemicals, trade, diesel, repair and maintenance services, oil and gas extraction and transportation. Sugarcane is the main feedstock for bioethanol production, and then, it holds the first position of commodity outputs. The inorganic chemicals are also related to a high economic activity because of its use as chemicals to the industrial process and also as fertilizer in the agricultural stages. Other 111 commodities—such as non-metallic mineral products, pesticides, services to companies, gasoline, information services, real estate services and metal products—also contributed to the ethanol production sector; however, lower economic effects were observed.

As Table 7.44 shows, the 1G-optimized scenario is related to the largest economic effect when the direct and indirect effects over commodity outputs are accounted for US$2.27 billion. The 1G-basic scenario holds the second position with an overall economic activity of US$1.90 billion. Although this scenario is related to a high sugarcane economic activity, there is no electricity surplus associated with ethanol production, decreasing the total economic activity of the scenario. The 1G2G scenario is related to the lowest overall impact on the economy (roughly US$1.72 billion), highlighting that a more efficient technology requires fewer resources from the economy to produce the same amount of ethanol compared with the other scenarios. In general, the contribution from indirect commodities—i.e., those required to produce commodities directly purchased by biorefineries—varies from 17 to 25 % depending on the scenario.

7.8 Social Assessment

The social assessment presented in this section is based on the previous economic results from the Input–Output Analysis using a hybrid model (Sect. 7.7). Consequently, the integrated sugarcane and ethanol production models described in Sect. 7.7 were assessed. In order to perform the social assessment—as explained in Chap. 6—data related to job creation per dollar output, for instance, available in previous studies (Guilhoto and Sesso 2005, 2010) based on the Brazilian System of National Accounts (IBGE 2009) were used. Education profile was determined by considering data on workers' degree profiles for each Brazilian economic sector according to the Ministry of Labor online database (MTE 2015). The number of occupational accidents of sectors was estimated based on data of accidents per thousand workers according to the Ministry of Social Security (MPS 2015).

Results in Fig. 7.32 show the number of direct and indirect jobs considering a US$1 billion change in ethanol final demand. The 1G-basic scenario is related to higher job creation levels mainly because of two factors. First, relatively more sugarcane is demanded for the same ethanol output. Second, the agricultural

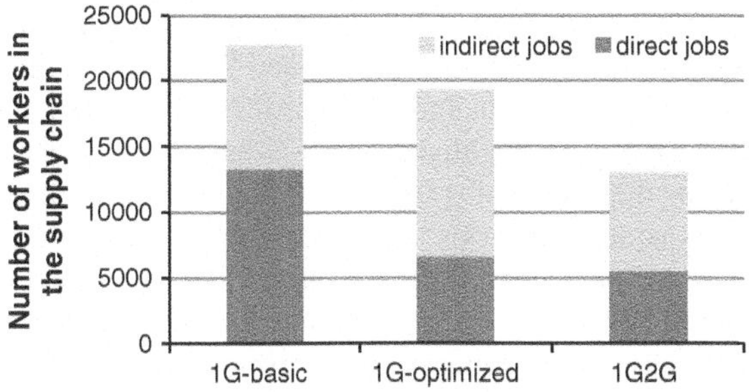

Fig. 7.32 Number of direct and indirect jobs of a US$1 billion change in ethanol final demand, considering three vertically integrated production models

scenario of 1G-basic is highly reliant upon manual operations—about 30 % of the sugarcane is manually harvested—then the number of workers will be higher than in other scenarios. As Fig. 7.32 shows, the worker profile is quite different among the scenarios. For the scenarios 1G2G and 1G-optimized, more indirect jobs are observed, indicating less participation of sugarcane sector—the number of jobs is importantly reduced due to the increase in mechanization levels for the 1G2G and 1G-optimized scenarios—and more contribution from indirect commodities such as trade, repair and maintenance services, services to companies, transportation, information services and other sectors.

The 1G2G scenario is related to the lowest number of occupational accidents (see Fig. 7.33). This effect is observed because sugarcane in 1G2G is more reliant on mechanical operations and also because of the higher ethanol yields in the 1G2G industrial phase when compared to the other scenarios. As a consequence, a lower level of accidents per thousand workers compared to 1G-basic and 1G-optimized scenarios is observed. For the 1G-optimized scenario, accidents were mostly indirect, mainly because of contributions from trade, repair and maintenance services, transportation, metal products, other products from extractive industry, rubber and plastic products and other sectors.

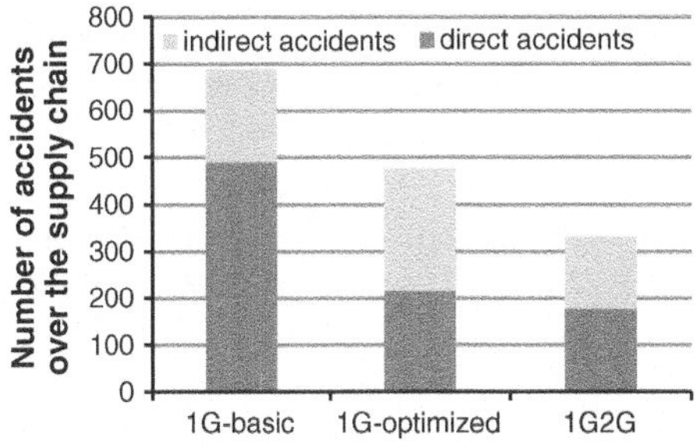

Fig. 7.33 Number of direct and indirect accidents over the supply chain associated with a US$1 billion ethanol final demand

A valuable outcome from the hybrid model is the possibility of describing the whole ethanol supply chain. Tables 7.45, 7.46 and 7.47 show wage, education and gender profiles, which can be obtained by considering the weighted contribution from each sector. Table 7.45 highlights that 1G-basic supply chain has a higher share of workers in the range of 1.01–1.5 minimum wages (about 64 % of all workers). As much as new scenarios take place, such as 1G-optimized and 1G2G, new wage profiles are observed: The range from 1.01 to 1.5 minimum wages is reduced to 52 % and 53 % of workers of the supply chain, respectively. Therefore, scenarios, 1G2G and 1G-optimized, are related to an improvement on average wage profile of the supply chain.

Regarding the education profile, some improvements associated with the 1G2G and 1G-optmized scenarios are clear (see Table 7.46). For the 1G-basic scenario, only 25 % of the supply chain workers have finished high school. Moreover, about 17 % of workers have not finished the elementary school. When it comes to 1G2G scenario, however, the supply chain education profile is significantly different: The share of workers who have finished high school increases from 25 to 35 % and those associated with the unfinished elementary school decreases from 17 to 11 %. This effect may be attributed to the lower participation of workers from the agricultural sector—most of the sugarcane workers, especially cutters, have up to the elementary school profile (Moraes et al. 2009)—and increase in workers from sectors related to higher education profile, such as repair and maintenance services whose workers have finished, in most cases, the high school level.

Another improvement related to 1G2G and 1G-optimized scenarios is the increase in the share of female workers over the supply chain (see Table 7.47). Although the ethanol supply chain is majorly configured by male workers, the overall picture may favor the participation of women in future ethanol production scenarios. Similar to the case of education profile, the 1G2G and 1G-optimized supply chains are more reliant upon indirect sectors, such as trade, which has a higher share of female workers.

The results of the social assessment aimed to improve the discussion in the social assessment of ethanol production technologies by using a quantitative approach based on the hybrid input–output LCA approach. In general, it is possible to

Table 7.45 Relative frequencies of wage categories over the supply chain considering workers of three vertically integrated production models (Brazilian minimum wage = 724 BRL = 315 US$, reference year: 2014)

	Number of minimum wages								
	Up to 1 (%)	1.01–1.5 (%)	1.51–2 (%)	2.01–3 (%)	3.01–4 (%)	4.01–5 (%)	5.01–7 (%)	7.01–10 (%)	Above 10.01 (%)
1G-basic	6	64	20	6	2	1	1	0	0
1G-opt.	8	52	27	7	2	1	1	1	1
1G2G	8	53	24	8	2	1	1	1	1

Table 7.46 Relative frequency of different education levels over the supply chain considering workers of three vertically integrated production models

	Education level								
	Illiterate (%)	Unfinished elementary school (%)	Finished elementary school (%)	Unfinished middle school (%)	Finished middle school (%)	Unfinish high school (%)	Finished high school (%)	Unfinished college (%)	Finished college (%)
1G-basic	3	17	13	15	14	7	25	2	2
1G-opt.	2	11	9	13	16	9	34	3	3
1G2G	2	10	9	13	16	9	35	3	4

Table 7.47 Gender profile of workers of the entire supply chain considering three vertically integrated production scenarios

	Male (%)	Female (%)
1G-basic	78.3	21.7
1G-optimized	71.3	28.7
1G2G	70.8	29.2

observe that sugarcane and biorefinery technological changes may affect the entire supply chain. Although a lower level of job creation may be observed, positive social effects may also occur such as lower accident levels, increase in the average wage, better education profile and more participation of female workers in the ethanol supply chain.

References

ACS (2013) Cost-saving measure to upgrade ethanol to butanol—a better alternative to gasoline. http://www.acs.org/content/acs/en/pressroom/newsreleases/2013/april/cost-saving-measure-to-upgrade-ethanol-to-butanol-a-better-alternative-to-gasoline.html. Accessed 19 Apr 2013

Afschar AS, Vaz Rossell CE, Schaller K (1990) Bacterial conversion of molasses to acetone and butanol. Appl Microbiol Biotechnol 34:168–171

Cardoso TF (2014) Avaliação socioeconômica e ambiental de sistemas de recolhimento e uso da palha de cana-de-açúcar. PhD thesis, School of Agricultural Engineering, University of Campinas

Cardoso TF, Cavalett O, Chagas MF et al (2013) Technical and economic assessment of trash recovery in the sugarcane bioenergy production system. Sci Agric 70:353–360

Cavalett O, Cunha MP, Chagas MF et al (2013) An exploratory economic analysis of sugarcane harvest extension using sweet sorghum in the Brazilian sugarcane industry. In: XXVIII ISSCT Congress, São Paulo, 24–27 June 2013

CNCIC (2014) China National Chemical Information Center. Octanol market study. http://chemroundtables.com/wp-content/uploads/2012/09/2012-China-Report-sample-revised1.pdf. Accessed 25 Oct 2014

Corn Refiners Association (2006) Corn oil. http://www.corn.org/wp-content/uploads/2009/12/CornOil.pdf. Accessed 12 June 2015

Dias MOS, Pereira LG, Junqueira TL et al (2014) Butanol production in a sugarcane biorefinery using ethanol as feedstock—Part I: integration to a first generation sugarcane distillery. Chem Eng Res Des 92(8):1441–1451

EPA (2010) EPA lifecycle analysis of greenhouse gas emissions from renewable fuels. http://www.epa.gov/otaq/renewablefuels/420f10006.pdf. EPA—Environment Protection Agency. Accessed June 2015

Ezeji T, Blaschek HP (2008) Fermentation of dried distillers' grains and soluble (DDGS) hydrolysates to solvents and value-added products by solventogenic clostridia. Bioresour Technol 99:5232–5242

Guilhoto JJM, Sesso UA (2005) Estimação da matriz insumo-produto a partir dedados preliminares das contas nacionais. Economia Aplicada 9(2):277–299

Guilhoto JJM, Sesso UA (2010) Estimação da matriz insumo-produto utilizandodados preliminares das contas nacionais: aplicação e análise de indicadores econômicos para o Brasil em

2005. Economia e Tecnologia 23:53–62. http://www.usp.br/nereus/wp-content/uploads/Metodologia-guilhoto-sesso-EA-2010.pdf. Accessed 22 Nov 2012

Hassuani SJ, Leal MRLV, Macedo IC (eds) (2005) Biomass power generation—sugarcane bagasse and trash. CTC and PNUD, Piracicaba

Humbird D, Davis R, Tao L (2011) Process design and economics for biochemical conversion of lignocellulosic biomass to ethanol. National Renewable Energy Laboratory—NREL. http://www.nrel.gov/biomass/pdfs/47764.pdf. Accessed 03 June 2014

IBGE (The Brazilian Institute of Geography and Statistics) (2009) System of national accounts. http://www.ibge.gov.br/home/estatistica/economia/matrizinsumo_produto/default.shtm. Accessed 15 Sep 2013

Junqueira TL (2015) Techno-economic feasibility analysis of process alternatives for ethanol production in Brazil. PhD thesis, School of Chemical Engineering, University of Campinas

Liu Z, Ying Y, Li F et al (2010) Butanol production by Clostridium beijerinckiiATCC 55025 from wheat bran. J Ind Microbiol Biotechnol 37:495–501

Mariano AP, Maciel Filho R (2012) Improvements in biobutanol fermentation and their impacts on distillation energy consumption and wastewater generation. Bio Energ Res 5(2):504–514

Mariano AP, Dias MOS, Junqueira TL et al (2013a) Butanol production in a first-generation Brazilian sugarcane biorefinery: technical aspects and economics of greenfield projects. Bioresour Technol 135:316–323

Mariano AP, Dias MOS, Junqueira TL et al (2013b) Utilization of pentoses from sugarcane biomass: techno-economics of biogas vs. butanol production. Bioresour Technol 142:390–399

May A (2012) Sistema Embrapa de Produção Agroindustrial de Sorgo Sacarino para Bioetanol Sistema BRS1G—Tecnologia Qualidade Embrapa. Documentos 139. Embrapa Milho e Sorgo. Sete Lagoas

Milanez AY, Nyko D, Valente MS et al (2014) A produção de etanol pela integração do milho-safrinha às usinas de cana-de-açúcar: avaliação ambiental, econômica e sugestões de política. Revista do BNDES 41:147–208

Milanez AY, Nyko D, Valente MS et al (2015) De promessa a realidade: como o etanol celulósico pode revolucionar a indústria da cana-de-açúcar—uma avaliação do potencial competitivo e sugestões de política pública. BNDES Setorial 41:237–294

Moraes MAFD, Figueiredo MG, Oliveira FCR (2009) Migração de Trabalhadores na Lavoura Canavieira Paulista: uma investigação dos impactos socioeconômicos nas cidades de Pedra Branca, Estado do Ceará, e de Leme, Estado de São Paulo. Revista de Economia Agrícola 56 (2):21–35

MPS (Ministry of Social Security) (2015) Anuário Estatístico da Previdência Social—AEAT. http://www.previdencia.gov.br/2015/01/estatisticas-anuario-estatistico-de-acidentes-do-trabalho-2013-ja-esta-disponivel-para-consutla/. Accessed 20 Jan 2015

MTE (Ministry of Labor and Employment) (2015) Programa de Disseminação de Estatísticas do trabalho—PDET. http://bi.mte.gov.br/bgcaged/login.php. Accessed 20 Jan 2015

Orbichem (2013a) Tecnon Orbichem. Prices on chemicals—n-Butanol. http://www.orbichem.com/userfiles/CNF%20Samples/nbt_13_11.pdf. Accessed 9 July 2015

Orbichem (2013b) Tecnon Orbichem. Prices on chemicals—2-Ethylhexanol. http://www.orbichem.com/userfiles/CNF%20Samples/2eh_13_11.pdf. Accessed 9 July 2015

Orbichem (2013c) Tecnon Orbichem. Prices on chemicals—Acetone. http://www.orbichem.com/userfiles/CNF%20Samples/act_13_11.pdf. Accessed 9 July 2015

Pacheco TF (2012) Tecnologia industrial. In: May A et al (eds) Sistema Embrapa de Produção Agroindustrial de Sorgo Sacarino para Bioetanol. Sistema BRS1G—Tecnologia Qualidade Embrapa. Documentos 139. Embrapa Milho e Sorgo, Sete Lagoas, pp 92–106

Parrella RAC, Schaffert RE (2012) Cultivares. In: May A et al (eds) Sistema Embrapa de Produção Agroindustrial de Sorgo Sacarino para Bioetanol. Sistema BRS1G—Tecnologia Qualidade Embrapa. Documentos 139. Embrapa Milho e Sorgo, Sete Lagoas, pp 14–22

Pereira LG, Dias MOS, Junqueira TL et al (2014) Butanol production in a sugarcane biorefinery using ethanol as feedstock—Part II: integration to a second generation sugarcane distillery. Chem Eng Res Des 92(8):1452–1462

Qureshi N, Saha BC, Hector RE et al (2010) Production of butanol (a biofuel) from agricultural residues: Part II—use of corn stover and switchgrass hydrolysates. Biomass Bioenerg 34:566–571

Riittonen T, Toukoniitty E, Madnani DK et al (2012) One-pot liquid-phase catalytic conversion of ethanol to 1-butanol over aluminium oxide—the effect of the active metal on the selectivity. J Catal 2:68–84

Roffler S, Blanch H, Wilke C (1987) Extractive fermentation of acetone and butanol: process design and economic evaluation. Biotechnol Prog 3:131–140

Tsuchida T, Kubo J, Yoshioka T et al (2008) Reaction of ethanol over hydroxyapatite affected by Ca/P ratio of catalyst. J Catal 259:183–189

Watanabe MDB, Chagas MF, Cavalett O et al (2015) Hybrid input-output life cycle assessment of first- and second-generation ethanol production technologies in Brazil. J Ind Ecol. doi:10.1111/jiec.12325 (online version)

Chapter 8
Use of VSB to Plan Research Programs and Public Policies

Tassia L. Junqueira, Bruna Moraes, Vera L.R. Gouveia, Mateus F. Chagas, Edvaldo R. Morais, Marcos D.B. Watanabe, Marcelo Zaiat and Antonio Bonomi

8.1 Introduction

The process of technological development often requires efforts in scientific and technological research. On the other hand, one can identify process bottlenecks and opportunities for using existing or new by-products for more sustainable applications. Many times, these requirements will require technological developments—that should be assessed to confirm the positive impacts derived from these innovations—and obtained new by-products could demand public policies to develop their market or improve positive impacts of their adoption.

Virtual Sugarcane Biorefinery (VSB) presents a useful framework for evaluating and assessing research programs, defining and evaluating the effects of public policies. This chapter describes two cases regarding the use of VSB for R&D evaluation—enzyme development and use and energy cane as a new variety for sugarcane biorefinery—and a public policy formulation for biogas production and use in the biorefinery context.

T.L. Junqueira · B. Moraes · V.L.R. Gouveia · M.F. Chagas · E.R. Morais
M.D.B. Watanabe · M. Zaiat · A. Bonomi (✉)
Brazilian Bioethanol Science and Technology Laboratory (CTBE),
Brazilian Center of Research in Energy and Materials (CNPEM), Campinas, SP, Brazil
e-mail: antonio.bonomi@bioetanol.org.br

M. Zaiat
University of São Paulo, São Carlos, SP, Brazil

A. Bonomi et al. (eds.), *Virtual Biorefinery*, Green Energy and Technology,
DOI 10.1007/978-3-319-26045-7_8

8.2 Case Study 1: 2G Ethanol

The capability of VSB to assess the development stage of new technologies and perform comparative assessments of biorefinery alternatives are some of the features that make VSB a tool with great potential to support planning of research programs in the bioenergy context. Based on the results obtained from the VSB, technological and sustainability bottlenecks are identified, which is useful to define research priorities. Given the importance of sugar-energy industry in the Brazilian economy and energy matrix, actions to accelerate the development of 2G ethanol production are desirable to increase the competitiveness and sustainability of this industry.

In order to improve the techno-economic feasibility of 2G ethanol production, two main aspects must be considered: feedstock and 2G process yield and costs. Since 2G ethanol production is based on the processing of lignocellulosic material, higher fiber content and higher agricultural productivities are desirable features for a bioenergy feedstock, and in doing so, energy cane stands out as a promising alternative. Regarding 2G process, enzymatic hydrolysis can be seen as the core step of biochemical route for ethanol production, thus impacting overall process yield and costs.

In this section, the assessment of integrated 1G2G process using energy cane either to complement or replace sugarcane and with advances on enzymatic hydrolysis was carried out in order to show the potential of further research on these important subjects.

8.2.1 Energy Cane

Energy cane is a more robust sugarcane variety with higher fiber content and productive potential and it has been envisioned as a promising alternative for production of 2G biofuels, biochemicals and energy (Matsuoka et al. 2014).

In spite of being already produced in small commercial scales, the high agricultural yields reported in the literature are, in most cases, based on the results from experimental areas. In large-scale production areas, it is difficult to deal with uncontrolled variables, such as climate conditions and pest infestations, which highly influence biomass productivity. Also in agricultural phase, the entire production system still needs to be established for energy cane, using equipment and management practices appropriate for this new biomass.

Season extension is also a possibility using energy cane since this biomass is not envisioned for sucrose production. The development of new harvesting equipment, which runs through preferred and compressed pathways, can make possible harvesting energy cane even during the rainy season.

There are still uncertainties regarding the juice extraction process due to higher fiber content of energy cane. It is expected that the extraction efficiency is slightly lower for same imbibition rates practiced for sugarcane.

In a previous study, the VSB was used to evaluate economic aspects of energy cane production and processing to obtain 1G and 2G ethanol in medium- and long-term scenarios (Milanez et al. 2015). The introduction of energy cane in this study was a suggestion of some specialists from sugar-energy sector and companies engaged in the research and development of 2G ethanol. For instance, GranBio—a Brazilian company of industrial biotechnology that operates the first commercial plant of cellulosic ethanol in the Southern Hemisphere—has an experimental station with an area of 60 ha of energy cane (GranBio 2015). GranBio experimental station is a partnership with the Agronomic Institute of Campinas, which has been developing the energy cane (also called "supercane") for six years (Toledo 2015).

Milanez et al. (2015) considered the introduction of energy cane along with other advances on 2G process to estimate the technological progress and production cost over time. In order to isolate the independent effect of energy cane as feedstock, the present study evaluates its use as an alternative to produce both 1G and 2G ethanol in the off-season period considering the same technology employed for sugarcane conversion to ethanol.

Besides promising results of preliminary studies, even with conservative assumptions, energy cane production and processing systems are not yet consolidated, and many questions remain to be answered. In view of this scenario, research programs combined with practical demonstrations are still required to clarify the questions about energy cane. Coordinated actions for research, combined with public policies to incentive the private sector participation, are necessary to turn energy cane into reality.

Scenario Description

In this case, the use of energy cane (EC) either to complement or replace sugarcane (SC) in an integrated 1G2G ethanol production process was assessed. For comparison, composition and agricultural productivity considered for energy cane and sugarcane in this analysis are shown in Table 8.1.

As a basis for comparison, the current 2G technology integrated to optimized autonomous distillery processing sugarcane (scenario 1G2G-Cur presented in Sect. 7.2.1) was considered. In this scenario, 2G ethanol is produced in the season

Table 8.1 Energy cane and sugarcane composition and productivity (Milanez et al. 2015)

Composition[a] (% wt)	Sugarcane[b]	Energy cane[c]
Water	70.3	66.8
Fibers	12.7	21.3
Sucrose	14.0	8.1
Reducing sugars	0.6	2.5
Others	2.4	1.3
Agricultural productivity (t ha^{-1})	80	200

[a]Does not include mineral impurities
[b]Considering only stalk composition
[c]Considering full composition (stalk + straw)

and off-season periods using bagasse and straw from sugarcane. Two other scenarios were defined: 1G2G (SC + EC) that considers production of 1G and 2G ethanol from sugarcane in the season and from energy cane in the off-season period and 1G2G (EC) that assumes production of 1G and 2G ethanol from energy cane in both season and off-season periods. Process flowsheet, conditions and yields for energy cane processing are mostly the same as those considered for sugarcane in the 1G2G-Cur scenario. Table 8.2 summarizes the operational characteristics for the three scenarios.

For scenario 1G2G-Cur, part of LCM (sugarcane bagasse and straw) is used in the season as feedstock for 2G process and as fuel in the CHP, while the remaining fraction is stored for off-season operation. Sugarcane reception, cleaning and juice treatment are not operational in the off-season, whereas process steps from juice concentration to ethanol dehydration and CHP operate with some idle capacity and 2G process runs at full capacity all year-round.

In scenario 1G2G (SC + EC), all sugarcane LCM is consumed in the season and energy cane is processed in the off-season. In this way, 1G ethanol is also produced in the off-season from energy cane juice, which is obtained using the same milling tandem (5 three-roll mills), but with lower efficiency due to higher fiber content of the energy cane. The energy cane bagasse is used to produce 2G ethanol and to

Table 8.2 Operational characteristics for integrated 1G2G ethanol production scenarios

Parameter	1G2G-Cur		1G2G (SC + EC)		1G2G (EC)	
Period[a]	Season	Off-season	Season	Off-season	Season	Off-season
Processing capacity (10^6 t $year^{-1}$)	SC: 4.0[b]	–	SC: 4.0[b]	EC: 1.7[c]	EC: 2.6[c]	EC: 1.7[c]
	Straw: 0.18[d]	–	Straw: 0.18[d]	–	–	–
1G feedstock	SC juice	–	SC juice	EC juice	EC juice	EC juice
2G feedstock	SC bagasse and straw	SC bagasse and straw	SC bagasse and straw	EC bagasse	EC bagasse	EC bagasse
Cleaning system	Dry cleaning	None	Dry cleaning	None	None	None
Extraction efficiency (%)	96 (5 mills)	–	96 (5 mills)	90 (5 mills)	80 (2 mills)	80 (2 mills)
Power demand (MW)[e]	28.7	2.3	30.4	22.9	21.9	21.9

[a]Season and off-season periods were defined equal to 200 and 130 days, respectively
[b]Includes only sugarcane stalks
[c]Includes energy cane stalks and straw
[d]Considering vegetal impurities and straw bales (dry basis)
[e]Average power demand in the period, considering—when applicable—dry cleaning, sugar extraction, and 1G and 2G processes

supply steam and electricity to the process in the off-season. The amount of energy cane processed in the off-season is defined in such a way that fibers fed in milling are the same as processed in the season. Therefore, 1G process is operational in the off-season but with some idle capacity, since the quantity of juice is significantly lower than that generated with sugarcane; similarly, 2G process does not operate at full capacity in the off-season, because an additional amount of fibers is provided by straw in the season.

In scenario 1G2G (EC), in which only energy cane is processed, specialists have proposed a milling tandem with 2 three-roll mills and extraction efficiency of 80 % (Milanez et al. 2015). The quantity of energy cane was defined so that the same amount of fibers of previous scenarios is fed to the extraction system. All plant operates at full capacity in the off-season period, since it is considered that energy cane can be delivered at the plant all year-round.

In all scenarios, backpressure turbines are employed since 2G ethanol production is favored; thus, steam is produced only to meet process requirements, considering a 20 % reduction in 2.5-bar steam demand, which would be achieved through thermal integration.

Techno-economic Analysis

Comparison between technical results is presented in Table 8.3 for the three scenarios evaluated. In general, 1G ethanol production (per tonne of cane) is significantly lower for EC, since it presents less sugars in its composition and reduced efficiency in milling was assumed. On the other hand, 2G ethanol production is larger for EC due to higher fiber content and lower steam consumption for 1G process when compared to SC.

Among the three scenarios, 1G2G (SC + EC) has the largest annual outputs for both ethanol and electricity, since it combines the processing of SC and EC. However, in terms of ethanol productivity per planted area, scenario 1G2G (EC) showed the greatest ethanol production, almost twice the amount produced in 1G2G-Cur that processes only sugarcane.

The techno-economic analysis was performed to assess the impact of energy cane use on the industrial stage. In order to perform the economic analysis, the

Table 8.3 Anhydrous ethanol and electricity production for the evaluated scenarios

Scenarios	1G2G-Cur	1G2G (SC + EC)	1G2G (EC)
1G anhydrous ethanol—SC (L t^{-1})	85.4	85.4	–
1G anhydrous ethanol—EC (L t^{-1})	–	56.8	50.5
2G anhydrous ethanol—SC (L t^{-1})	22.1	23.5	–
2G anhydrous ethanol—EC (L t^{-1})	–	27.3	32.9
1G anhydrous ethanol (10^6 L $year^{-1}$)	341.4	436.9	215.3
2G anhydrous ethanol (10^6 L $year^{-1}$)	88.6	139.1	140.3
Overall anhydrous ethanol (10^6 L $year^{-1}$)	430.0	576.7	355.6
Ethanol productivity (m^3 ha^{-1})	8.6	10.7	16.7
Electricity (GWh $year^{-1}$)	309.0	420.6	306.7

Table 8.4 Summary of CAPEX associated with the main 1G2G industrial areas, considering different scenarios

CAPEX (US$ million)	1G2G-Cur	1G2G (SC + EC)	1G2G (EC)
1G + interface areas			
Administrative infrastructure, engineering and utilities	86.6	99.3	69.4
Sugarcane reception, juice extraction and evaporation	70.5	73.5	45.1
Ethanol production (fermentation, distillation/dehydration)	93.6	102.9	53.7
Combined heat and power (steam and electricity generation)	109.2	125.4	84.2
2G areas			
Pretreatment and C5 liquor separation/evaporation	75.9	102.9	80.4
Separated C5 fermentation	11.2	15.8	14.8
Enzymatic hydrolysis and C6 liquor separation	34.7	58.0	47.9
Yeast propagation	0.7	0.7	0.7
Total	482.4	578.5	396.2

capital expenditures (CAPEX) were calculated and they are presented in Table 8.4. The difference between the estimated values is mainly due to the extraction area, which is less significant for the 1G2G (EC) scenario, since a relatively small equipment is needed to extract the energy cane juice. Another variable is the amount of processed LCM, which impacts the CHP and pretreatment areas; this effect is more evident on scenario 1G2G (SC + EC).

The biomass production costs of both conventional sugarcane (SC) and energy cane (EC) were calculated according to the CanaSoft model. The results in Table 8.5 show that energy cane production cost (roughly US$17 per tonne) is clearly lower due to higher productivity per hectare when compared to the figures observed in conventional sugarcane. Straw remained with the cost of roughly US $36 per dry tonne. These results of biomass production costs will play a key role in the economic results from cash flow analysis and ethanol cost that will be presented.

The comparison of integrated 1G2G ethanol production scenarios using both conventional sugarcane and energy cane is presented in Fig. 8.1. The integrated

Table 8.5 Summary of biomass production costs

Biomass production cost	1G2G-Cur	1G2G (SC + EC)	1G2G (EC)
Sugarcane (US$ t^{-1})	28.50	28.52	–
Straw (US$ $t^{-1}{}_{dry}$)	36.38	36.38	–
Energy cane (US$ t^{-1})	–	18.63	17.07

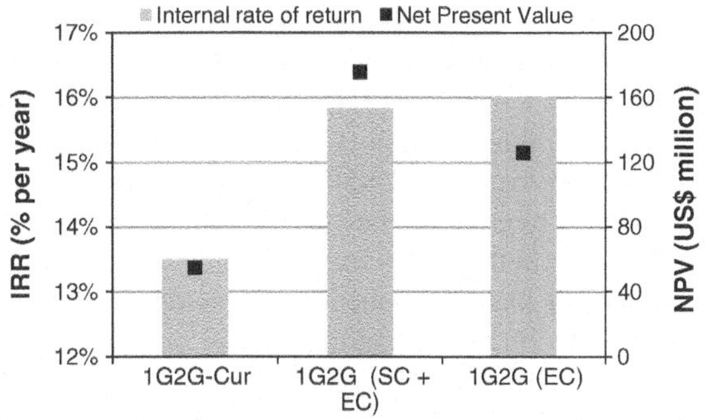

Fig. 8.1 Results of the economic analysis considering different integrated 1G2G scenarios

1G2G plant using conventional sugarcane during season and energy cane during the off-season (1G2G SC + EC) leads to higher net present value (NPV)—roughly US $175 million—in comparison with the 1G2G-Cur and 1G2G plant using energy cane all year-round (1G2G EC), which means that more value would be added to the plant when 1G2G (SC + EC) project is executed. In regard to the internal rate of return (IRR), both 1G2G (SC + EC) and 1G2G (EC) presented very close figures: 15.8 and 16.0 % per year, respectively.

These IRRs are above the minimum acceptable rate of return (12 % per year), which indicates that these scenarios are good alternatives for 1G2G ethanol and electricity production considering the economic point of view. The 1G2G-Cur scenario, on the other hand, was related to lower IRR and NPV values—13.5 % per year and US$55 million, respectively. These results indicate that 1G2G-Cur is economically viable; however, conventional sugarcane has lower productivity and higher production costs when compared to EC. The introduction of energy cane significantly reduces biomass production costs, thus positively impacting the operating costs in the industrial stage. Moreover, the additional revenues from electricity and 2G ethanol production during the off-season period contribute to improve the economic results of 1G2G (SC + EC) and 1G2G (EC) scenarios.

Figure 8.2 shows the main differences of 1G2G ethanol costs as well as the allocation of such value to both 1G and 2G ethanol production (allocation

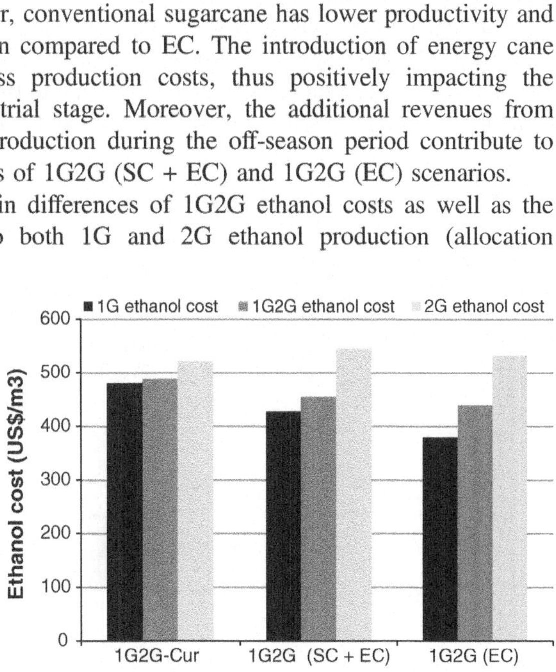

Fig. 8.2 Average 1G2G ethanol costs and allocation to 1G and 2G ethanol

procedure was discussed in Chap. 6). The average ethanol production cost obtained from both 1G and 2G ethanol (1G2G cost) is gradually reduced when observing values from left to right (Fig. 8.2). These results denote the influence of biomass cost reduction on the total 1G2G ethanol costs for 1G2G-Cur, 1G2G (SC + EC) and 1G2G (EC) scenarios: US$0.489, 0.456, 0.440 per liter, respectively.

In order to estimate the 1G ethanol costs, mirror scenarios were built for 1G2G technology. In the mirror scenarios, investments and industrial processes associated with the 2G ethanol production are excluded in order to represent the conditions by which only 1G ethanol and electricity would be produced in equivalent industrial facilities. Therefore, the 1G ethanol costs associated with 1G2G-Cur, 1G2G (SC + EC) and 1G2G (EC) scenarios were estimated: US$0.480, 0.427 and 0.380 per liter of ethanol. In Fig. 8.2, the 1G ethanol costs are reduced from left to right mainly due to biomass cost reduction. It is important to highlight that the 1G ethanol cost reduction was higher when compared to the 1G2G ethanol cost reduction. This result was expected because the 1G ethanol cost is strongly dependent on biomass cost—about 60 % of total 1G ethanol cost.

The 2G ethanol cost was then determined according to the allocation criteria discussed in Chap. 6. It is clear that the reduction of 1G ethanol cost caused impacts on the 2G ethanol costs. As much as the 1G ethanol cost is associated with significant decreasing costs, more capital and operating costs will be allocated to the 2G ethanol. As Fig. 8.2 shows, values associated with 1G2G-Cur, 1G2G (SC + EC) and 1G2G (EC) were of US$0.522, 0.545 and 0.532 per liter, respectively. The results show that the higher the difference between 1G and 1G2G ethanol costs, the higher the cost allocation to 2G ethanol.

With regard to 2G ethanol cost breakdown, results in Fig. 8.3 show the expenses on capital, biomass, enzyme, maintenance, labor and other costs. Once again, the capital cost was the main cost element representing from 35 to 40 % of the total 2G ethanol costs. In the case of 1G2G (SC + EC) and 1G2G (EC), higher capital costs

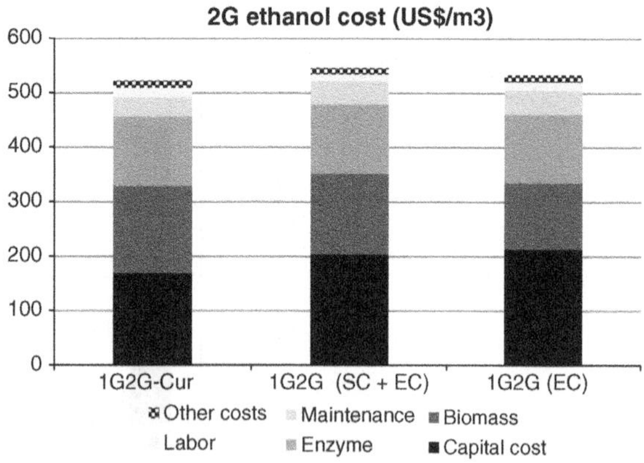

Fig. 8.3 Breakdown of 2G production costs for scenarios with and without energy cane

are observed due to higher allocation of investment to the 2G ethanol, as a consequence of 1G ethanol cost reduction. On the other hand, it is clear that biomass production costs are reduced as much as the participation of energy cane in the total biomass input increases. For instance, the biomass cost for the 1G2G (EC) scenario is approximately US$0.121 per liter and about US$0.148 per liter for the 1G2G (CC + EC) scenario. In the case of 1G2G-Cur, which does not make use of energy cane, the contribution of biomass to the total 2G ethanol cost was around US$0.160 per liter. Observing the overall picture, it is clear that 2G ethanol costs of the different scenarios were very similar, varying from US$0.522 to 0.545 per liter.

8.2.2 Enzymes

The use of enzymes to perform hydrolysis allows carrying out the process in milder conditions, when compared to acid hydrolysis, with higher yields and lower sugar degradation (Mussatto et al. 2010). However, residence time on enzymatic hydrolysis is relatively long (few days), which increases the susceptibility to microbial contamination (Binod et al. 2011). Several factors affect enzymatic hydrolysis, such as substrates (concentration and quality), cellulase activity and operating conditions, e.g., temperature and pH (Taherzadeh and Karimi 2007).

Operating hydrolysis with high initial substrate concentration is desirable in terms of energy and economic viability since more concentrated hydrolysates are generated and smaller equipment is required; however, product inhibition and high viscosity affect hydrolysis efficiency and lead to high electricity consumption for mixing (Binod et al. 2011).

Besides, the contribution of enzymes to the 2G ethanol production cost has been pointed out to be controversial, since there is little information available about the enzyme costs, especially in protein basis. Enzyme costs are usually reported in terms of volume of ethanol produced, which is dependent on other factors, such as enzyme loading and overall ethanol yield (Klein-Marcuschamer et al. 2012).

Research toward optimization of this operation and improvement of cellulase activity has been carried out aiming at increasing hydrolysis yield and rate, and enzyme robustness under industrial operating conditions as well as reducing product and substrate inhibition. Optimization of enzyme cocktails through the use of genetic techniques and search for new enzymes are some efforts around this issue (U.S. DOE 2015; Sun and Cheng 2002).

Taking into account that there are several challenges to be overcome related to enzymatic hydrolysis, a scenario considering improvements in this process operation—use of higher solid content and superior hydrolysis rate and yield—was defined to measure the potential gains that would be achieved. For this scenario, operating conditions and yields for current 2G technology were assumed (see Sect. 7.2.1.1), except for enzymatic hydrolysis that was based on advanced 2G technology. Table 8.6 presents the comparison between current and advanced 2G technologies for enzymatic hydrolysis.

Table 8.6 Process conditions and yields for enzymatic hydrolysis (EH)

Technology	Current EH	Advanced EH
Temperature (°C)	50	65[a]
Residence time (h)	48	36
Solid content (%)	15	25
Cellulose conversion to glucose (%)	60	80
Xylan conversion to xylose (%)	60	80

[a]Microorganisms resistant to higher temperature may allow higher solid content in enzymatic hydrolysis

The acquisition of enzymes can be through either on-site production or purchase from supplier, since both methods will probably coexist and compete with each other. When the enzyme production is colocated with a 2G plant, steam and electricity are provided by the facility as well as the carbon source; in addition, the whole fermentation broth can be sent to hydrolysis, reducing costs related to concentration and transportation. On the other hand, external suppliers benefit from economies of scale and are constantly researching ways to improve enzyme performance and reduce costs.

A comparison between current technology without modifications (scenario 1G2G-Cur) and the scenario with advanced enzymatic hydrolysis (1G2G-AdvEH) is shown in Table 8.7.

Improvement on enzymatic hydrolysis caused an increase of 30 % in 2G ethanol production and a decrease of 13 % in electricity generation. These changes are not only due to the incremental production of 2G ethanol related to higher yields, but also due to the increase in solid content on enzymatic hydrolysis that allowed the production of more concentrated C6 liquor. Consequently, there is a reduction in steam consumption which, in turn, increases lignocellulosic material availability for pretreatment and decreases the amount of electricity generated in the backpressure turbines.

For 1G2G-Cur scenario, the enzyme contribution to the 2G ethanol production cost was considered to be US\$0.13 L^{-1} based on Milanez et al. (2015). If the enzyme production costs were US\$4.24 kg^{-1} protein as calculated by Humbird et al. (2011) for on-site production, the amount of enzyme added would be 16 mg per g of cellulose. In the 1G2G-AdvEH scenario, it was assumed that the development of enzymes (e.g., increase in activity) and advances on enzymatic hydrolysis would allow achieving higher yields with lower residence times by using

Table 8.7 Technical results for 1G2G-Cur and 1G2G-AdvEH scenarios

Scenarios	1G2G-Cur	1G2G-AdvEH
Total anhydrous ethanol production (L TC^{-1})	107.5	114.1
1G anhydrous ethanol production (L TC^{-1})	85.4	85.4
2G anhydrous ethanol production (L TC^{-1})	22.1	28.8
Electricity (kWh TC^{-1})	77.4	67.2

the same enzyme/cellulose proportion. As a result, the contribution of enzyme to the 2G ethanol production cost is US\$0.106 L^{-1}, around 20 % less than that assumed for 1G2G-Cur scenario.

In order to achieve the enzyme cost predicted for advanced 2G technology (US \$0.06 L^{-1}), further improvements need to be accomplished, such as:

- Increase in overall 2G ethanol production, due to advances in other process steps (e.g., pretreatment and fermentation);
- Reduction in enzyme production cost and/or enzyme recycle.

In order to assess the impacts of advanced enzymatic hydrolysis on the techno-economic results of the different scenarios, estimations on the CAPEX were calculated (Table 8.8). It is important to highlight that improvements on enzymatic hydrolysis (1G2G-AdvEH) slightly reduce the CAPEX, especially because of smaller hydrolysis reactors as a consequence of lower residence times and higher solid contents when compared to the baseline scenario (1G2G-Cur). The investment in cogeneration system also presented a small reduction in the 1G2G-AdvEH scenario, since less lignocellulosic material is burned because of the lower steam consumption.

Techno-economic Analysis

The results of the techno-economic analysis are presented in Fig. 8.4. As it shows, both 1G2G-Cur and 1G2G-AdvEH proved to be economically feasible, presenting internal rates of return above the minimum acceptable rate of return (12 % per year) and positive NPV. The 1G2G-AdvEH scenario, however, is related to better economic results for both IRR and NPV: 15.4 % per year and US\$123 million, respectively, when compared to the 1G2G-Cur whose results were roughly 13.6 % per year and US\$57 million. As expected, advanced enzymatic hydrolysis causes a positive impact on the 1G2G-AdvEH scenario mainly due to the extra revenues from the additional 2G ethanol production as well as the decreased capital costs

Table 8.8 Summary of CAPEX associated with the main industrial areas for scenarios 1G2G-Cur and 1G2G-AdvEH

CAPEX (US\$ million)	1G2G-Cur	1G2G-AdvEH
1G + interface areas		
Administrative infrastructure, engineering and utilities	86.6	84.3
Sugarcane reception, juice extraction and evaporation	70.5	67.3
Ethanol production (fermentation, distillation/dehydration)	93.6	97.0
Combined heat and power (steam and electricity generation)	109.2	106.2
2G areas		
Pretreatment and C5 liquor separation/evaporation	75.9	76.7
Separated C5 fermentation	11.2	11.9
Enzymatic hydrolysis and C6 liquor separation	34.7	20.5
Yeast propagation	0.7	0.7
Total	482.4	464.6

Fig. 8.4 Results of the economic analysis considering different hydrolysis technologies

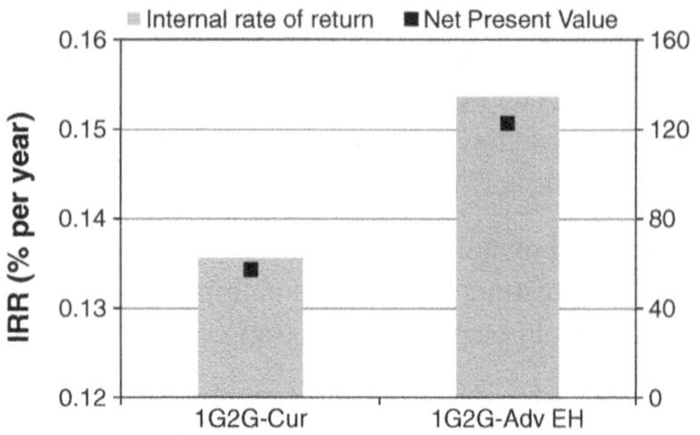

associated with lower investment in hydrolysis reactors and cogeneration system, as previously discussed.

The effects of advanced enzymatic hydrolysis are clearer when observing the breakdown of 2G ethanol costs. As depicted in Fig. 8.5, considerable reductions were achieved when shifting from 1G2G-Cur to 1G2G-AdvEH scenario. As initially assumed, hydrolysis technology in 1G2G-AdvEH scenario would lead to 20 % enzyme cost reduction when compared to 1G2G-Cur. Considering the importance of capital cost to the 2G ethanol, its cost per liter decreased by 30 % due to a lower investment in hydrolysis reactors (and cogeneration system) as well as due to higher ethanol yield achieved using more advanced enzyme technology. The increased 2G ethanol yield of 1G2G-AdvEH has also contributed to decrease the costs per liter associated with biomass, maintenance (also reduced due to lower CAPEX), labor and others. The overall effect of the new enzyme technology on the 2G ethanol production was a cost reduction of about 21 % considering the decline from US$0.516 to US$0.405 per liter. This result clearly proves the importance of the improvement on enzyme technology for 2G ethanol production, as well as the importance of having a simulation tool to economically assess the new technology.

Fig. 8.5 Second-generation ethanol cost breakdown considering different hydrolysis technologies

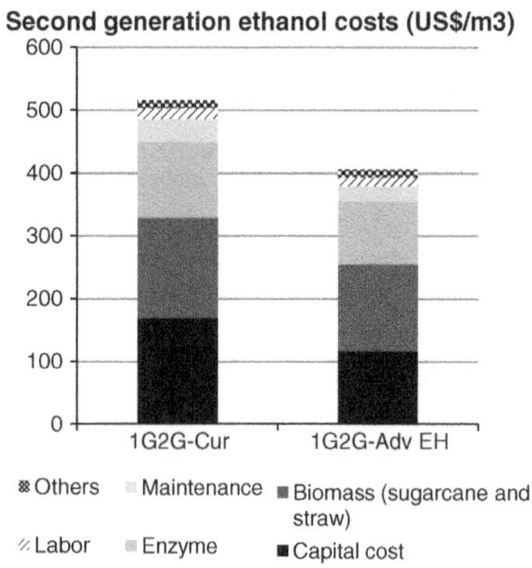

8.3 Case Study 2: Vinasse—Methane Use Alternatives

Aiming to justify the implementation of public policy mechanisms to accelerate the integration and expansion of biogas production concept within the sugarcane biorefineries in Brazil, the VSB has been used to assess different scenarios involving such premise. The proposal consists of the use and recovery of the by-products from the biogas production process from vinasse with technical, economic, social and environmental benefits.

The politically motivated incentives in Brazil within the biogas field are scarce, contemporary and mostly statewide character. The main efforts have been performed in the electric energy generation, following the trend of increasing the renewable sources in the energy matrix (Secretaria de Energia 2012). The nationwide regulation officializing the biomethane was also an important step (ANP 2015), although further public policies are still required to spread such conception widely through the private sector and to enhance biogas use diversification. The sugarcane ethanol chain is a suitable field to strength the biogas concept (Moraes et al. 2015). The valorization of vinasse to biogas can be an alternative to optimize the energy potential and sustainability of bioethanol production, and through this significant sector, the biogas initiatives may be improved and political directions may be provided as well.

This section brings together arguments based on an empirical approach supplemented by well-founded theoretical concepts, drawn up into the VSB, raising inquiries to support the implementation of research programs and public policies toward a more sustainable biorefinery model.

8.3.1 Biorefinery Assessment: Biogas Production in Greenfield and Brownfield Systems

Considering the characteristic of the VSB, this framework is mostly indicated in this current case to provide useful information that could base future prospects. By exemplifying such contextualization, assessments on biogas production within sugarcane biorefineries were performed considering both greenfield and brownfield scenarios. At first, the greenfield context was taken into account. Five different scenarios including biogas production and utilization were assessed in the VSB and compared with a baseline scenario (1G-Anx-Op) in which biogas production was not considered. In a second stage, the biogas scenarios were evaluated considering incremental cash flow analysis within the brownfield context. All the assessed scenarios consisted of 1G annexed distilleries using 50 % of the juice for sugar production and the other 50 %, together with molasses, for ethanol production. For electricity production, 65-bar boilers were employed, using bagasse and 50 % of the available straw as fuel. The main considerations adopted for such distilleries are presented in Table 8.9.

Table 8.9 Main considerations of the 1G annexed distillery in which biogas scenarios were based for both greenfield and brownfield contexts

Parameter	Value
Season period (days)	200
Sugarcane processed (TC per season)	4.0×10^6
Area of sugarcane (hectare)	52.6×10^3
Annual anhydrous ethanol production (m^3 per season)	214.8×10^3
Annual sugar production (t per season)	205.5×10^3
Annual vinasse production (m^3 per season)	1.8×10^6
Vinasse yield ($m^3_{vinasse}\ m^{-3}_{ethanol}$)	8.5

Regarding the biogas cases in the greenfield context, one scenario including standard biodigestion process (St-BG-Cog) was assessed together with four optimized scenarios, in which the inclusion of a technological package composed of biogas/biomethane and fluid fertilizer production units from vinasse was considered, as outlined in Fig. 8.6. For the incremental cash flow analysis, only the biogas scenarios including the technological package (optimized scenarios) were considered as they were applied in brownfield distilleries. An overview of all the major characteristics of the selected scenarios is presented in Table 8.10.

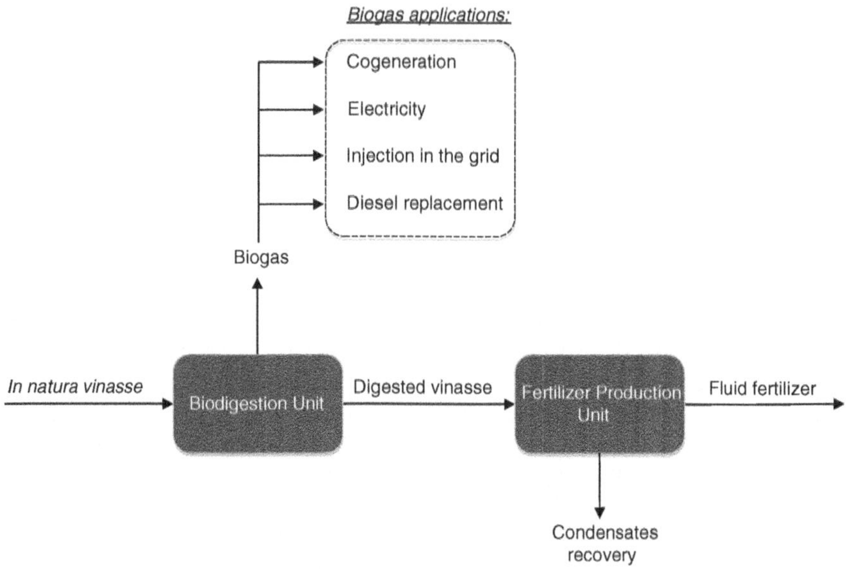

Fig. 8.6 Simplified diagram of the technological package of the optimized scenarios involving the production of biogas and fluid organic fertilizer from the vinasse

Table 8.10 Overview of the assessed scenarios based on 1G sugarcane biorefineries

[a]Scenarios	Annexed distillery	Biogas production	Biogas use	Fluid fertilizer production	[b]Context
1G-Anx-Op	Optimized	No	No	No	G
St-BG-Cog	Optimized	Standard	Cogeneration	No	G
Op-BG-Cog	Optimized	Optimized	Cogeneration	Yes	G/B
Op-BG-Elec	Optimized	Optimized	Electricity generation	Yes	G/B
Op-BM-Grid	Optimized	Optimized	[c]Sale to the natural gas grid	Yes	G/B
Op-BM-Diesel	Optimized	Optimized	[c]Replacement of diesel	Yes	G/B

[a]Acronym: *Anx* annex, *Op* optimized, *St* standard, *BG* biogas, *Cog* cogeneration, *Elec* electricity, *BM* biomethane

[b]G and B denote greenfield and brownfield assessments, respectively

[c]Uses in which biogas must be converted into biomethane

Power generation from biogas was calculated according to Moraes et al. (2014), based on the different biogas/biomethane applications, specified as follows:

- Cogeneration: combustion efficiency in the industrial boilers equivalent to 30 % (Siemens, model SGT 100, 4700 kW; Munich, Germany);
- Electricity generation: stationary internal combustion engine with 38 % efficiency (Caterpillar, Inc., model DM 5234, 50 Hz, 1500 rpm, 400 V; Peoria, IL, USA);
- Biomethane sale: purification of biogas to biomethane to be injected in the natural gas grid;
- Diesel replacement: energy equivalence of biomethane to diesel based on their lower heating value (LHV).

In the case of diesel replacement, hybrid heavy-duty biomethane–diesel engines were adopted, considering a parcel of diesel supplementation for the diesel engines converted to biomethane use (Moraes et al. 2014; Knight and Young 2009). Therefore, replacement of 70 % of diesel used in the agricultural operations and transport of sugarcane, straw, inputs and vinasse was considered (Soares 2013). The fluid fertilizer production was performed through the concentration of digested vinasse by evaporation to 25 °Bx. Ammonia was added to digested vinasse after the concentration step in order to achieve the desired N:K ratio for the agricultural application in sugarcane ratoons. The application doses of conventional inorganic fertilizer or fluid fertilizer produced from digested vinasse were defined according to recommendations for Brazilian sugarcane areas. A summary of the main parameters adopted for the biodigestion process and biogas uses is displayed in Table 8.11.

Table 8.11 Main parameters adopted for the biodigestion process according to each biogas production scenario

Parameter	St-BG-Cog	Op-BG-Cog	Op-BG-Elec	Op-BM-Grid	Op-BM-Diesel
Total nitrogen of in natura vinasse (kg m^{-3})	0.395	0.395	0.395	0.395	0.395
COD of in natura vinasse (kg m^{-3})	33.6	33.6	33.6	33.6	33.6
COD removal efficiency (%)	0.72	0.85	0.85	0.85	0.85
In natura vinasse volumetric flow rate (m^3 h^{-1})	382.7	382.7	382.7	382.7	382.7
CH_4 production per COD removed (Nm^3_{CH4} kg^{-1}_{COD})	0.29	0.32	0.32	0.32	0.32
Volumetric fraction of CH_4 in biogas (m^3_{CH4} m^{-3}_{biogas})	0.60	0.76	0.76	0.76	0.76
Volumetric fraction of CH_4 in biomethane (m^3_{CH4} $m^{-3}_{biomethane}$)	n.a.	n.a.	n.a.	0.965	0.965
Lower heating value (LHV) of biogas (10^3 kJ Nm^{-3})	21.5	27.3	27.3	30.5[a]	30.5[a]

[a]Relative to biomethane (96.5 % CH_4 v/v)

8.3.2 *Techno-economic Analysis*

Investment and operating costs of the biogas production unit were based on Procknor (2008) considering an internal circulation (IC) reactor (Driessen and Yspeert 1999) and including a treatment system for H_2S removal from biogas. Other economic parameters for the cash flow analysis were described in Chap. 6. The market price of biomethane (US$0.42 Nm^{-3}) for the current assessment was based on the equivalence with natural gas in terms of the lower heating value (LHV). The price of natural gas was based on the Petrobras price for distributors, exempt from taxes and charges (IBP 2014).

8.3.2.1 Greenfield Scenarios

Investment data of the different biorefinery cases are exposed in Table 8.12, accounting for the additional value regarding technological package considered in

Table 8.12 Investment values of the whole sugarcane biorefinery (greenfield context), discretizing the additional value relative to the technological package, according to the different scenarios

Investment (US$ million)	1G-Anx-Op	St-BG-Cog	Op-BG-Cog	Op-BG-Elec	Op-BM-Grid	Op-BM-Diesel
Base	424.84	430.80	430.07	424.63	420.45	420.45
Additional	n.a.	14.50[a]	30.27[b]	31.25[b]	31.48[b]	31.50[b]
Total	424.84	445.30	460.34	455.88	451.93	451.94

[a]Additional value in this scenario refers only to the biodigestion plant (without fertilizer production unit) and biogas used for cogeneration
[b]These additional investments were considered in the incremental cash flow analysis (brownfield context)

the optimized scenarios. Such additional value in the standard scenario St-BG-Cog concerns only to the biodigestion plant without the fertilizer production unit and biogas applied to the cogeneration system. Compared to the baseline scenario (1G-Anx-St), the inclusion of the cogeneration with biogas use (St-BG-Cog and Op-BG-Cog) caused a slight increase in the base investment due to the required increment in the generation and distribution system of steam and electricity. The technological package inclusion in the optimized scenarios, in turn, represented about 7 % of the total investment and promoted some decrease in the base investment (except for the cogeneration case Op-BG-Cog), especially in the biomethane scenarios (Op-BM-Grid and Op-BM-Diesel), in which savings in the electricity generation and distribution system were achieved. This fact was due to the lower generated energy/power caused by the steam use in the fertilizer production unit.

Figure 8.7 presents the main economic results obtained considering the greenfield context. The techno-economic analysis of the biogas scenarios compared to the baseline scenario (1G-Anx-St) revealed that the standard biogas technology aiming cogeneration (St-BG-Cog) leads to slightly lower NPV and IRR values—US$123 million and 15.5 % per year, respectively. This result was expected because the standard biodigestion technology is poorly explored, not achieving the real biogas potential, and the electricity from biogas is the only product as well. Other products, however, could be obtained from vinasse biodigestion such as those products presented in the optimized scenarios. Compared to the baseline and standard cogeneration scenarios (1G-Anx-St and St-BG-Cog), all the optimized ones promoted the reduction of production costs associated with the biomass (Table 8.13), indicating greater cost competitiveness. For instance, the Op-BG-Cog scenario improves the economic results compared with both 1G-Anx-Op and St-BG-Cog because of two main factors: First, the efficiency of the biogas-producing reactors is improved from 72 to 85 %, and second, two products

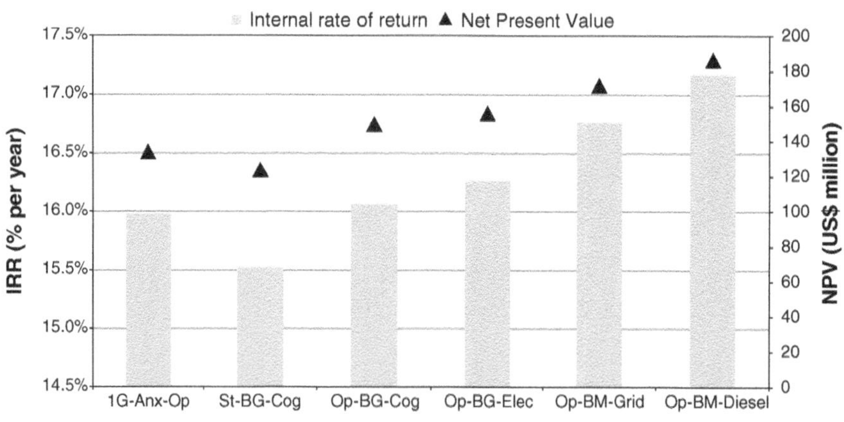

Fig. 8.7 Economic results for the greenfield context

Table 8.13 Production costs resulting from the techno-economic assessment according to the different scenarios in the greenfield context

Production costs	1G-Anx-Op	St-BG-Cog	Op-BG-Cog	Op-BG-Elec	Op-BM-Grid	Op-BM-Diesel
Sugarcane (US$ TC^{-1})	27.90	27.90	25.04	25.04	25.04	23.22
Straw (US$ $t^{-1}_{dry\ straw}$)	36.38	36.38	36.38	36.38	36.38	33.54
Sugar (US$ t^{-1})	337.64	340.75	328.24	326.33	320.87	312.61
Ethanol (US$ m^{-3})	449.57	453.71	437.06	434.51	427.24	416.25
Electricity (US$ MWh^{-1})	44.57	44.98	43.33	43.08	42.36	41.27
Biomethane (US$ Nm^{-3})	n.a.	n.a.	n.a.	n.a.	0.31	0.30

are obtained: bioelectricity and fluid fertilizer. As a consequence, revenues in Op-BG-Cog are higher and improve both NPV and IRR: US$148 million and 16.1 % per year, respectively. The Op-BG-Elec scenario also produces bioelectricity and fluid fertilizer. In this scenario, however, the efficiency of biogas conversion into electricity is higher than that in the Op-BG-Cog due to better technology associated with combustion efficiency: For electric generation in a stationary internal combustion engine, an electricity generation with 38 % efficiency was considered; on the other hand, the combustion efficiency in the industrial boilers for cogeneration was just 30 %.

The best economic results were associated with the biomethane production and its application, as occurred in Op-BM-Grid and Op-BM-Diesel scenarios. In the Op-BM-Grid, biomethane is produced from biogas and then exported to the natural gas grid—whose average distance from the industrial plant was assumed to be 3 km. Therefore, the additional revenues from biomethane exports to the grid improved both NPV and IRR—US$173 million and 16.8 % per year, respectively. However, it is important to notice that location of the sugarcane mill relative to the grid is a key factor for obtaining such positive revenues. Biomethane sales have the same logistic issues of natural gas because of the development of a dedicated network to the distribution of biomethane. The distribution costs for longer distances might be prohibitive, considering the uncertainties involved in the immaturity of this biomethane application. In the case of Op-BM-Diesel, a fraction of diesel used in sugarcane agricultural operations and transport is replaced by biomethane, which significantly reduces biomass production costs (Table 8.13). The main impact of such reduction was on the industrial operating costs associated with biomass, making this the most competitive scenario among the optimized ones. In the Op-BM-Diesel, the highest NPV and IRR values were observed: US$189 million and 17.2 % per year, respectively.

8.3.2.2 Brownfield Scenarios: Incremental Cash Flow Analysis

Results obtained from incremental cash flow analysis (Fig. 8.8) show economic viability of vinasse biodigestion technologies that could be introduced in brownfield industrial plants. Similar to the previous assessment, the best economic results were observed in the biomethane scenarios, providing both higher NPV and IRR values: US$23.3 million and 22.2 % per year for Op-BM-Grid, and US$35.8 million and 25.1 % for Op-BM-Diesel. These values were much lower in the biogas scenarios, achieving NPV and IRR of US$7.7 million and 15.7 %, respectively, for Op-BG-Cog, and US$9.5 million and 16.3 %, respectively, for Op-BG-Elec. Such results reinforce that adding value to biogas can improve the economic aspect within the sugarcane biorefinery concept.

The analysis considering brownfield conditions is more suitable for the current sugarcane biorefinery scenario, in which the inclusion of biogas concept has been gaining strength. The existing sugarcane mills in Brazil are focusing on improving their economic results, searching for innovative and profitable technologies. In this

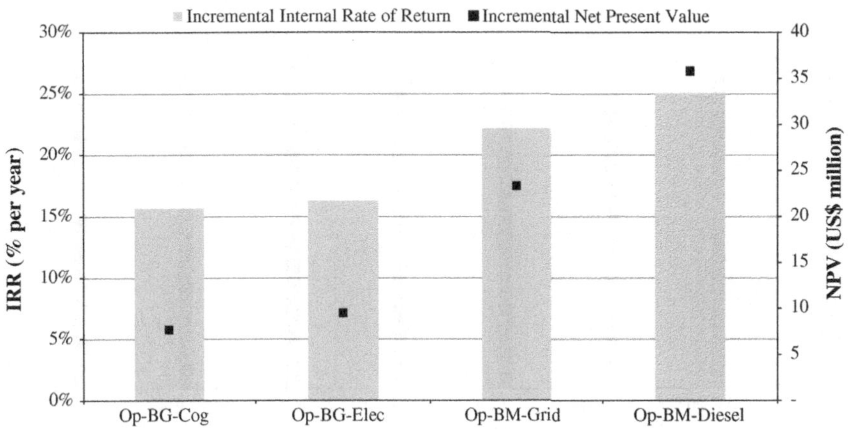

Fig. 8.8 Economic results obtained from the incremental cash flow analysis (brownfield context)

context, the proposed technological package arises as an attractive alternative, adding value to the most voluminous waste of the bioethanol production sector. However, the momentum provided by public policies is still essential to effectively insert the biogas concept in the sugarcane biorefinery model.

8.3.3 Environmental Assessment

Environmental impact assessment for ethanol production in different assessed scenarios is evaluated using a cradle to gate Life Cycle Assessment, considering the greenfield context.

The comparative environmental impacts of ethanol production for selected impact categories and for the selected vinasse biodigestion scenarios are presented in Fig. 8.9. These scores give the relative environmental impacts of ethanol production including sugarcane production and transportation processes, industrial conversion in the biorefinery and all the previous life cycle stages. Ethanol distribution and use were not included in this comparative assessment, since these stages will not differ among the evaluated scenarios.

Scenario with basic biodigestion (St-BG-Cog) presented environmental impacts slightly worse than equivalent annexed plant without vinasse biodigestion (1G-Anx-Op). This fact occurs mainly due to the need of sodium hydroxide to produce sodium bicarbonate used to control pH in biodigestion process. The large amount of surplus electricity, produced with biogas used in boilers, does not compensate the extra environmental impacts caused by soda usage. In addition, this scenario presents no agricultural benefit, since only the organic matter is removed from vinasse and its application in sugarcane fields remains the same from scenario without vinasse biodigestion.

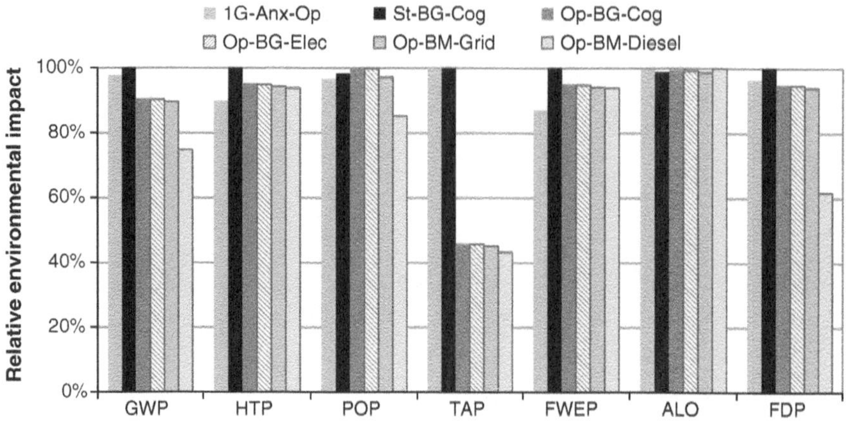

Fig. 8.9 Comparative environmental scores for ethanol in different scenarios (greenfield context) (**Note** *GWP* climate change; *HTP* human toxicity; *TAP* terrestrial acidification; *FWEP* freshwater eutrophication; *ALO* agricultural land occupation; *FDP* fossil depletion)

Optimization of vinasse application area and mineral fertilizer replacement reduces environmental scores of ethanol from optimized biodigestion scenarios (Op-BG-Cog, Op-BG-Elec, Op-BM-Grid, Op-BM-Diesel) when compared to 1G-Anx-Op in climate change (GWP), terrestrial acidification (TAP) and fossil depletion (FDP) categories. For human toxicity (HTP) and freshwater eutrophication (FWEP) categories, industrial inputs required for biodigestion (sodium hydroxide) and fertilizer production (ammonia and sulfuric acid) lead to higher impacts of ethanol production.

In general, differences between the optimized biodigestion scenarios are very small, except for Op-BM-Diesel scenario. Small differences in ethanol environmental scores are related to different biogas uses, which reflect on allocation factors that were calculated based on economic criteria. The best environmental results were observed in scenario with diesel replacement, since the agricultural scores for this scenario are benefited by partial replacement of the fossil fuel.

Breakdown of the environmental impacts of ethanol production is presented in Fig. 8.10 for climate change category. Results indicate that agricultural phase has a very high impact on the ethanol production chain, and reduction in environmental impacts related to fertilizer and diesel replacement on the agricultural phase helps to reduce environmental impacts. On the other hand, industrial inputs used in biodigestion and fertilizer production processes bring additional impacts to ethanol life cycle.

Due to the importance of the agricultural processes in the ethanol life cycle, the impacts on climate change category of sugarcane and straw production and transportation are detailed in Fig. 8.11 for selected scenarios. It is important to highlight that reduction in mineral fertilizer consumption decreases 22 % of the GHG emissions in sugarcane production. In the scenario with fertilizer and diesel

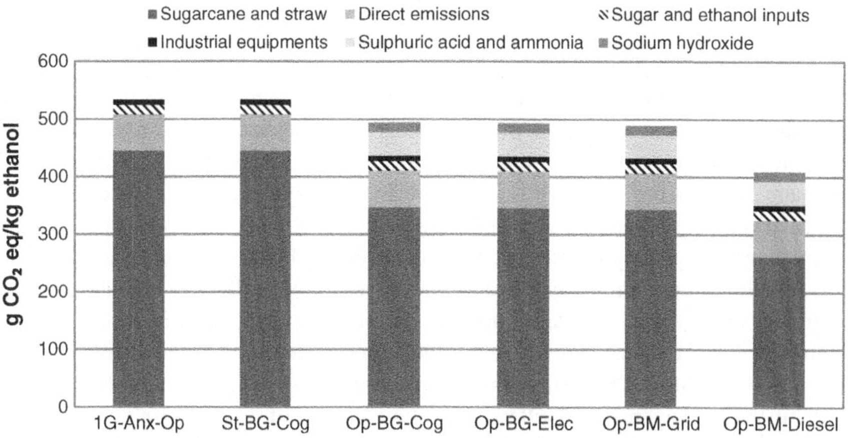

Fig. 8.10 Breakdown of the environmental impacts on climate change for ethanol production in different scenarios (greenfield context)

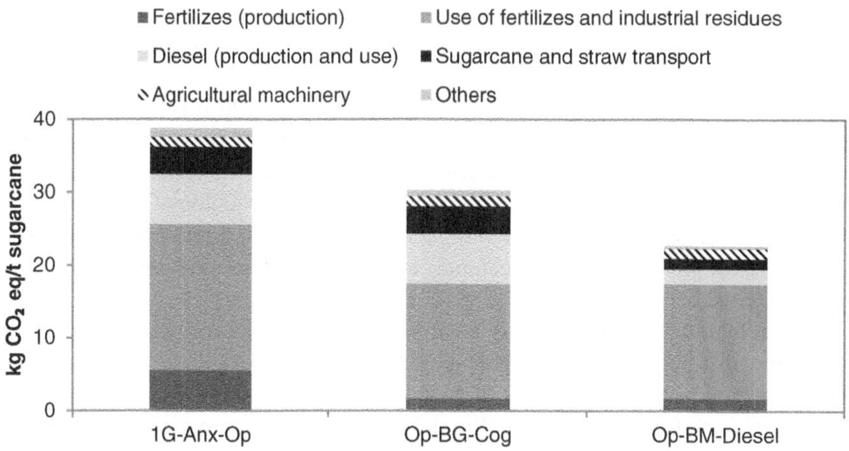

Fig. 8.11 Breakdown of the environmental impacts on climate change for sugarcane production in selected scenarios (greenfield context)

replacement, reduction is even greater, reaching 42 % in relation to traditional sugarcane production system.

It is worthwhile to mention that there are no differences in the impacts obtained for all the scenarios considering greenfield and brownfield systems.

8.3.4 Overall Balance: Uncertainties and Perspectives

Although the economic prospects for biomethane production within the sugarcane biorefinery scenario are more positive than the assessed biogas uses (for both greenfield and brownfield contexts), there are many uncertainties regarding biomethane applications, involving technical and economic aspects. There is some information in the literature regarding the technical viability of hybrid heavy-duty biomethane–diesel engines in the sugarcane ethanol sector (Knight and Young 2009), although no commercial example was yet found in Brazil. The same occurs with the biomethane injected into the natural gas grid, in which the current examples in the country are only small farm properties in the South region close to the grid (Cardoso 2015), generating around only 1 % of the potential biogas production from a sugarcane distillery with the capacity of 4 MTC per season.

The environmental feature should also receive more attention aside the economic perspective in order to stimulate such biogas conception. The commercialization of carbon credits could provide extra revenues to the distilleries adopting the clean development mechanism (CDM) and certified emission reduction (CER) (Salomon et al. 2011; Moraes et al. 2014). However, the carbon credit market still has low notability in Brazil, experiencing the lowest sale price at the last auction in 2012. Compared to the first auction (in 2007), the price devaluation represented a drop of 80 %, closing in 3.3 euro per tonne of CO_2 equivalent (BM&F BOVESPA 2012). This fact highlights the necessity of the combined efforts of government, scientific community and environmental agencies in order to add value and enhance the impacts (technical, economic and environmental) of the technology. The creation of national programs focused on the mitigation of emissions would be an important step, e.g., by providing incentive funds for carbon capture and remuneration of environmental services.

Through an overall balance, it is notable the surplus bioenergy generation in the sugarcane distilleries as a favorable key factor to improve their economic and environmental aspects, by means of biogas production from the waste generated during the manufacturing process. The different alternatives for biogas/biomethane applications could positively impact the scenario of the sugarcane biorefineries, either to bioelectricity generation or cogeneration, or to the biomethane sale into the natural gas grid or use as an alternative fuel. Such potential uses should be chosen according to the context of the biorefineries and the current political and economic national scenarios. For instance, it seems more economically attractive to sell the biomethane to the natural gas grid than to the electricity grid considering the present Brazilian energy matrix. The current relatively lower price paid to new energy sources through public auctions for renewable energy in Brazil compared to traditional ones contributes to marginalization of bioelectricity (ABRADEE 2015). However, the distance to the gas grid might be an impeditive factor, and in this case, the other uses seem to be more adequate. Even so, if there were some incentives to a mandatory market where the distributors are required to purchase a percentage of biogas energy in the energy matrix, the scenarios could also be different. These simple examples highlight the importance of creation of research programs and effective public policies regarding biogas use alternatives, offering a wider range of choices for the biorefineries and encouraging them to adopt such more sustainable concept.

References

ABRADEE (2015) Leilões de energia. In: Associação Brasileira de Distribuidores de Energia Elétrica – ABRADEE. Available at http://www.abradee.com.br/setor-eletrico/leiloes-de-energia. Accessed Jun 2015

ANP (2015) Resolução ANP Nº 8 de 30/01/2015. In: Agência Nacional do Petróleo, Gás Natual e Biocombustíveis. Available at http://www.anp.gov.br. Accessed Feb 2015

Binod P, Janu KU, Sindhu R et al (2011) Hydrolysis of lignocellulosic biomass for bioethanol production. In: Pandey A, Larroche C, Ricke SC, Dussap CG, Gnansounou E (eds) Biofuels. Academic Press, Amsterdam, pp 229–250

BM&F BOVESPA (2012) Leilões de créditos de carbono nº 001/2012. Available at http://www.bmfbovespa.com.br/consulta-leiloes/leiloes-de-credito-de-carbono-login.aspx?idioma=pt-br. Accessed Jun 2015

Cardoso MT (2015) SCGAS – Companhia de Gás de Santa Catarina – aposta no biometano para ampliar a rede para o interior. In: Página Sustentável. Available at http://www.paginasustentavel.com.br/index.php?option=com_content&view=article&id=2437:scgas-aposta-no-biometano-para-ampliar-a-rede-para-o-interior&catid=1:energia&Itemid=9. Accessed Jun 2015

Driessen W, Yspeert P (1999) Anaerobic treatment of low medium and high strength effluent in the agro industry. Water Sci Technol 40(8):221–228. doi:10.1016/S0273-1223(99)00629-0

GranBio (2015) Cana-energia. http://www.granbio.com.br/conteudos/cana-energia. Accessed Jun 2015

Humbird D, Davis R, Tao L (2011) Process Design and Economics for Biochemical Conversion of Lignocellulosic Biomass to Ethanol. National Renewable Energy Laboratory – NREL. Available at http://www.nrel.gov/biomass/pdfs/47764.pdf. Accessed Jun 2015

IBP (2014) Preco do Gás Natural. In: Instituto Brasileiro de Petróleo, Gás e Biocombustíveis – IBP. Available at http://200.189.102.61/SIEE/dashboard/ PrecoDoGasNatural. Accessed Jun 2015

Klein-Marcuschamer D, Oleskowicz-Popiel P, Simmons BA et al (2012) The challenge of enzyme cost in the production of lignocellulosic biofuels. Biotechnol Bioeng 109:1083–1087. doi:10.1002/bit.24370

Knight VM, Young CEF (2009) Análise de custo-benefício da substituição do diesel por gás natural veicular em ônibus na Região Metropolitana de São Paulo. Revista de Economia Mackenzie 7(3):24–36

Matsuoka S, Kennedy AJ, Santos EGD et al (2014) Energy cane: its concept, development, characteristics, and prospects. Adv Botany 2014:1–13. doi:10.1155/2014/597275

Milanez AY, Nyko D, Valente MS et al (2015) De promessa a realidade: como o etanol celulósico pode revolucionar a indústria da cana-de-açúcar – uma avaliação do potencial competitivo e sugestões de política pública. BNDES Setorial 41:237–294

Moraes BS, Junqueira TL, Pavanello LG et al (2014) Anaerobic digestion of vinasse from sugarcane biorefineries in Brazil from energy, environmental, and economic perspectives: profit or expense? Appl Energ 113:825–835

Moraes BS, Zaiat M, Bonomi A (2015) Anaerobic digestion of vinasse from sugarcane ethanol production in Brazil: challenges and perspectives. Renew Sust Energ Rev 44:888–903. doi:10.1016/j.rser.2015.01.023

Mussatto SI, Dragone G, Guimarães PMR et al (2010) Technological trends, global market, and challenges of bio-ethanol production. Biotechnol Adv 28(6):817–830. doi:10.1016/j.biotechadv.2010.07.001

Procknor C (2008) Energia elétrica a partir da vinhaça. Revista da Sociedade dos Técnicos Açucareiros e Alcooleiros do Brasil – STAB. Available at http://www.unica.com.br/convidados/25641156920337715081/energia. Accessed Mar 2014

Salomon KR, Lora EES, Rocha MH et al (2011) Cost calculations for biogas from vinasse biodigestion and its energy utilization. Coop Sugar 42(9):23–31

Secretaria de Energia (2012) Plano Paulista de Energia – PPE. In: Secretaria de Energia do Estado de São Paulo. Available at http://www.energia.sp.gov.br/ a2sitebox/arquivos/documentos/491.pdf. Accessed Jun 2015

Soares PA (2013) Use of biomethane in diesel engines. Dedini S.A. Personal Communication

Sun Y, Cheng J (2002) Hydrolysis of lignocellulosic materials for ethanol production: a review. Bioresour Technol 83(1):1–11. doi:10.1016/S0960-8524(01)00212-7

Taherzadeh MJ, Karimi K (2007) Enzymatic-based hydrolysis processes for ethanol from lignocellulosic materials: a review. Bioresources 2:707–738

Toledo M (2015) Desenvolvida após melhoramentos genéticos, 'supercana' visa energia. http://www1.folha.uol.com.br/mercado/2015/05/1627014-com-alto-indice - de-fibras-e-biomassa-supercana-visa-a-producao-de-energia.shtml. Accessed Jun 2015

U.S. DOE (2015) Lignocellulosic Biomass for Advanced Biofuels and Bioproducts: Workshop Report. In: U.S. Department of Energy—Office of Science. http://genomicscience.energy.gov/biofuels/lignocellulose. Accessed Jun 201

Chapter 9
Final Remarks

Antonio Bonomi, Otávio Cavalett, Marcelo Pereira da Cunha and Marco A.P. Lima

This book presents the development and use of a simulation tool, the so-called Virtual Sugarcane Biorefinery (VSB), for the assessment of technical, economic, social, and environmental aspects of biorefinery alternatives. The integrated analysis, i.e., consideration of entire sugarcane chain, analyzed within the three aspects of sustainability, can be considered as a special feature of this effort. The framework for linking all these aspects is, in fact, the main innovative aspect of the tool. Another integration aspect that is also important to mention is that the knowledge provided by the chemical process design and its simulation helped to better describe and update methods, models, and upstream process to the bioenergy sector context.

Within this context, appropriate analyzing tools and design methods, as well as optimization procedures coupled with the VSB, can be used to properly assess and optimize sustainability's triple bottom line, which assembles economic, social, and environmental performance metrics. This assessment and optimization could be further improved with more detailed models for each one of the sugarcane life cycle phases and unit operations. An optimization strategy would allow having a better guideline for assessing the sustainability attractiveness of different perspectives of technology configurations and better define their commercial implementation strategy.

A. Bonomi (✉) · O. Cavalett
Brazilian Bioethanol Science and Technology Laboratory (CTBE),
Brazilian Center of Research in Energy and Materials (CNPEM),
Campinas, SP, Brazil
e-mail: antonio.bonomi@bioetanol.org.br

O. Cavalett
e-mail: otavio.cavalett@bioetanol.org.br

M.P. da Cunha
Institute of Economics (IE), University of Campinas (UNICAMP),
Campinas, SP, Brazil
e-mail: marcelocunha@eco.unicamp.br

M.A.P. Lima
Institute of Physics Gleb Wataghin (IFGW), University of Campinas (UNICAMP),
Campinas, SP, Brazil
e-mail: maplima@ifi.unicamp.br

A. Bonomi et al. (eds.), *Virtual Biorefinery*, Green Energy and Technology,
DOI 10.1007/978-3-319-26045-7_9

Regarding methodological issues, it will be fundamental to constantly updating VSB with improved methodologies as the scientific knowledge in the area evolves. For example, better metrics for evaluating climate change impacts, considering the most recent IPCC report shall be included in the environmental assessment. Social metrics will also gain robustness, for example, as the Social Life Cycle Assessment gain more case studies and the method become more disseminated and improved.

Although the focus of the VSB is the sugarcane sector, as demonstrated in Chaps. 3–5, and 7, it can be adapted to evaluate other biomass such as sweet sorghum and energy cane, beyond others. The models within VSB have the possibility of being customized to other production chains using the same integration and sustainability assessment framework. The construction of the tool is presented in details in this book, and it is expected that it serves as an inspiration and pave the way for similar approaches to be constructed and applied in other economic sectors.

Regarding the dissemination of the tool, more effort will be spent to develop simplified user-friendly versions of it. This initiative could generate several tailor-made versions for specific sectors or industries, which would allow the tool to be more attractive and practical for intended users. In addition, these consolidated versions could be made available in the Internet and, therefore, increasing the tool's range of use all over the world. Other possibility is the use of the VSB as an open facility at the Brazilian Center of Research in Energy and Materials (CNPEM). Within this arrangement, users from all over the world can have full support from the VSB team for adapting and using the tool in their own research and/or evaluation. This is considered as the best option for users to get experienced with the VSB use.

VSB was conceived and developed to promote the development of sugarcane industry, approximating the key players for this purpose, including R&D institutions and enterprises, from public and private initiatives. As shown in Chap. 8, the tool has a great potential to be used for evaluating the progress of research and development programs, identifying main technical, economic, social, and environmental bottlenecks of the process under assessment. In fact, this has been the main use of the VSB so far and the success of the tool for this purpose has been demonstrated in several publications of the VSB team (e.g., Jonker et al. 2015; Cardoso et al. 2013; Dias et al. 2014; Moraes et al. 2014; Cavalett et al. 2012).

Furthermore, the tool is also designed for development plans assessment and promotion of better and scientific-based public policies in the bioenergy sector. As the tool gains more robustness and acceptance, it can be used for defining and evaluating public policies in Brazil (e.g., to support sugarcane and corn ethanol flexible plants—Milanez et al. 2014—and for second-generation ethanol production—Milanez et al. 2015) and also for other regions and countries of the world.

For the use of the VSB to fulfill these last two purposes, it was of the upmost importance the idea of conceiving and developing the VSB in a National Laboratory in the Brazilian context, providing the tool with the required impartiality and distance to individual or companies interests.

References

Cardoso TF, Cavalett O, Chagas MF, Moraes ER, Carvalho JLN, Franco HCJ, Galdos MV, Scarpari F, Braunbeck OA, Cortez LAB, Bonomi A (2013) Technical and economic assessment of trash recovery in the sugarcane bioenergy production system. Scientia Agrícola 70:353–360

Cavalett O, Junqueira T, Dias MOS et al (2012) Environmental and economic assessment of sugarcane first generation biorefineries in Brazil. Clean Technol Envir 14(3):399–410. doi:10.1007/s10098-011-0424-7

Dias MOS, Pereira LG, Junqueira TL, Pavanello LG, Chagas MF, Cavalett O, Maciel Filho R, Bonomi A (2014) Butanol production in a sugarcane biorefinery using ethanol as feedstock—Part I: integration to a first generation sugarcane distillery. Chem Eng Res Des 92 (8):1441–1451

Jonker JGG, Van Der Hilst F, Junginger HM, Cavalett O, Chagas MF, Faaij APC (2015) Outlook for ethanol production costs in Brazil up to 2030, for different biomass crops and industrial technologies. Appl Energy 147:593–610

Milanez AY, Nyko D, Valente MS, Xavier CEO, Kulay LA, Donke ACG, Matsuura MISF, Ramos NP, Morandi MAB, Bonomi A, Capitani DHD, Chagas MF, Cavalett O, Gouveia VLR (2014) A produção de etanol pela integração do milho-safrinha às usinas de cana-de-açúcar: avaliação ambiental, econômica e sugestões de política. Revista do BNDES 41:147–207

Milanez AY, Nyko D, Valente MS, Sousa LC, Bonomi A, Jesus CDF, Watanabe MDB, Chagas MF, Rezende MCAF, Cavalett O, Junqueira TL, Gouveia VLR (2015) De promessa a realidade: como o etanol celulósico pode revolucionar a indústria da cana-de-açúcar - uma avaliação do potencial competitivo e sugestões de política pública. BNDES Setorial 41:237–294

Moraes BS, Junqueira TL, Pavanello LG et al (2014) Anaerobic digestion of vinasse from sugarcane biorefineries in Brazil from energy, environmental, and economic perspectives: profit or expense? Appl Energ 113:825–835

The manufacturer's authorised representative in the EU is Springer Nature Customer Service Centre GmbH, Europaplatz 3, 69115 Heidelberg, Germany. If you have any concerns regarding our products, please contact ProductSafety@springernature.com

Printed and bound by CPI Group (UK) Ltd, Croydon, CR0 4YY

29/07/2026

02180577-0004